Composites Science and Technology

Composites Science and Technology (CST) book series publishes the latest developments in the field of composite science and technology. It aims to publish cutting edge research monographs (both edited and authored volumes) comprehensively covering topics shown below:

- Composites from agricultural biomass/natural fibres include conventional composites-Plywood/MDF/Fiberboard
- Fabrication of Composites/conventional composites from biomass and natural fibers
- Utilization of biomass in polymer composites
- Wood, and Wood based materials
- Chemistry and biology of Composites and Biocomposites
- Modelling of damage of Composites and Biocomposites
- Failure Analysis of Composites and Biocomposites
- Structural Health Monitoring of Composites and Biocomposites
- Durability of Composites and Biocomposites
- Biodegradability of Composites and Biocomposites
- Thermal properties of Composites and Biocomposites
- Flammability of Composites and Biocomposites
- Tribology of Composites and Biocomposites
- Bionanocomposites and Nanocomposites
- Applications of Composites, and Biocomposites

To submit a proposal for a research monograph or have further inquries, please contact springer editor, Ramesh Premnath (ramesh.premnath@springer.com).

More information about this series at http://www.springer.com/series/16333

Mohamed Thariq Hameed Sultan ·
Azwan Iskandar Azmi · Mohd Shukry Abd Majid ·
Mohd Ridzuan Mohd Jamir · Naheed Saba
Editors

Machining and Machinability of Fiber Reinforced Polymer Composites

Springer

Editors
Mohamed Thariq Hameed Sultan
Faculty of Engineering
Department of Aerospace Engineering
Universiti Putra Malaysia
Serdang, Selangor, Malaysia

Azwan Iskandar Azmi
Faculty of Mechanical Engineering
Technology
Universiti Malaysia Perlis
Arau, Malaysia

Mohd Shukry Abd Majid
Faculty of Mechanical Engineering
Technology
Universiti Malaysia Perlis
Arau, Malaysia

Mohd Ridzuan Mohd Jamir
Faculty of Mechanical Engineering
Technology
Universiti Malaysia Perlis
Arau, Malaysia

Naheed Saba
Lab Biocomposite Tech
INTROP, Universiti Putra Malaysia
Serdang, Selangor, Malaysia

ISSN 2662-1819 ISSN 2662-1827 (electronic)
Composites Science and Technology
ISBN 978-981-33-4152-4 ISBN 978-981-33-4153-1 (eBook)
https://doi.org/10.1007/978-981-33-4153-1

This Springer imprint is published by the registered company Springer Nature Singapore Pte Ltd.
The registered company address is: 152 Beach Road, #21-01/04 Gateway East, Singapore 189721, Singapore

Preface

This book covers recent development and practices of machining of fiber reinforced polymer composites. These particular types of composite materials are often manufactured to near-net shape with certain machining and cutting processes, which are inevitable to meet tight-dimensional tolerance and intricate shape requirements. Improper selection of machining parameters and tooling would lead to severe damage on the materials which has the likelihood to deteriorate their mechanical performance during in-life services. This book elaborates state-of-the-art machining techniques and approaches carried out by prominent researchers worldwide on fiber reinforced polymer composites. A sound knowledge of modern research techniques and applications of machining fiber reinforced polymer composites is furnished for related industries, namely, those in aerospace, automotive, marines and construction fields.

The book explores the influence of drill geometry design in alleviating the onset delamination damage of fiber reinforced polymer composites. A combination of chisel and cutting edges, helix and point angles, diameter and length plays a very crucial role in producing minimum delamination induced damage on fiber reinforced polymer composite components. Likewise, the proper settings of drilling process parameters, namely, cutting speed, feed rate, depth-of-cut and material removal rate for minimisation of delamination damage are deliberated in this book. Further analytical as well as experimental investigations to explore the critical thrust force, which attributed to the delamination behaviour during drilling of fiber reinforced polymer composite, are covered. The book continues to illustrate some innovative numerical and analytical studies to comprehend the mechanics of machining of fiber reinforced polymer composites. Examples are given on these topics so that they complement some of the experimental work carried out by researchers worldwide. Challenges in evaluating the amount of heat dissipated through workpiece, tool and chips during machining of these composite materials using an iterative inverse heat conduction technique are comprehensively discussed in this book. Finally, this book also provides a recent knowledge of tribological aspect of machining of fiber reinforced polymer composites, apart from the novel method of cutting composites using non-traditional processes such as electro-discharge machining, laser processing and abrasive waterjet machining.

As a final note, we assure the readers that the information provided in this book aims to assist researchers, scientists and practitioners to understand the best practices on machining of fiber reinforced polymer composites to meet the essential requirements and demand of composite industries. We acknowledge and are thankful to all authors who have fully committed to contribute and share their research finding in this edited book. Lastly, we are grateful to Springer team for continuous support at every stage of this book production in ensuring it is published in a timely manner.

Serdang, Malaysia — Mohamed Thariq Hameed Sultan
Perlis, Malaysia — Azwan Iskandar Azmi
Perlis, Malaysia — Mohd Ridzuan Mohd Jamir
Perlis, Malaysia — Mohd Shukry Abd Majid
Kuala Lumpur, Malaysia — Naheed Saba

Contents

Editors and Contributors

About the Editors

Prof. Ir. Ts. Dr. Mohamed Thariq Hameed Sultan is a Professional Engineer (PEng) registered under the Board of Engineers Malaysia (BEM), a Professional Technologist (PTech) registered under the Malaysian Board of Technologists, and also a Chartered Engineer (CEng) registered with the Institution of Mechanical Engineers United Kingdom, currently attached to the Universiti Putra Malaysia as the Head of the Biocomposite Technology Laboratory, Institute of Tropical Forestry and Forest Products (INTROP), UPM Serdang, Selangor, Malaysia. Being the Head of the Biocomposite Technology Laboratory, he is also appointed as an Independent Scientific Advisor to Aerospace Malaysia Innovation Centre (AMIC) based in Cyberjaya, Selangor, Malaysia. He received his Ph.D. from the University of Sheffield, United Kingdom. He has about 10 years of experience in teaching as well as in research. His area of research interests includes hybrid composites, advance materials, structural health monitoring, and impact studies. So far, he has published more than 100 international journal papers and received many awards locally and internationally. In December 2017, he was awarded a Leaders in Innovation Fellowship (LIF) by the Royal Academy of Engineering (Raeng), United Kingdom. He is also the Honourable Secretary of the Malaysian Society of Structural Health Monitoring (MSSHM) based in UPM Serdang, Selangor, Malaysia. Currently, he is also attached to the Institution of Engineers Malaysia (IEM) as the Deputy Chairman in the Engineering Education Technical Division (E2TD).

Dr. Azwan Iskandar Azmi received his bachelor's degree in Mechanical Engineering from Purdue University, USA in 1999 and Master's in Advanced Manufacturing Technology from Universiti Teknologi Malaysia in 2003. In 2013, he completed his doctoral study at The University of Auckland, New Zealand in Mechanical Engineering with specialisation in the area of fiber reinforced composite machining. He is currently serving the Faculty of Mechanical Engineering Technology, Universiti Malaysia Perlis (UniMAP) as Associate Prof. Dr. Azwan is a Professional Engineer (PEng) registered under the Board of Engineers Malaysia

(BEM), a Professional Technologist (PTech) registered under the Malaysian Board of Technologists (MBOT), and also a Chartered Engineer (CEng) registered with the Institution of Mechanical Engineers (IMechE) in the United Kingdom. He has more than 10 years of experience in teaching, research, and industries. His area of research interests includes machining and machinability of carbon, glass, and their hybrid composites. Currently, the research also covers the drilling and milling of lignocellulosic reinforced polymer composites. His interest extends on the machinability study of difficult-to-cut metal alloys such as titanium, Inconel, and nickel-titanium (NiTi) alloys. So far, he has more than 50 research articles in renowned international journals, proceedings and review papers. He has been awarded with research funds from Ministry of Higher Education Malaysia (MOHE) and Ministry of Science Technology and Innovation (MOSTI). Due to the research outputs, Dr. Azwan has served as technical reviewer for a number of reputable ISI ranked journals and international conferences.

Dr. Mohd Shukry Abd Majid received his Bachelor of Engineering in Mechanical Engineering from University Manchester Institute of Science and Technology (UMIST) in 2001. Upon his return to Malaysia, he worked as a research and development (R&D) engineer at a semiconductor industry before joining Universiti Malaysia Perlis (UniMAP) as a lecturer in 2004. He completed his M.Sc. in Mechanical Systems Engineering from the University of Liverpool in 2005 and his Ph.D. in Composite Engineering from Newcastle University, United Kingdom in 2011. Currently, he is serving Universiti Malaysia Perlis as an Associate Professor at Faculty of Mechanical Engineering Technology. An advocate of interdisciplinary research, his research interests lie in the strength of material's area with emphasis on the composite piping, looking at the performance of composite structures, NDE's of composites and natural fiber/green composites, hybrid reinforced/filled polymer composites, lignocellulosic reinforced/filled polymer and biodegradable composites. His aptitude for high-quality research of international standing has been supported by his 188 Scopus indexed publications with 47 (32 Q1) publications in ISI-ranked Journals having a cumulative impact factor (CIF) of 148.948. He received his professional engineer qualification (Ir.) from Board of Engineer Malaysia (BEM) in Mac 2016 and has been a Chartered Engineer (CEng) from the Engineering Council, United Kingdom since 2014. Dr. Mohd Shukry has been honoured with numerous local and international awards for his achievements. He is the first recipient from Technical University Network (MTUN) to have been awarded Malaysia's Research Star Award 2017, as one of Malaysia's most promising and influential researchers by the Ministry of Higher Education Malaysia (MOHE).

Dr. Mohd Ridzuan Mohd Jamir is a Professional Engineer (PEng) registered under the Board of Engineers Malaysia (BEM), a Professional Technologist (PTech) registered under the Malaysian Board of Technologists, and also a Chartered Engineer (CEng) registered with the Institution of Mechanical Engineers United Kingdom, currently attached to the Universiti Malaysia Perlis as Associate Professor at the

Faculty of Mechanical Engineering Technology. He obtained Diploma in Mechanical Engineering from Universiti Teknologi Malaysia (UTM) in 2006. He graduated in Bachelor of Engineering (Mechanical) and Master of Engineering (Mechanical) from Universiti Teknologi Malaysia (UTM) in the year of 2009 and 2010, respectively. He also worked as Quality Assurance Engineer at Venture Pintarmas Sdn Bhd, Johor Baharu in 2010 before joined Universiti Malaysia Perlis (UniMAP) in the same year. He received his Ph.D. in Mechanical Engineering from Universiti Malaysia Perlis (UniMAP) in 2016. He has about 10 years of experience in teaching as well as in research. An advocate of interdisciplinary research, his research interests includes the strength of material's area with emphasis on the natural fiber composite, looking at the performance of composite structures and tribological properties, hybrid reinforced/filled polymer composites, lignocellulosic reinforced/filled polymer and biodegradable composites. He has published more than 100 international journal research paper with Scopus index and ISI ranked journal publications. In addition, he is also a regular reviewer for high-impacts ISI ranked journals.

Dr. Naheed Saba completed her Ph.D. (Nanocomposites) with Distinction from Laboratory of Biocomposites Technology, Institute of Tropical Forestry and Forest Products (INTROP), Universiti Putra Malaysia, Serdang, Selangor, Malaysia in 2017. She completed her master's in chemistry and also completed her postgraduate diploma in environment and sustainable development from India. She has published over 50 scientific Peer review Articles in International Journal and 4–5 Articles come under top cited articles in Construction and Building Materials, and Polymers Journal during 2016–2019. She edited six books from Elsevier and also published more than 25 book chapters in Springer, Elsevier, and Wiley publication. She attended few international conferences and presented research papers. Her research interest areas are nanocellulosic materials, fire-retardant materials, natural fibere reinforced polymer composites, biocomposites, hybrid composites, and nanocomposites. She is also recipient of International Graduate Research Fellowship and Graduate on Time (GOT) award from Universiti Putra Malaysia, Malaysia. Presently she is Editor-in-Chief of The Journal of Composites and Advance Materials (ISSN 2716-8018) is a peer-reviewed bi-annual journal. She also serves as reviewer for several international journals such as Cellulose, Constructions and Building Materials, Composite Part A, Composite Part B, Journal of Polymers and The Environment, Journal of Energy Storage, Journal of Elastomers and Plastics, Journal of Materials Research and Technology, BioResources, and Carbohydrate Polymers. Her H-index is 17 for Scopus database and 20 for Google Scholar.

Contributors

Fahad Almaskari Department of Aerospace Engineering, Khalifa University of Science and Technology, Main Campus, Abu Dhabi, United Arab Emirates

Mohd Hazim Mohamad Amini Faculty of Bioengineering and Technology, Universiti Malaysia Kelantan, Jeli Campus, Jeli, Kelantan, Malaysia

Azwan Iskandar Azmi Faculty of Mechanical Engineering Technology, Universiti Malaysia Perlis, Pauh Putra Campus, Pauh, Perlis, Malaysia

M. Roy Choudhury Department of Mechanical Engineering, National Institute of Technology Meghalaya, Shillong, India

L. Crouzeix Institut Clément Ader, UMR 5312, CNRS, Toulouse, France

K. Debnath Department of Mechanical Engineering, National Institute of Technology Meghalaya, Shillong, India

H. Dutta Department of Mechanical Engineering, National Institute of Technology Meghalaya, Shillong, India

Munish Kumar Gupta Key Laboratory of High Efficiency and Clean Mechanical Manufacture, School of Mechanical Engineering, Shandong University, Jinan, People's Republic of China

Farrukh Hafeez DIAC, University of Birmingham Dubai, Dubai, United Arab Emirates

M. H. Hassan School of Mechanical Engineering, Universiti Sains Malaysia, Nibong Tebal, Pulau Pinang, Malaysia

Robert Heinemann Department of Mechanical, Aerospaceand Civil Engineering, The University of Manchester, Manchester, UK

Ozden Isbilir Mechanical Engineering Department, Engineering Faculty, Karabuk University, Karabuk, Turkey

Sikiru Oluwarotimi Ismail Department of Engineering, School of Physics, Engineering and Computer Science, Centre for Engineering Research, University of Hertfordshire, Hatfield, Hertfordshire, UK

R. Izamshah Advanced Manufacturing Center (AMC), Universiti Teknikal Malaysia Melaka (UTeM), Durian Tunggal, Melaka, Malaysia

M. F. Jaafar Advanced Manufacturing Center (AMC), Universiti Teknikal Malaysia Melaka (UTeM), Durian Tunggal, Melaka, Malaysia

M. S. Kasim Advanced Manufacturing Center (AMC), Universiti Teknikal Malaysia Melaka (UTeM), Durian Tunggal, Melaka, Malaysia

D. Lamouche Safran Aircraft Engines (Villaroche). Rond-Point René Ravaud, Moissy-Cramayel, France

Tan Chye Lih Faculty of Mechanical Engineering Technology, Universiti Malaysia Perlis (UniMAP), Arau, Perlis, Malaysia

R. N. Mahapatra Department of Mechanical Engineering, National Institute of Technology Meghalaya, Shillong, India

Paul Tarisai Mativenga School of Mechanical, Aerospace and Civil Engineering, University of Manchester, Manchester, UK

Mozammel Mia Department of Mechanical Engineering, Imperial College London, South Kensington, London, UK

Mazlan Mohamed Faculty of Bioengineering and Technology, Universiti Malaysia Kelantan, Jeli Campus, Jeli, Kelantan, Malaysia

Misbah Niamat Mechanical Engineering Department, Muhammad Nawaz Sharif University of Engineering and Technology, Multan, Pakistan

Danil Yu Pimenov Department of Automated Mechanical Engineering, South Ural State University, Chelyabinsk, Russia

G. Surya Rao Department of Mechanical Engineering, National Institute of Technology Meghalaya, Shillong, India

Shoaib Sarfraz Manufacturing Department, School of Aerospace, Transport and Manufacturing, Cranfield University, Cranfield, Bedfordshire, UK

D. K. Sarma Department of Mechanical Engineering, National Institute of Technology Meghalaya, Shillong, India

Essam Shehab Mechanical and Aerospace Department, School of Engineering and Digital Sciences, Nazarbayev University, Nur-Sultan, Kazakhstan

Jamal Sheikh-Ahmad Department of Mechanical Engineering, Khalifa University of Science and Technology, SAN Campus, Abu Dhabi, United Arab Emirates

Norshah Aizat Shuaib Faculty of Engineering Technology, Universiti Malaysia Perlis, Padang Besar, Malaysia

Sharizal Ahmad Sobri Faculty of Bioengineering and Technology, Universiti Malaysia Kelantan, Jeli Campus, Jeli, Kelantan, Malaysia

X. Sourd Institut Clément Ader, UMR 5312, CNRS, Toulouse, France;
Safran Aircraft Engines (Villaroche). Rond-Point René Ravaud, Moissy-Cramayel, France

S. A. Sundi Faculty of Mechanical & Manufacturing Engineering Technology, Universiti Teknikal Malaysia Melaka (UTeM), Durian Tunggal, Melaka, Malaysia; Advanced Manufacturing Center (AMC), Universiti Teknikal Malaysia Melaka (UTeM), Durian Tunggal, Melaka, Malaysia

David Whitehead Department of Mechanical, Aerospaceand Civil Engineering, The University of Manchester, Manchester, UK

Irina Ming Ming Wong Faculty of Mechanical Engineering Technology, Universiti Malaysia Perlis, Pauh Putra Campus, Pauh, Perlis, Malaysia

Hariz Zain Faculty of Engineering Technology, Universiti Malaysia Perlis, Padang Besar, Malaysia

R. Zitoune Institut Clément Ader, UMR 5312, CNRS, Toulouse, France

Influence of Drill Geometry Design on Drilling-Induced Damage Reduction in Fiber-Reinforced Polymeric Composites

Sikiru Oluwarotimi Ismail

Abstract Drilling is an extensively used manufacturing process for boring different and widely used fiber-reinforced polymeric (FRP) composite materials, among various machining operations. This process is inevitable for assembling/coupling of parts of systems. Despite of good inherent properties of the FRP composite materials, they are not easy to drill, due to the dissimilar properties of their constituents (mainly fiber/reinforcement and matrix). More than a few drilling-induced damage (DID) on FRP composites include delamination, surface roughness, fiber-pull out/uncut, among others. They severely affect the quality, structural integrity and applications of the drilled composite components. The most rampant among these damage is delamination; either peel-up or push-out type. Importantly, these damage are frequent and attributed mainly to the geometry design of the drill bits used. It is highly germane to consider and further study the influence of the drill geometry design (DGD) on reduction of DID on FRP composite components and improve the quality of the drilled holes. Therefore, this present chapter focuses on a current status/trend in the drilling of FRP composites and comprehensively reports optimum drill geometry designs (DGDs) for different FRP composites. It was evident that a combination of an efficient drill geometry (chisel and cutting edges, helix and point angles, diameter, length, material, among others) design and suitable selected drilling process parameters (cutting speed, feed rate, depth-of-cut, material removal rate, among others) produced minimum DID on FRP composite components. This knowledge is required to guide drill designers, manufacturers, machinists and researchers in their search for high performance drilling phenomenon.

Keywords Drill geometry design (DGD) · Drilling · Fiber-reinforced polymeric (FRP) composites · Drilling-induced damage (DID) · Delamination

---

S. O. Ismail (✉)
Department of Engineering, School of Physics, Engineering and Computer Science, Centre for Engineering Research, University of Hertfordshire, Hatfield, Hertfordshire AL10 9AB, UK
e-mail: s.ismail3@herts.ac.uk

M. T. Hameed Sultan et al. (eds.), *Machining and Machinability of Fiber Reinforced Polymer Composites*, Composites Science and Technology,
https://doi.org/10.1007/978-981-33-4153-1_1

1 Introduction

A minimum of two different materials make up a composite material. These materials are known as fiber and matrix; reinforcement and binder, respectively. They improve performance and reduce production costs of the composite materials. For many decades, fiber-reinforced polymeric (FRP) composite are the right materials for many industrial applications, such as automotive and aviation components, marine as well as oil and gas industries. The wide application of FRP composite materials depends on their outstanding mechanical properties (Callister 1999). However, drilling operation remains unavoidable for FRP composite components in the process of completing assembly. Mechanical couplings for joining FRP components with screws and pins are commonly used in various industrial applications. Whenever holes are needed for assembling of FRP components, drilling operation is unavoidably necessary.

Fibers are harder to break or drill than resin and the interfacial bond or strength within the fiber and the resin is less than the strength of the reinforcing fibers, according to Zitoune and Collombet (2007). Therefore, in FRP composite drilling, delamination damage is the most critical and rampant, followed by fiber pull-out among several drilling-induced damage (DID), as reported (Zemann et al. 2015). Other DID include surface roughness, de-bonding, fiber uncut, cracking, matrix melting and cratering, among others. These damage decrease the bearing capacity of FRP composite parts and shorten the service life of the assembled parts. DID are difficult to repair. Also, they lead to poor quality of holes, which accounts for 60% of all rejected components, mainly caused by delamination. The rejects and refusal of these parts are very expensive (Tsao and Hocheng 2005). DID lead to severe reduction in properties of FRP composites, especially on mechanical behaviours and load-carrying capacity of the composite parts.

Moreover, the aforementioned DID on FRP composite materials have been attributed to many factors. These factors include, but are not limited to, chosen drilling process parameters, the nature of the FRP composite samples and importantly the drill geometry design (DGD). Drill is a complex cutting tool that rotates when creating hole on a material or producing chips by shear deformation. It contains both cutting lip(s) and flutes. Flutes support the removal of chips and passage of cutting fluid (Ismail et al. 2017a). It is a mostly used cutting tool (Ismail and Dhakal 2017). The geometries of drills are complex, as shown in Fig. 1.

Today, there are different drills available in the drilling world. These include hole saw, core, digger, step, twist and special, to mention but a few. They are used to perform different drilling purposes. Twist drill is the most common type and used. A nearly 250 million twist drills are consumed by American manufacturing industries yearly (DeGarmo et al. 2003). The GDD greatly determines the quality of drilled holes. A well-designed drill produces a less damaged component. This is because both the concentrated and uniformly distributed loads depend on the DGD, as reported (Ismail et al. 2017b; Ojo et al. 2017). The concentrated and uniformly distributed loads occur at initially entrance of the drill and during maximum engagement of drill

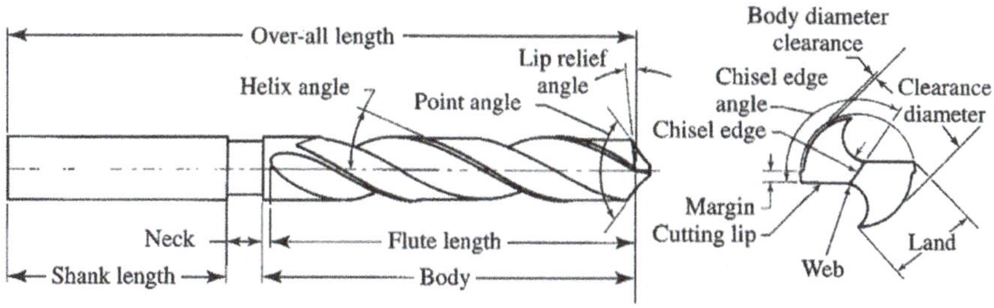

Fig. 1 Parts of a typical drill bit, showing complexity of its geometry design (Ismail et al. 2017a)

in the FRP composite materials during drilling operation, respectively. These two forces contribute to the type of damage on the FRP composite materials.

For instance, the values of point, helix and lip relief angles determine the amount of torque and thrust force developed from the cutting force and feed rate, respectively during drilling operation. A strong connection exists between the feed rate and thrust force. Precisely, thrust force depends on feed rate. Feed rate often increases with thrust forces during drilling process. It has been reported that the major factor that responsible for the occurrence of delamination was thrust force (Ojo et al. 2017; Ismail et al. 2017a, b). Delamination damage occurs whenever the drilling force exceeds the threshold value. It is rampant in multi-layer or hybrid FRP composite laminates, as illustrated in Fig. 2.

A further elaboration coupled with an extensive discussion on drilling-induce delamination will be provided in the later sub-chapters.

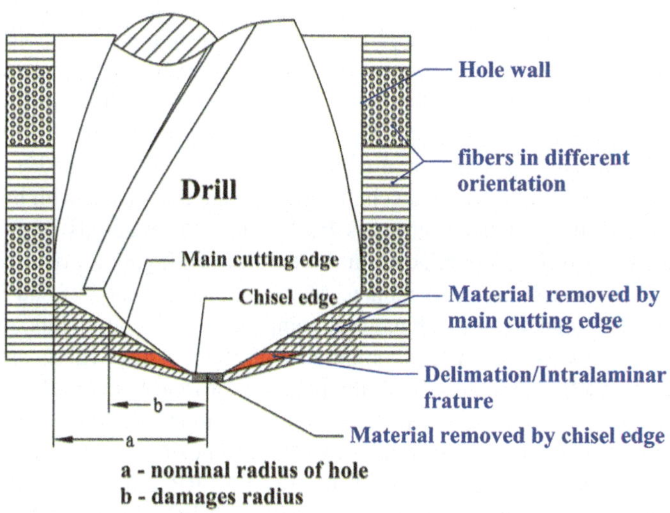

Fig. 2 Diagram of removed materials and delamination/inter-laminar fracture damage caused by different cutting edges (Jia et al. 2016)

2 Composite Materials

The making of material structures has evolved from metallic to composite materials. Presently, composite materials are widely used, because of their outstanding physical and mechanical properties (Sardiñas et al. 2006). Composite can be simply defined as a combination of two or more classes of materials to form a single material with an improved property for a required application. Composites are either metal matrix, ceramic matrix or polymer matrix composite type. According to Goni et al. (2000) and Soldani et al. (2011), polymer matrix composites (PMCs) have been extensively used to manufacture numerous engineering components by many companies. These applications are rampant in automobile, marine, power/energy, telecommunication, security, military, sport/game, aerospace industries, to mention but a few. The widely use of composite materials can be attributed to their lightweight, high strength, exceptional corrosion and acoustic properties, among other outstanding properties, especially when compared with several metallic and alloy materials.

Furthermore, composite materials provide an extensive range of potential automotive applications. Composite automotive parts include, but are not limited to, intake manifolds, door linings, brakes, dash boards, interior finish, suspension, bumper systems, bonnets, steering, wheel covers and body panels. For instance, the structural application of carbon FRP composites proved better than many engineering materials (Das 2001), due to their supreme strength-to-weight ratio. Moving forward, the concept of light weighting is essential for high performance in automobile racing. The recent advances in design and development of various innovative composite materials has increased their applications in automotive industries. Nowadays, components of various cars and aircrafts are products of various types of composite materials. Therefore, both cars and aircrafts become lighter. Consequently, the rate of fuel consumption is reduced, making application of composite materials more economical when compared with other conventional and older engineering materials. It has been reported that composites offer 50–70% reduction in automotive mass, when compared with the use of steel (25–35%) and aluminum (40–55%), with specific strength 2–6 times higher than that of metals (Basavarajappa et al. 2008).

Notwithstanding, there are many challenges that are associated with both manufacturing and drilling of heterogeneous FRP composite materials. Soldani et al. (2011) stated that research conducted in the field of drilling is not only expensive, also it requires enough time and money. Also, it has a tendency of causing fewer dangers against human health with natural (hemp, jute, sisal, cotton, flax, to mention but a few) FRP composites, when compared with the synthetic (glass and carbon, among others) FRP composites. With the presence of long fibers in the eliminated chip of some FRP composite materials, machining of long FRP composites must be wisely conducted. Industrial machining of FRP composites attracts some complications, due to the presence of long fibers in some FRP composite materials and hence resulting to the risk of damage on both tools (drills) and materials/workpiece (composites) during the drilling process (Mkaddem and El Mansori 2009). Therefore, the drilling parameters, process design and drill geometry must be properly selected for a particular FRP composite in a bid to obtain an optimised drilling operation.

3 Drilling, Drill Geometry and Types

Drilling process is one of the hole-creating operations that are widely used in many machining industries. Among the manufacturing processes, drilling is a frequently used process (Sanjay 2006). Drills are solid, rigid, multi-point rotary cutting tools with a range of cutting angles, edges and various shanks, as earlier depicted in Fig. 1.

There are several drill geometries (Fig. 3). Each of these drills is used for various drilling operations. It depends on the nature of FRP composite materials, desired quality of drilled holes, chosen drilling parameters and conditions.

Drills of various geometries have been designed, manufactured and used to create holes of different diameters on numerous materials, including FRP composites. These drills include, but are not limited to, twist drill, one-shot, brad-centre, reamer, step, among others, as listed in Fig. 3 and subsequently discussed.

Several twist drills have been used in many decades ago. They were designed to accommodate a few specific point geometries. Therefore, there are limitations to the extent they could be used to obtain damage-free drilling. They were originally designed for a specific application; drilling of metals and few alloys that were available. These conventional DGDs include, but are not limited to, planar, cylindrical, conical, ellipsoidal or hyperboloidal. Both inputs and outputs of a drilling process depend on the geometry of a drilling tools, commonly referred to as drills. Therefore, a geometric modelling of a drill plays an important role in its design.

In addition, drilling performance is a function of the drill geometry shape as well as dimensions. They both determine the drilling performance: drill wear, drilling forces (thrust and torque), drilling dynamics and resultantly, the quality if the produced holes (Xiong et al. 2009). This makes the DGD attracts a great importance and uncompromised consideration. An inefficiently designed drill usually produces a poor distribution of forces along the cutting plane, resulting into bad performance and undesirable cutting ability. It therefore increases the machining and/or manufacturing costs. A loss of cutting ability increases drill wear rate, which adversely increases the energy consumption, as rubbing effect replaces an efficient cutting or drilling. Both drilling thrust (along the drilling axis) and torque are the two main drill performance characteristics. These characteristics are determined by the drill point geometry (Paul et al. 2005).

3.1 Twist Drill

Twist drill is the widest used drilling tool for many drilling processes. They possess several parts (angles, lips, edges, grooves, plains and lands) that work together to

Fig. 3 Various drill geometries used for carbon FRP composite drilling (Feito et al. 2016)

achieve drilling operation (Fig. 1). The complexity of drill bit geometries, such as twist drill needs careful consideration during design and choice for operation. They are different types of twist drills. They are used according to their appropriateness for a particular drilling activity. Twist drills can be classified based on the shapes of their shanks (straight, square, taper and bit), flutes (double/two, three, four and straight), angles (helix, clearance and point of different angles) and land (sub-land), to mention but a few. Twist drills are also categorised based on the material used: high speed steel (HSS), cobalt, carbide, diamond, tungsten carbide (WC), polycrystalline diamond (PCD), among others. Also, there are coated (titanium and tin) and uncoated twist drills. These names are combined to give drill a descriptive label, such as diamond-coated carbide twist drill, titanium coated HSS twist drill, among others. Drilling dynamics and twist drill wear depend on its size and shape. Machine tools require power, but much when cutting tools (including drills) are not well-designed (Ismail et al. 2017a). Therefore, one of the effective ways to prevent rapid wear of twist drill wear as well as increase drilling efficiency is by selecting clearly defined geometry of the primary cutting edge. In a bid to design a new drill for a best performance, its flute profiles have been generally designed to combine simulation analysis for forward and backward manoeuvres (Fetecau et al. 2009).

3.2 One-Shot Drill

One-shot drill possesses a primary or main, secondary cutting edges and two points, produced by both the main and secondary cutting edges. The structure of one-shot drill expands the drilling process in stages, as described in Fig. 4 by Jia et al. (2016). Stages 1–4 show how long the edge of the chisel pushes the last layer of the FRP before drilling takes place. After that, the removal of the remaining material at stage 2 is carried out by the primary cutting edge. Stage 3 then reveals a secondary cutting edge with a small angle associated with material removal, and finally, the secondary

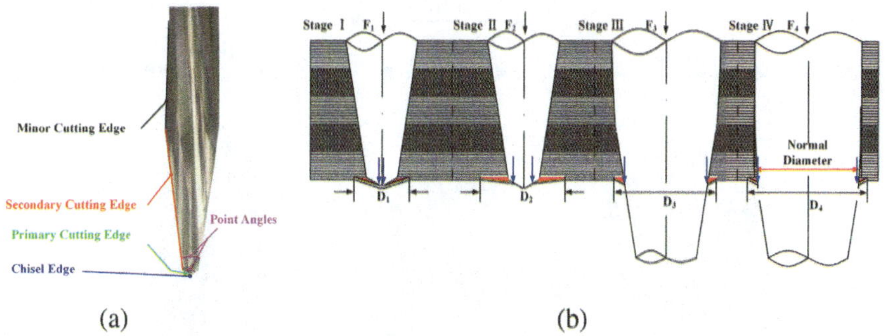

Fig. 4 **a** Typical one-shot drill, depicting its parts and **b** four stages of its drilling process (Jia et al. 2016)

cutting edge gradually penetrates and the cutting edge begins milling close to the end of the drilling operation, at final step 4. To reduce the impact of DID at exit point of the drill, the direction of cutting in the last stage 4 is expected to be reversed towards the drilling point. A periodic saw-tooth structure the drill cutting edges to achieve the purpose of reversing the cutting direction was then proposed.

3.3 Brad-Centre, Reamer and Step Drills

Feito et al. (2015) comparatively investigated the effect of three uncoated special 6 mm drill geometries (Fig. 5) on drilling of woven carbon FRP composite materials, using different cutting conditions.

The results depicted that the two classes of delamination increased with feed rates for both step and brad drills, while it was nearly constant for the reamer drill, especially with exit delamination almost recorded delamination-free drilling (Fig. 6). The step drill recorded highest entry delamination, followed by brad drill and reamer drill exhibited lowest entry delamination damage. Conversely, brad drill recorded highest exit delamination, followed by step drill, while reamer drill still maintained least delamination damage response. Therefore, an optimal result was obtained with reamer drill geometry. This can be traced to the lowest values of both drilling forces (thrust and torque) during drilling evolution, as detailed in another similar and more comprehensive study (Feito et al. 2016). The results of both developed drilling forces are presented in Fig. 7. These forces determine the possibility of occurrence of both types of delamination, because delamination extent, area or factor often increased mainly with thrust force (Feito et al. 2016; Ojo et al. 2017; Ismail and Dhakal 2017;

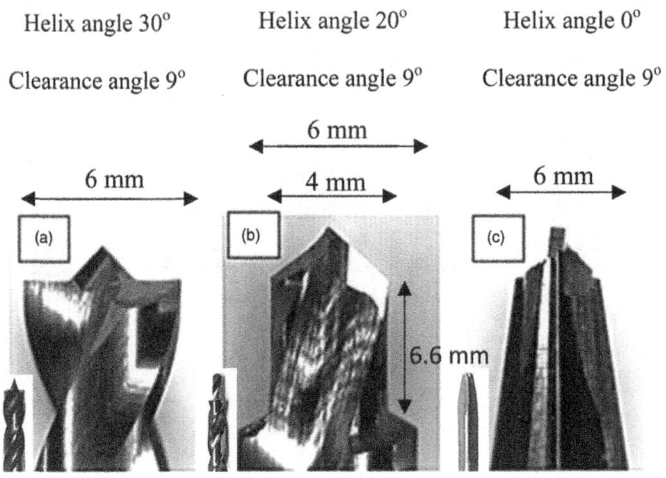

Fig. 5 **a** A brad-centre, **b** step and **c** reamer drill geometries/dimensions (Feito et al. 2015, 2016)

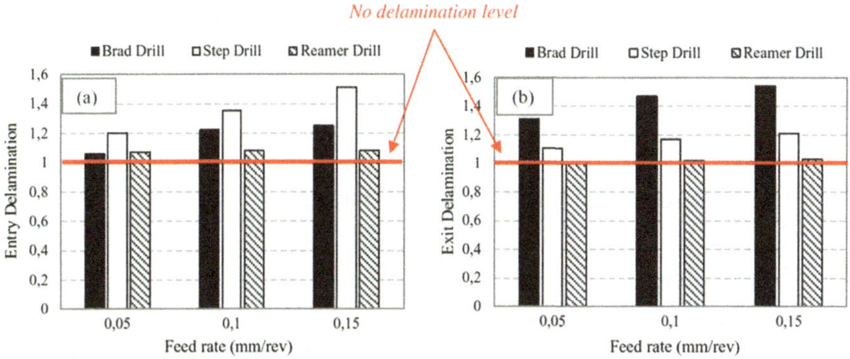

Fig. 6 Influence of drill geometries on both **a** entry/peel-up and **b** exit/push-out delamination (Feito et al. 2015)

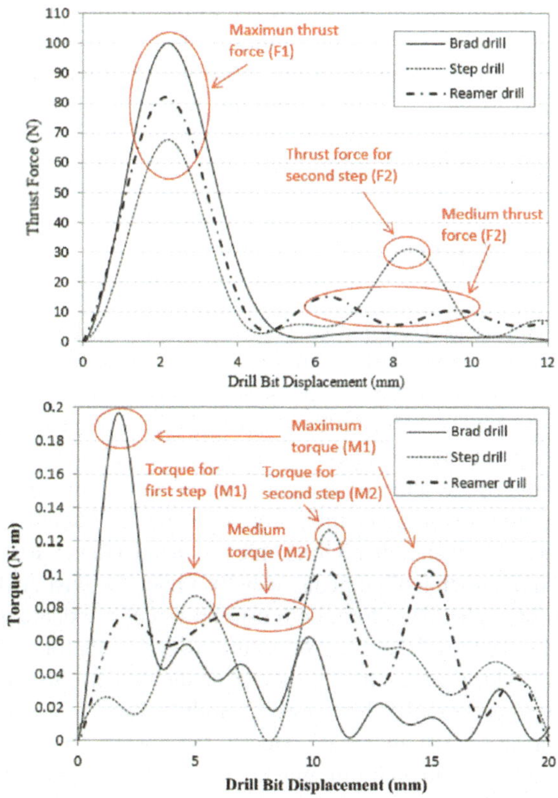

Fig. 7 Effect of drill geometries (brad, step and reamer) on thrust force and torque during drilling of carbon FRP composite materials (Feito et al. 2016)

Ismail et al. 2017a, b). In summary, it was stated that combination of a correct geometry of drill tip and selection of feed rate was an indispensable factor required to decrease drilling-induced delamination.

4 Other Important Drill Geometry Designs

4.1 Diameters

Dharan and Won (2000) stated that an increase in drill diameter led to an increase in both thrust force and torque developed when drilling carbon FRP composite laminates of fiber volume fraction and thickness of 0.63 and 9.9 mm, respectively. These results were similar to that of Zitoune et al. (2010), when drill diameters of 4 and 6 mm recorded lower thrust forces and torques than drill diameter of 8 mm when drilling carbon FRP/aluminium composite. This observation was attributed to higher chisel edge length and steep rise in chip cross-sectional area of a bigger diameter of 8 mm. Conversely, Lee et al. (2008) reported a remarkable increase in thrust force when drill diameter of 8 mm was increased.

4.2 Angles

Chen (1997) recorded a decrease in torque developed when point angle of a twist drill was increased during drilling of carbon FRP composite laminates. This occurred, because an increase in point angle of a twist drill leads to an increase in the drill orthogonal rake angle. This occurred at each point on the main cutting edge. However, the point angle increased with thrust force. Hence, in an attempt to achieve a reduced thrust force and consequently, a decreased delamination damage on carbon FRP composite during drilling process, a choice of smaller point angle is recommended. In addition, a higher helix angle produced a lower drilling forces (thrust and torque). Similar to that of point angle, helix angle of a twist drill increases with its orthogonal rake angle. Therefore, both drilling forces are decreased. In addition, a larger drill chisel edge rake angle produced a lower drilling forces, as reported. A decrease in the drill chisel edge length produced an increase in chisel edge rake angle, consequently, it reduced the drilling forces. Both drilling forces increased when the drill web thickness was increased. Hence, the correct choice of a smaller chisel edge web thickness and web-thinning are required to decrease the thrust force and achieve an optimal drilling process.

In addition, Shyha et al. (2009) compared the effects of two step drill geometries (point angles of 118° and 140°) on thrust force developed during experimental drilling of carbon FRP laminate. It was observed that a lower thrust force was obtained with a higher point angle of 140°. This was attributed to a smaller interaction occurred between the step drill chisel edge and the carbon FRP laminate. Additionally, influences of two helix angles (24° and 30°) of step drills were further investigated. From the results obtained, it was observed that difference in their helix angles attracted insignificant response in terms of DID recorded. Similarly, highest force and minimum delamination damage were recorded with a higher 120° point angle twist drill, when compared with a smaller point angle 85° twist, a special step, a dagger and a brad drills, under the same drilling parameters and conditions (Durão et al. 2010). Conversely, Karnik et al. (2008) stated that drill point angle has a significant influence on drilling-induced delamination. Hence, lower point angles of twist drills produced smaller delamination factor. This implies that thrust force reduced with the lower point angles. This was in agreement with similar results reported by Tsao (2008) and Palanikumar et al. (2008). Tsao (2008) stated that a decrease in twist drill point angle led to a significant increase in the drill relief angle and consequently, it reduced thrust force and its associated delamination damage. And, Palanikumar et al. (2008) studied the effects of three point of angles (85°, 115° and 130°) of twist drills when drilling glass FRP composite laminates. From the experimental results, it was observed that twist drill with the smallest point angle of 85° recorded an optimal results, when compared with others. Consequently, it exhibited a lowest value of drilling-induced delamination response.

5 General Drilling of FRP Composite Materials

FRP composites are quite different from metals. Fundamentally, metals are homogeneous materials and FRP composites are heterogeneous, with isotropic and anisotropic properties, respectively. With the heterogeneous nature of FRP composites due to the dissimilar properties of their constituents, drilling of composite structures has remained a major challenge. This is traced to the different responses of various fibers/fillers (reinforcements) and matrices (binders) under same cutting forces (torque and thrust) and interfacial drill-composite temperature. A combination of more dissimilar constituents in a single composite material increases the possibility and degree of the drilling problem. It requires a more robust drill tool and a better drilling process. For instance, when a carbon reinforced composite needs to be fastened to metal parts of the automobile, it is inevitably required that a hole needs to be produced in the material stack, which comprises the composite and metal. A report from the Abrasive Technology (2001) showed that polycrystalline diamond (PCD)—a stronger material can be used as a cutting tool (drill) material. The presence of diamond in PCD drill increases its resistance against a severe abrasive nature of many synthetic carbon FRP composite materials. Nevertheless, PCD cannot withstand required high cutting forces for metal-matrix composites (MMCs) drilling, particularly with titanium metal.

More also, Veniali et al. (1995) studied the drilling responses of an aramid FRP plastics, a smear occurrence on the drilling tool was reported. The possibility of

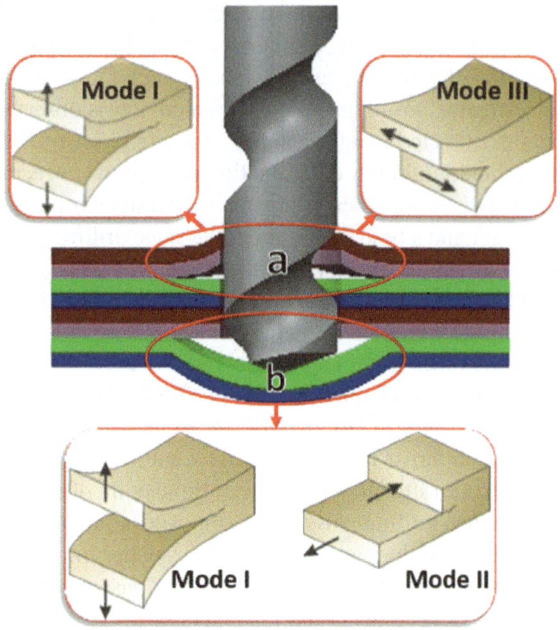

Fig. 8 Delamination-induced damage, showing **a** peel-up or entry and **b** push-out or exit types and their associated modes/mechanisms of deformation (Ojo et al. 2017)

controlling both entry and exit delamination-induced damage (Fig. 8) was later concluded, by having a good relationship between machining/drilling parameters. The entry and exit delamination phenomena are technically referred to as a peel-up and push-out type, respectively. Consequently, this relationship produced forces and torque, which determined the degree of the DID on the FRP composites.

6 Drilling-Induced Delamination Damage

Jain and Yang (1994) modelled the delamination zone of FRP composite laminates, caused by both critical feed rates and thrust forces. Concerning drill geometry designs (DGDs), chisel edge breadth has been reported to be most significant factor which determined the measure of thrust force and therefore delamination damage. To further establish this finding, a diamond-impregnated tubular drill was designed and experimentally tested. The results showed that diamond-impregnated tubular drill tool produced a minimal thrust with a higher hole quality when compared with commonly used twist drills. Further investigation on drill geometries (chisel edge, cutting lip and point angles) and their influences on critical thrust force, above which delamination occurs has been carried out by Ismail et al. (2017b), as illustrated in Fig. 9 and later explained in Sect. 7 (Fig. 17).

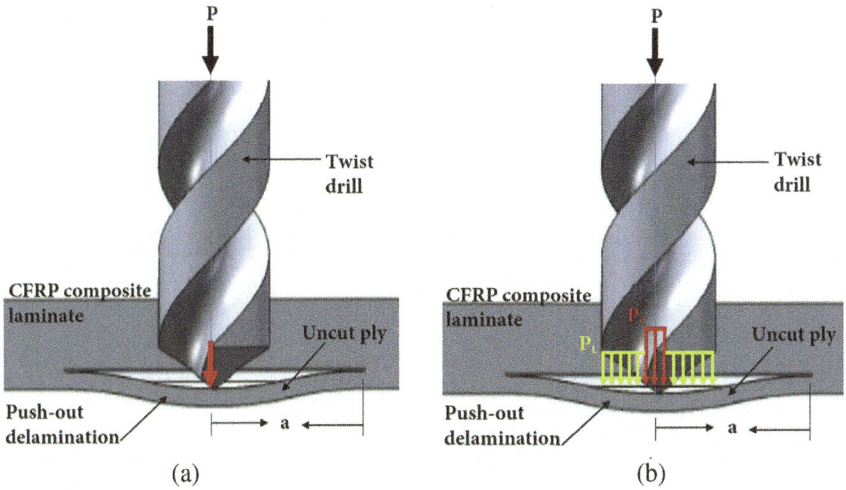

Fig. 9 Delamination phenomenon, caused due to the **a** concentrated load and **b** distributed loads on chisel edge and cutting lips (Ismail et al. 2017b)

Furthermore, Lazar and Paul (2011) experimental analysed drilling of FRP composites to determine the cutting loads distribution along thickness of the workpiece and drill radius. The loads were axially and tangentially distributed. Three different drills were used to analyse the thrust and torque curves. An occurrence of maximum loads on the plies in contact with the drill tip was observed. The initiation of an exit delamination was attributed to the period when the drill tip exited the workpiece. However, cases where the initial interlaminar crack extended outside the limits of the future hole to produce a quantifiable delamination or an occurrence further propagation as the tool continued its movement towards leaving the workpiece was not studied. Capello (2004) hypothesised two main variances in the pattern of delamination. Based on this analysis and in a bid to counter the hypothesised mode of delamination, a new simple prototyped device was designed and built. The effectiveness of this device was verified. From the results obtained, the possibility of significantly decrease of delamination through the proposed device was observed and established.

Additionally, a detailed analysis of influences of some drill geometries, namely; saw, candle stick, core and step drills on delamination has been analytically reported (Hocheng and Tsao 2003), as shown in Figs. 10a–d, respectively, and their results were compared with a standard conventional twist drill (Fig. 10e). Various critical thrust forces of all the drill types were predicted. This study on occurrence of delamination is necessary to establish drilling parameters and drill geometry governing delamination, predict and establish a suitable and an optimal relationship. From the analytical results obtained, the physical influences of various drill geometry were mathematically represented. Therefore, it was concluded that critical thrust forces of the various drill types can be reduced. When ratio between radius, c of saw drill

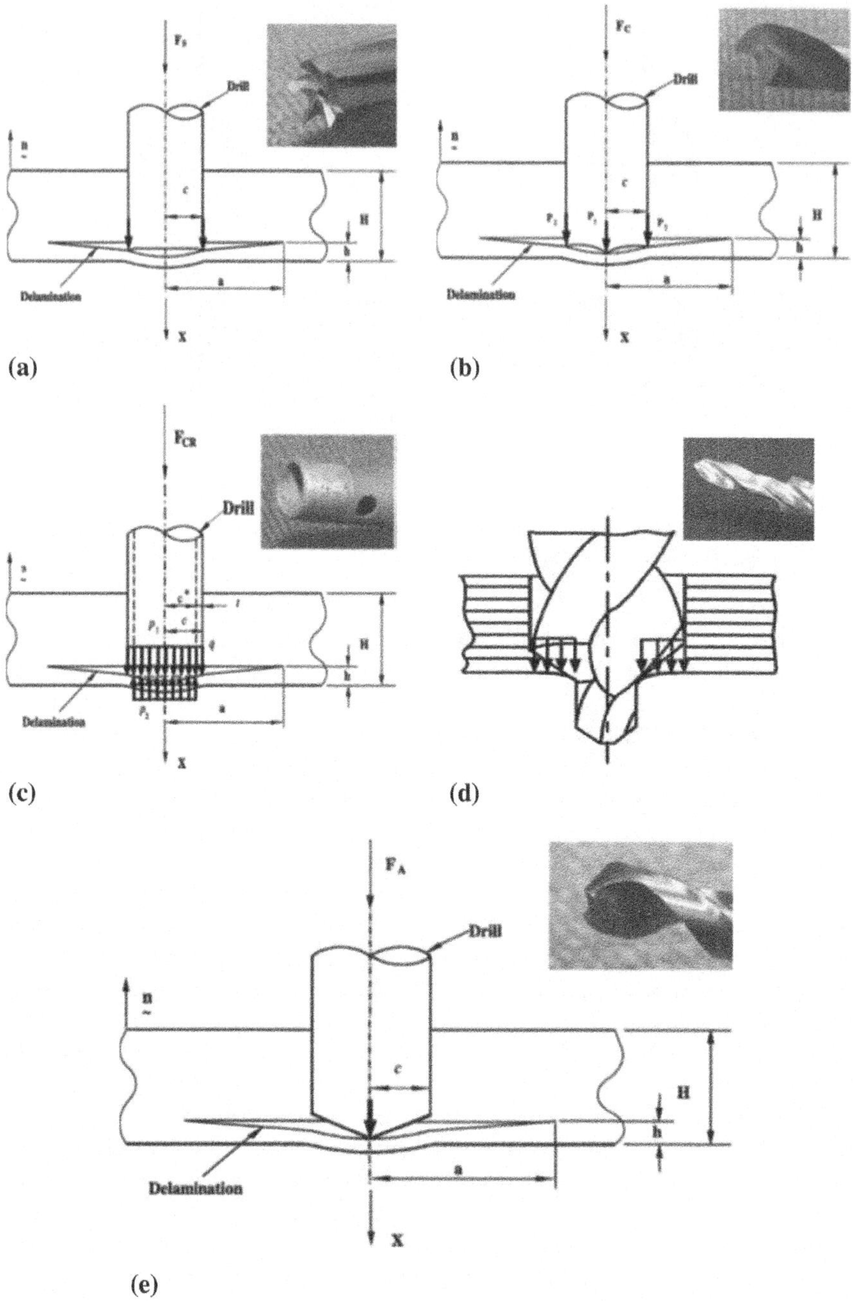

Fig. 10 Analysis of push-out delamination in **a** saw, **b** candle stick, **c** core, **d** step drills and **e** twist drill types (Hocheng and Tsao 2003)

and the extent of the elliptical delaminated region, a along the major or longer axis equals zero ($s = \frac{c}{a} = 0$), saw drill reduces to the twist drill (Fig. 10a). A very high critical thrust force is recorded whenever s approaches 1. Similarly, when ratio between peripheral circular force, P_2 and central concentrated force, P_1 equals zero ($\alpha = \frac{P_2}{P_1} = 0$), candle stick drill reduces to the twist drill (Fig. 10b), but candle stick drill reduces to the saw drill when α equals infinity ($\alpha = \frac{P_2}{P_1} = \infty$). In addition, when s equals zero ($s = \frac{c}{a} = 0$) and ratio between thickness, t and radius, c of core drill equals 1 ($\beta = \frac{t}{c} = 1$), core drill (Fig. 10c) reduces to the twist drill. Lastly, step drill (Fig. 10d) reduces to the twist drill (Fig. 10e) when $i = \xi = 0$ and $s = 0$, where $i = 1 - n$, n represents the number of consecutive increase of secondary cutting lips during drilling operation and ξ denotes the separated thickness of the push-out delaminated laminar traced to value of n. However, step drill behaves like core drill type when $\beta = \frac{t}{c} = 1$ and $i = \xi = 0$. These were similarly reported in another study carried out by the same researchers (Hocheng and Tsao 2006).

Fernandes and Cook (2006) studied the effects of a ‘one shot’ drill on force and torque developed during drilling of carbon FRP composite material. Also, the influences of tool/drill wear and workpiece thickness on both thrust force and torque were investigated. The drilling process was categorised into various phases and related to associate DID responses. From the results obtained, it was evident that a five-step process with a subsequent step related to various drilling and reaming processes physically from a drilling operation with one-shot drill. An extension and prediction of the tool life as well as enhancement of productivity and quality of drilled holes are established with the empirical model. Similarly, Tsao and Hocheng (2003) investigated into the possibility of achieving a delamination-free drilling, using an analytical method to establish a relationship between chisel edge length and drill diameter. The formulation was based a linear elastic fracture mechanics (LEFM) of FRP composites. A set of holes was obtained, and their optimum set of preliminary hole diameters related to the chisel edge lengths were computed. It was established that drilling of FRP composite laminates with medium to large holes and void of delamination-induced damage are achievable, provided the ratio of the drill chisel edge length is effectively controlled.

7 The Influence of Tool Geometry

The influence of tool geometry determines the shape and angle of the projected part of the cutting tool. It affects the type of machining process, materials, efficiency and quality savings of finished parts and tool life.

The factory machining process attracts inclined cutting. Nevertheless, orthogonal cutting has been predominant in the simulation environment of numerical machining studies, because of its ease and usefulness to acquire data about challenging variable measurements. Soldani et al. (2011) supported this notion during experimental studies, stated that an interpretation of results is often proved to be difficult, due to the complexity that may arise from a cutting process as a result of anisotropic nature of the composite materials. Despite this limitation, it is important to further draw necessary information about the morphology of the chip, sub-surface damage, the effect of the orientation of the fibers and the role of the shape (geometry) of the tool/drill.

Therefore, Nayak et al. (2005) studied additional development on orthogonal cutting of FRP composite materials and extensively worked on the effects of cutting-edge radius, fiber orientation, slope angle and depth of cut. Accurate validation of several numerical models in many scientific publications have been successful using the experimental results obtained from their work.

7.1 Effect of Drill Bit Geometry

The geometry of drill significantly determines the quality and integrity of the holes produced on FRP composite materials during drilling operation. Some researchers have undertaken some studies on the effects of drill geometry on DID responses when cutting FRP composites. For instance, Niketh and Samuel (2016) initiated the surface finishing of a drill flute and edge in a sustainable titanium alloy machining to decrease both friction and torque. Feito et al. (2018) illustrated through experimental investigation that drilling with a step drill produced a lower thrust force and de-bonding or separation in carbon FRP composite materials. Girot and Géhin (2002) found that the shape of a drill decreased the surface temperature of the drill bit used and the adhesion of aluminum during dry drilling of Al2024 alloys. This results may not be too far from the drilling of abrasive carbon FRP composite materials. A one-time shot drill that possessed a continuous saw tooth structure has been adopted to alter the drilling conditions of holes and reduce DID impact on the material used (Jia et al. 2016).

Moving forward, the point geometry of a twist drill has been successfully optimised by Paul et al. (2005). The aim of the study was to minimise drilling forces (thrust and torque) during drilling process. A set of drill grinding parameters obtained from a point geometry parameterisation of three different drills. The drill point geometries included conical, helical and racon. The essential properties of their profiles were maintained. Comparatively, these various kinds of drill profiles recorded the optimal results. Therefore, optimised conical point profiles were manufactured and used to perform experiments for the purpose of obtaining data for validation of the already developed analytical models for drill designed geometry. The simulation results established a conical point as an optimal drill design, as a noticeable decrease in both drilling forces was observed. A decrease of nearly 40% in each of the drilling forces. Additionally, a significant decrease in thrust force was recorded with racon drill, but with a minimal enhancement in torque. The optimised helical drill geometry exhibited a great enhancement above the threshold helical point geometry by decreasing each of the drilling forces above 40%.

Furthermore, modification of a drill point design with plane rake faces has been investigated (Wang and Zhang 2008). Fundamental chisel edge and the lip geometries were modified in an attempt to design a new twist drill. Later, the developed models were validated with two sets of experiments to determine the drilling forces (thrust and torque) as well as the drill-life. From the results obtained, it was evident that up to 46.9% and 13.2% of thrust force and torque can be reduced with an average of 23.8% and and maximum of 24.9%, respectively, when compared with the conventional or traditional twist drills, considering same drilling parameters. The modified drills were more superior to the traditional twist drills, as obtained from their drill-life tests. A better performance of the modified drills was observed, because the workpiece contained pieces of some broken conventional drills, under same drilling parameters and conditions.

Effects of two different geometrically designed drills and cutting parameters on power, delamination and specific cutting pressure have been studied (Davim and Reis 2003). Both helical flute-straight shank and brad-spur drills with diameter of 5 mm each were used experimentally and statistically to study drilling of carbon fiber reinforced plastics. The woven and 0°/90° oriented carbon FRP composite laminates were manufactured by autoclaves. The results from the experimental investigation showed that feed rate and cutting velocity had a greatest influence on the delamination of the material at entrance and exit, respectively for both drill types physically and statistically. The entrance/peel-up delamination was higher than the exit/push-out type for both drill types. However, helical flute-straight shank drill exhibited greater peel-up and push-out delamination-induced damage on the carbon FRP composite than the brad-spur drill. In addition, the helical flute-straight shank drill produced lower specific cutting pressure and power than the brad-spur drill, under the same feed rate and cutting speed, as considered cutting parameters. Both drills established that feed rate has a greater influence on the power than cutting speed, both experimentally and statistically. Grilo et al. (2013) conducted a related study, but with different drill tools: spur, four-flute and helicoidal drills, as shown in Fig. 11, respectively. From the results obtained, spur drill produced an optimal and delamination-free drilling, with a greater production rate as well as feed rate and spindle speed of 2025 mm/min and 6750 rpm, respectively.

7.2 Further Effects of Drill Geometry on Carbon FRP Composite Drilling

An efficient drilling of carbon FRP composites stacked with aluminium has been investigated and reported by Garrick (2007), using polycrystalline diamond (PCD) drill. Though, after 200 holes, a wear land was observed on the cutting edge of the drill, leaving it with an option or re-sharpening. Also, the operating range of tungsten carbide was observed bigger than the PCD drill. This could be solved with an

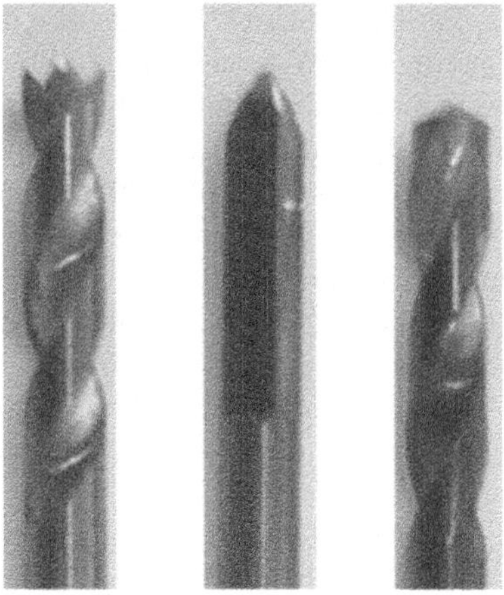

Fig. 11 Different types of drill geometries: **a** spur, **b** four-flute and **c** helicoidal (Grilo et al. 2013)

improved cutting edge by modifying it with k-land. This report did not assess delamination in the carbon reinforced composite and other DID responses. Hence, this report was not a holistic assessment of the new 86 series vein drill. This assessment is necessary to design and optimise the new drill for a better performance. However, Park et al. (2011) took over this challenge and assessed the drilling process of carbon FRP/titanium composite, using PCD drill. Both confocal and scanning electrons microscopes (SEM) were used to monitor the wear mechanism and progression on the drill surface. The performance of the drill was decreased, due to the formation of a micro-chipping or fracture on the drill cutting edges, close to its margin. In addition, the brittle nature of the PCD drill attracted a major chipping on its cutting edges during drilling of titanium. In comparison, PCD recorded a better performance than the tungsten drill type.

Gaitonde et al. (2008) suggested combination of low feed rate values and point angle to reduce DID response on carbon fiber reinforced plastic composites. Importantly, there is need to study the delamination process together with the drilling process in order to come up with a more detailed relationship that will help to understand the delamination mechanism in relationship to the drilling parameters and process. Moreover, the drill wear process is an important variable that could be used to improve the performance of the drill. In addition, the use of other drills apart from the twist drill is recommended.

Tsao (2008) used three different step-core drills: step-core-twist, step-core-saw and step-core-candlestick types (Fig. 12) to report the influence of drilling parameters (diameter ratio, feed rate and spindle speed) on drilling-induced delamination of composite materials. The result showed that the three drilling parameters have a significant effect on drilling-induced delamination. The optimal result of combination of drilling parameters was observed with all the step-core drill types, using a higher diameter ratio, lowest feed rate and highest spindle speed of 0.74 mm/mm, 8 mm/min

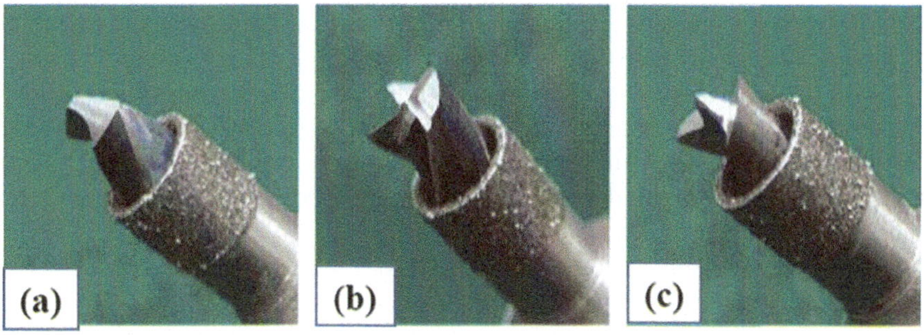

Fig. 12 Step-core drills, showing **a** twist, **b** saw and **c** candlestick types (Tsao 2008)

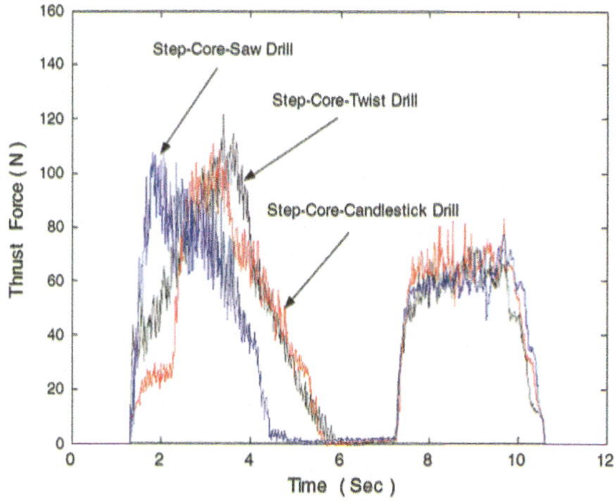

Fig. 13 Drilling evolution with different step-core drill types, showing thrust force–time relationship at diameter ratio, feed rate and spindle speed of 0.74 mm/mm, 16 mm/min and 800 rpm, respectively (Tsao 2008)

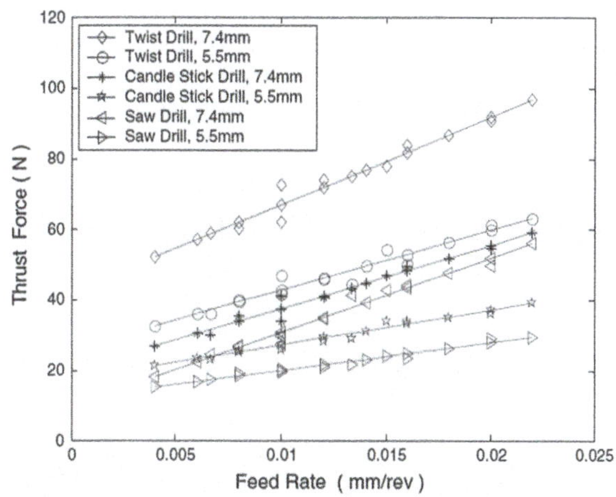

Fig. 14 Effects of the drill diameters and feed rates on thrust forces developed by different step-core drill types (Tsao 2008)

and 1200 rpm, respectively, as depicted in Figs. 13 and 14. These results were similar to that of Ismail et al. (2016).

These optimal results can be attributed to the following reasons: a lowest feed rate developed a lowest thrust force, which consequently produced a lowest value of delamination response. While, combination of a higher diameter ratio and highest spindle speed probably produced a highest cutting speed that was required for an optimum drilling phenomenon.

8 Numerical Simulation and Analytical Analysis

In addition to the aforementioned and discussed simulation and analytical results of DGD, Miao et al. (2009) studied the influences of drill design geometry on the deposition (interface) in a set of solid diamond-coated carbide twist drills, as shown in Fig. 15. They were modelled as a two-fluted drill, using a commercially available computer-aided design (CAD) software. The normal and residual stresses developed as a result of the mismatched thermal expansion coefficients during deposition process was simulated with aid of a finite element analysis (FEA), as illustrated in Fig. 16. The model generated was used to design different drill geometries and determine the deposition stresses.

In addition, it was concluded that the model was capable of design various drills with various geometric parameters. Also, the edge radius recorded the highest significant effects on the interface stresses, among the micro-level drill geometry. Other macro-level drill geometries, include point angle, helix and web thickness have tendency of influencing the web angle located at the drill cutting tip from 10° to 20° variances. But, with minor influence on the interface residual magnitudes. This research considered the twist drill with diamond-coated carbide and did not take into consideration the wear process of the tool as well as drilling process of material workpiece, especially carbon FRP composite materials. Although, the influences of coating thickness on interface residual stresses as well as induced mechanical

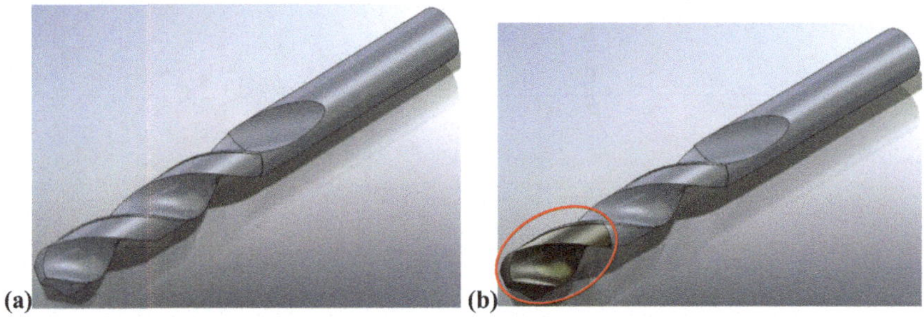

Fig. 15 Typical twist drill, showing **a** a solid CAD model and **b** diamond coated head (Miao et al. 2009)

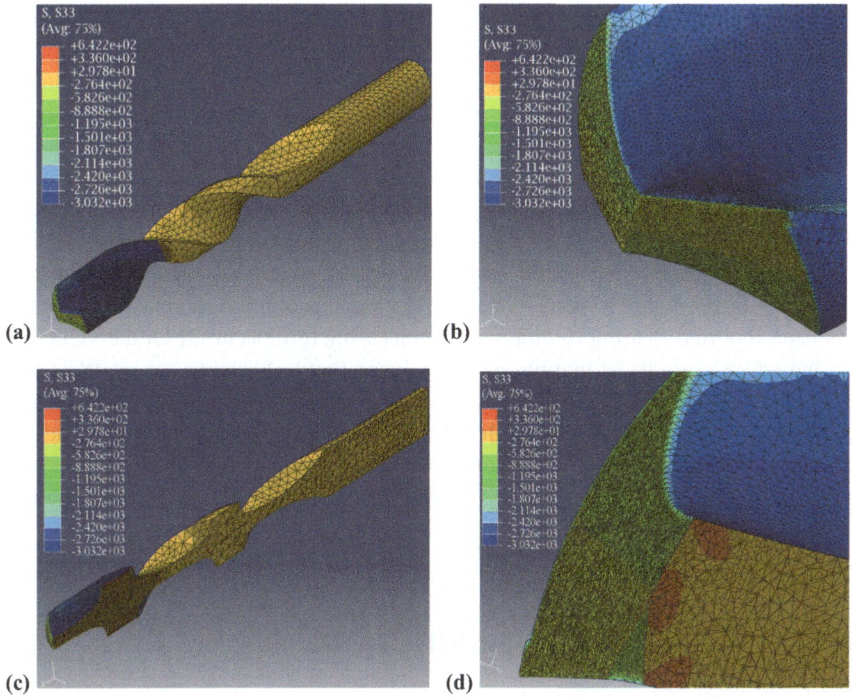

Fig. 16 Longitudinal normal stress distribution in a diamond-coated carbide twist drill, showing **a** a complete simulation, **b** a near drill tip, **c** and **d** their sectional views, respectively (Miao et al. 2009)

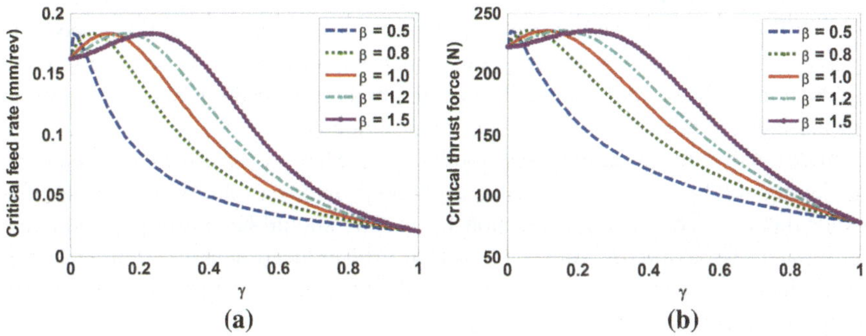

Fig. 17 Influence of chisel edge load on **a** critical feed rate and **b** lowest critical thrust force for various point angles (Ismail et al. 2017b)

and thermal loads on the deposition residual stress during drilling operation were proposed and recommended for future works.

Moving forward, Ismail et al. (2017b) analytically analysed the influenced of drill chisel edge and point angle ratios on the lowest critical thrust force. It was reported that carbon FRP composite laminates can be drilled without delamination damage through a combination of correct selection and effective monitoring of the drill geometrical parameters; with higher feed rate and thrust force as well as reduced chisel edge and point angle. These are illustrated in Figs. 17a, b, where γ denotes chisel edge ratio and β represents ratio between arbitrary and reference point angles.

The above extensive reviews showed that considerable research have been conducted to comprehend the relationship among DGD, delamination damage and the drilling process. However, little or no work has been done to understand and establish models of relationship among drilling mechanisms (drilling variables), drilling-induced delamination, DGD and the wear process. Moreover, although it has been reported that micro-chipping or micro-fracture occurred on the PCD drill cutting edges close to its margin, which reduced the performance of the PCD drill (Park et al. 2011). Currently, there is no reported study to attempt designing a new PCD drill tool (geometry) that will overcome this problem. Therefore, a deeper knowledge towards further optimisation of the drilling process, resistance to wear and delamination mechanism of FRP composites, using PCD drill is highly required, hence proposed or recommended.

9 Recommended Design for Effective Damage Reduction

To reduce the impact of drill-induced damage at exit point of a drill, it is necessary to reverse the cutting direction towards the drilling point. There must be a correct selection of values of drill geometries, such as diameter, angles (point, helix and relief), flutes (two, three or four), shanks (straight, square and taper) and length (web thickness, lip, cutting edge and flute) for a particular drilling of FRP composite. Also, good drill tip geometry and selection of appropriate drilling parameters, especially feed rates are important in reducing delamination. A decrease in feed rate leads to a reduction in thrust force and consequently, both entry/peel-up and exit/push-out drilling-induced delamination are reduced or eliminated.

Additionally, the intermittent saw-tooth structure has proven effective in changing the cutting conditions at the tip of the drill and provides a solution for drilling that avoids damage. The material selection for tooling should be carefully considered. Tool materials have a significant impact on drill-induced delamination and other DID, which have a significant impact on drilling life and structural strength of FRP composite materials.

10 Conclusions and Future Perspectives

10.1 Conclusions

Machining, especially drilling of FRP composite materials remains an inevitable process in a manufacturing industry, where coupling, assembly or mounting of components is necessary. Hence, this chapter has extensively and comprehensively elucidated the effects of various drill geometries or designs on drilling-induced damage, mainly delamination reduction in different FRP composite laminates.

It was evident that drill geometries significantly determined quality of drilled holes in terms of drilling-induce delamination, among other factors such as drilling parameters (feed rate, cutting speed, depth of cut and rate of material removal), tooling materials (HSS, PCD, carbide-coated and uncoated types) and conditions. It was established that delamination increased with drill diameter, because an increased in drill diameter led to an increase in both thrust force and torque developed when drilling carbon FRP composite laminates.

Furthermore, the drill point angle increased with thrust force, because a decrease in twist drill point angle led to a significant increase in the drill relief angle and consequently, it reduced thrust force and its associated delamination damage. Conversely, a higher helix angle produced a lower drilling forces (thrust and torque). While, a larger drill chisel edge rake angle exhibited lower drilling forces. A correct choice of a smaller chisel edge web thickness and web-thinning are both required to decrease the thrust force and achieve an optimal or delamination-free drilling process.

Therefore, the aforementioned results are very relevant to select a correct drill geometry for an efficient and optimal drilling of FRP composite material. These results are highly relevant to all drill designers, developers/manufacturers and researchers, as quest to optimise the drilling of FRP composite materials continues.

10.2 Future Perspectives

Drilling of FRP composite remains an indispensable machining process, provided holes are required for assembly of components of a system. Also, design and development of an innovative FRP composite materials for several applications in various manufacturing industries have increased the need for continuous improvement in the quality of drills. The efficiency of a drill significantly depends on its geometry design. Therefore, design and manufacturing of more novel or special drilling tools to effectively drill both newly and future developed FRP composite, especially natural/hybrid types remain a continuous and increasing exercises.

Furthermore, the use of computer numerical controlled (CNC) drilling centres has reduced some challenges associated with drilling of abrasive, anisotropic and heterogeneous FRP composite materials. For instance, several and different geometries of drills can be loaded into CNC drilling machine to perform various drilling processes within a single operation. Also, Special drills with through-holes at their centres have been designed, developed and utilised through the help of CNC drilling machines. The centre holes accommodate coolants, such as compressed/cold air for an effective cooling during drilling operation. There is a tendency of significant improvement on these technologies in the nearest future, especially with the current advent of smart manufacturing and robots in manufacturing technology (robotic arm drilling).

However, moving from subtractive, traditional or conventional manufacturing to additive manufacturing by some engineering industries may reduce the application of drills in the next decade, as drills are fast replacing with print nozzles in additive layer manufacturing technology.

References

Abrasive Technology (2001) Products for the precision engineering industry, drilling with polycrystalline diamond. https://assets.abrasive-tech.com/literature/pcddrill.pdf

Basavarajappa S, Chandramohan G, Paulo DJ (2008) Some studies on drilling of hybrid metal matrix composites based on Taguchi techniques. J Mater Process Technol 196:332–338

Callister (1999) An introduction, materials science and engineering. New York, Wiley Inc

Capello E (2004) Workpiece damping and its effect on delamination damage in drilling thin composite laminates. J Mater Process Technol 148:186–195

Chen WC (1997) Some experimental investigations in the drilling of carbon fibre reinforced plastic (CFRP) composite laminates. Int J Mach Tools Manuf 37(8):1097–1108

Das S (2001) The cost of automotive polymer composites: a review and assessment of DOE's lightweight materials composites research. Tech Rep. https://doi.org/10.2172/777656

Davim JP, Reis P (2003) Drilling carbon fiber reinforced plastics manufactured by autoclave-experimental and statistical study. Mater Des 24(5):315–324

DeGarmo EP, Black JT, Kohser RA (2003) Materials and processes in manufacturing, 9th edn. New York, Wiley

Dharan CKH, Won MS (2000) Machining parameters for an intelligent machining system for composite laminates. Int J Mach Tools Manuf 40(3):415–426

Durão LMP, Gonçalves DJS, Tavares JMRS, deAlbuquerque VHC, Vieira AA, Marques TA (2010) Drilling tool geometry evaluation for reinforced composite laminates. Compos Struct 92(7):1545–1550

Feito N, Díaz-Álvarez A, Cantero JL, Rodriguez-Millan M, Miguelez H (2016) Experimental analysis of special tool geometries when drilling woven and multidirectional CFRPs. J Reinf Plast Compos 35(1):33–55

Feito N, Díaz-Álvarez J, Cantero JL, Miguelez MH (2015) Influence of special tool geometry in drilling woven CFRPs materials. Procedia Eng 132:632–638

Feito N, Díaz-Álvarez J, López-Puente J, Miguelez MH (2018) Experimental and numerical analysis of step drill bit performance when drilling woven CFRPs. Compos Struct 184:1147–1155

Fernandes M, Cook C (2006) Drilling of carbon composites using a one shot drill bit. Part I: Five stage representation of drilling and factors affecting maximum force and torque. Int J Mach Tools Manuf 46:70–75

Fetecau C, Stan F, Oancea N (2009) Toroidal grinding method for curved cutting-edge twist drills. J Mater Process Technol 209:3460–3468

Gaitonde VN, Karnik SR, Campos Rubio AJ, Correia E, Abrão AM, Davim JP (2008) Analysis of parametric influence on delamination in high-speed drilling of carbon fibre reinforced plastic composites. J Mater Process Technol 203:431–438

Garrick R (2007) Drilling advanced aircraft structures with PCD (Poly-crystalline diamond) drills. SAE Int 2688–3627
Girot F, Géhin D (2002) Dry drilling of aluminum alloys for aeronautics. Mécanique Indus 3(4):301–313
Goni J, Mitxelena I, Coleto J (2000) Development of low-cost metal matrix composites for commercial applications. J Mater Sci Technol 16(7–8):743–746
Grilo TJ, Paulo RMF, Silva CRM, Davim JP (2013) Experimental delamination analyses of CFRPs using different drill geometries. Compos B Eng 45:1344–1350
Hocheng H, Tsao CC (2003) Comprehensive analysis of delamination in drilling of composite materials with various drill bits. J Mater Process Technol 140(1–3):335–339
Hocheng H, Tsao CC (2006) Effects of special drill bits on drilling-induced delamination of composite materials. Int J Mach Tools Manuf 46(12–13):1403–1416
Ismail SO, Dhakal HN, Popov I, Beaugrand J (2016) Comprehensive study on machinability of sustainable and conventional fibre reinforced polymer composites. Eng Sci Technol Int J 19(4):2043–2052
Ismail SO, Dhakal HN (2017) Philosophical study on composites and their drilling techniques. In: Thakur YK, Thakur MK (eds) Functional biopolymers. Springer series on polymer and composite materials, Chapter 9. Switzerland, Springer Nature, pp 239–280
Ismail SO, Dhakal HN, Dimla E, Popov I (2017a) Recent advances in twist drill design for composite machining: a critical review. Proc Inst Mech Eng Part B: J Eng Manuf 231(14):2527–2542
Ismail SO, Ojo SO, Dhakal HN (2017b) Thermo-mechanical modelling of FRP cross-ply composite laminates drilling: delamination damage analysis. Compos B Eng 108:45–52
Jain S, Yang DCH (1994) Delamination-free drilling of composite laminates. Trans ASME J Eng Indus 116:475–481
Jia Z, Fu R, Niu B, Qian B, Bai Y, Wang F (2016) Novel drill structure for damage reduction in drilling CFRP composites. Int J Mach Tools Manuf 110:55–65
Karnik SR, Gaitonde VN, Rubio JC, Correia AE, Abrao AM, Davim JP (2008) Delamination analysis in high speed drilling of carbon fibre reinforced plastics (CFRP) using artificial neural network model. Mater Des 29:1768–1776
Lazar M, Paul X (2011) Experimental analysis of drilling fibre reinforced composites. Int J Mach Tools Manuf 51:937–946
Lee SC, Jeong ST, Park JN, Kim SJ, Cho GJ (2008) Study on drilling characteristics and mechanical properties of CFRP composites. Acta Mech Solida Sin 21(4):364–368
Miao C, Qin F, Sthur G, Chou K, Thompson R (2009) Integrated design and analysis of diamond-coated drills. Comput-Aided Design Appl 6(2):195–205
Mkaddem A, El Mansori M (2009) Finite element analysis when machining UG reinforced PMCs plates: chip formation, crack propagation and induced damage. J Mater Design 30(8):33295–33302
Nayak D, Bhatnagar N, Mahajan P (2005) Machining studies of uni-directional glass fibre reinforced plastic (UD-GFRP) composites part 1: effect of geometrical and process parameters. Mach Sci Technol 9:481–501
Niketh S, Samuel GL (2016) Surface texturing for tribology enhancement and its application on drill tool for the sustainable machining of titanium alloy. J Clean Prod 167:253–270
Ojo SO, Ismail SO, Paggi M, Dhakal HN (2017) A new analytical critical thrust force model for delamination analysis of laminated composites during drilling operation. Compos B Eng 124:207–217
Palanikumar K, Campos Rubio J, Abrao AM, Correia AE, Davim JP (2008) Influence of drill point angle in high speed drilling of glass fiber reinforced plastics. J Compos Mater 42(24):2585–2597
Park K, Beal A, Kim D, Kwon P, Lantrip J (2011) Tool wear in drilling of composite/titanium stacks using carbide and polycrystalline diamond tools. Wear 271:2826–2835
Paul SG, Kapoor RE, Devor (2005) Chisel edge and cutting lip shape optimization for improved twist drill point design. Int J Mach Tools Manuf 45:421–431

Sanjay C (2006) Comparative analysis of various methods of surface roughness in drilling. J Advan Manuf Syst 5(1):75–87
Sardiñas RQ, Reis P, Davim JP (2006) Multi-objective optimization of cutting parameters for drilling laminate composite materials by using genetic algorithms. Compos Sci Technol 66(15):3083–3088
Shyha IS, Aspinwall DK, Soo SL, Bradley S (2009) Drill geometry and operating effects when cutting small diameter holes in CFRP. Int J Mach Tools Manuf 49(12–13):1008–1014
Soldani X, Santiuste C, Muñoz-Sánchez A, Miguélez MH (2011) Influence of tool geometry and numerical parameters when modelling orthogonal cutting of LFRP composites. Compos Part A Appl Sci Manuf 42:1205–1216
Tsao CC (2008) Investigation into the effects of drilling parameters on delamination by various step-core drills. J Mater Process Technol 206:405–411
Tsao CC, Hocheng H (2003) The effects of chisel length and associated pilot hole on delamination when drilling composite materials. Int J Mach Tools Manuf 43:1087–1092
Tsao CC, Hocheng H (2005) Computerized tomography and C-scan for measuring delamination in the drilling of composite materials using various drills. Int J Mach Tools Manuf 45(11):1282–1287
Veniali F, DiLlio A, Tagliaferri V (1995) An experimental study of the drilling of aramid composites. Trans ASME J Energy Res Technol 117:271–278
Wang J, Zhang Q (2008) A study of high-performance plane rake faced twist drills. Part I: Geometrical analysis and experimental investigation. Int J Mach Tools Manuf 48, 1276–1285
Xiong L, Fang N, Shi H (2009) A new methodology for designing a curve-edged twist drill with an arbitrarily given distribution of the cutting angles along the tool cutting edge. Int J Mach Tools Manuf 49:667–677
Zemann R, Sacherl J, Hake W, Bleicher F (2015) New measurement processes to define the quality of machined fibre reinforced polymers. Energy Procedia 100:636–645
Zitoune R, Collombet F (2007) Numerical prediction of the thrust force responsible of delamination during the drilling of the long-fibre composite structures. Compos Part A: Appl Sci Manuf 38(3):858–866
Zitoune R, Krishnaraj V, Collombet F (2010) Study of drilling of composite material and aluminium stack. Compos Struct 92(5):1246–1255

Thrust Force Analyses in Drilling FRP Composites

Tan Chye Lih and Azwan Iskandar Azmi

Abstract Whenever fiber-reinforced polymer (FRP) composites are drilled to produce a hole, it is necessary to obtain the desired dimensional requirements with minimal surface and sub-surface damage. A number of complications arise in this process, such as multiphase laminated materials (hard reinforced fibers within soft polymer matrix) and complex cutting edges of the drill bit. According to existing research, changes in the drilling parameters' setting and tool geometry play a critical role in influencing the thrust force and size of delamination zone. Thus, the workpiece's delamination responses can be minimised by identifying proper drilling parameters and tool geometry to reduce the effect of thrust force on uncut layers. Analytical as well as experimental investigations have explored the thrust force and delamination behaviour during composite drilling, to compute the critical thrust force at the onset of delamination during drilling. All studies cited in this review indicate that delamination damage can be avoided if the applied thrust force is lower than the critical thrust force value. A good agreement between the estimated critical thrust force and the measured thrust force was evident in certain studies. Considering this, the critical thrust force value can be a reliable benchmark or reference for industrial practice in reducing delamination damage for better assembly performance of the drilled FRP composites. Lastly, a general review of new approaches to reduce the drilling thrust force for reaching delamination free drilling for FRP composite laminates has been attempted.

Keywords Critical thrust force · Delamination · Drilling · FRP composites · Thrust force

---

T. C. Lih · A. I. Azmi (✉)
Faculty of Mechanical Engineering Technology, Universiti Malaysia Perlis (UniMAP), Pauh Putra Campus, 02600 Arau, Perlis, Malaysia
e-mail: azwaniskandar@unimap.edu.my

M. T. Hameed Sultan et al. (eds.), *Machining and Machinability of Fiber Reinforced Polymer Composites*, Composites Science and Technology,
https://doi.org/10.1007/978-981-33-4153-1_2

1 Introduction

Composite materials refer to a combination of two or more reinforcements within a matrix, insoluble in one another, to obtain the advantages of both materials while simultaneously mitigating undesired properties. The concept of composites was based on the existence of natural composites, such as wood (cellulose fiber and lignin) and bone (bone tissue and collagen). The oldest composite materials, which combined straw fibers with clay to increase the performance in construction, were found in Ancient Egypt. The fibers mentioned in the earlier definition act as a reinforcing phase, and another phase is embedded in these materials known as the matrix phase. The role of the reinforcement materials is to provide the composite primary mechanical strength and reinforce the matrix in any direction. The matrix holds and redistributes the load to reinforcement as well as provides a favourable environment resistance (Harris 1999). Figure 1 shows the stress–strain relationships for a FRP composite material. The high modulus fiber is strong but brittle, while the low modulus fiber and matrix are soft but ductile. The properties of FRP composites can be tailored by modifying the favourable characteristics of fibers and matrix system, Table 1. Thus, the advanced fiber-reinforced polymer (FRP) composites have received considerable attention as engineering materials because of their particular mechanical and physical properties. This is especially true in structural applications

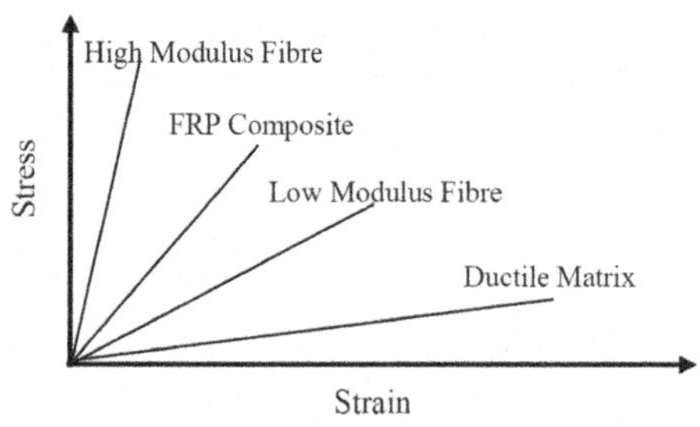

Fig. 1 Stress–strain relationships for hybrid composite material and its constituents

Table 1 Mechanical properties of fiber reinforcements and resin (Azrin Hani et al. 2015; Jawaid and Abdul Khalil 2011; Suppakul and Bandyopadhyay 2002)

No.	fiber/Resin	Density (g/cm^3)	Tensile strength (MPa)	Tensile modulus (GPa)	Elongation (%)
1	E–glass fibers	2.5	1900	69	4.8
2	Carbon fibers	1.61	3530	230	1.5
3	Kevlar (29) fibers	1.45	2900	70	3.6
4	Epoxy resin	1.15	93.15	2.59	5.7
5	Polyphenylene sulphide (PPS)	1.33	65.56	3.09	2.0

nowadays that require highly specific strength, are lightweight, have a better corrosion resistance, and are resilient against environmental attack over metallic materials. Therefore, the development and application of FRP composites is occurring at an increasingly fast pace to replace conventional metallic materials in various application parts. Within the composite materials group, synthetic FRP composites have been widely used in various manufacturing applications, such as aerospace components (e.g. tails, wings and fuselages), transportation (e.g. racing car bodies), military equipment, sporting goods (e.g. bicycle frames and badminton rackets), electronic items (e.g. printed wiring boards), and various other industrial applications.

Generally, the FRP composite materials are well advanced in producing near-net shape composite components by reducing the machining needs to a minimum. However, the secondary/finishing process, particularly drilling operation, is obligatory to aid subsequent assembly requirements in application. The high-performance drilling process with minimal surface and sub-surface damage is required to produce the quality and durability of the composite structures. Unfortunately, FRP composites have poor machinability and their cutting mechanisms have been considered distinct from the cutting behaviour of homogeneous materials such as metals. The several undesirable defects such as fiber pull-out, delamination, fiber matrix debonding, and fiber bridging induced by drilling processes drastically reduce the assembly structure's strength, Fig. 2 (Altin Karataş and Gökkaya 2018). Any major defects arising in the drilled parts may lead to rejection and significant loss to the industry.

Among drilling-induced damages, delamination has been recognised as a critical damage in drilling operation as it can reduce the structural integrity, cause poor assembly tolerance, and hence, can deteriorate their in-service performance. As indicated in previous studies (Geng et al. 2019; Wong et al. 1982), repairing a delamination defect can take 5–6 h, for fastening a hole in the airplane assembly. Therefore, addressing ways to improve delamination damage on drilled FRP composite is imperative. Generally, delamination damage is regarded as an adhesive or cohesive failure at the interface of a laminate. It occurs mainly due to drilling thrust force at the cutting zone exceeds the inter-laminar bond strength of the uncut thickness (Rahme et al. 2015). Apparently, the thrust force developed during the drilling of FRP composites

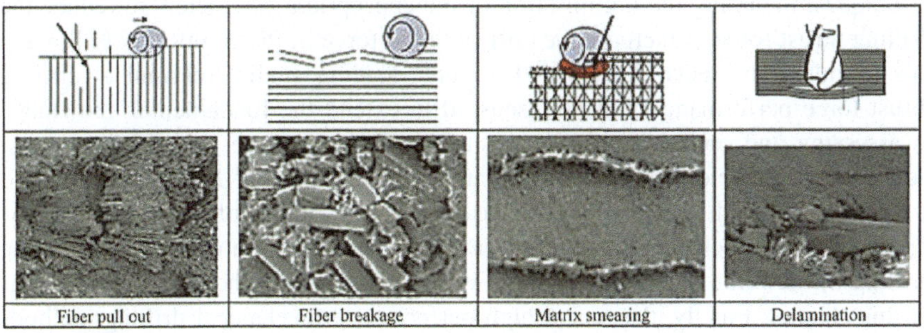

Fig. 2 Typical material damage mechanisms in FRP composites as a result of machining forces (Altin Karataş and Gökkaya 2018)

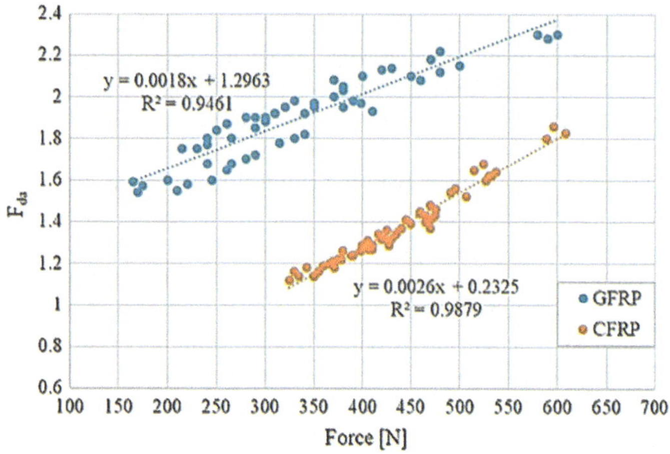

Fig. 3 The relationship of drilling thrust force and delamination factor at hole exit (Sorrentino et al. 2018a, b)

is the key factor that induces serious delamination around the hole, especially the push-out delamination. Figure 3 shows the relationship between drilling thrust force and the push-out delamination factor (Sorrentino et al. 2018a, b). It can be concluded that the size of drilling-induced delamination is highly relevant to the thrust force: higher the thrust force, greater will be the delamination factor. It is believed that a delamination-free drilling process can only be achieved when the thrust force does not exceed the inter-laminar bonding strength of the composite structure, namely "critical thrust force" (Zhenchao et al. 2014).

With regard to this, the majority of published works have reported that the settings of input process parameters, tool geometry, and properties of workpiece material have a greater influence on the thrust force during the drilling process, as shown in Fig. 4 (Abrão et al. 2007; Palanikumar and Muniaraj 2014; Tsao and Hocheng 2008b; Zhenchao et al. 2014). Therefore, in order to improve the structural integrity of drilled holes, many research works have attempted various delamination suppression techniques, including the optimisation of the appropriate parameters to reduce the drilling thrust force or increase the critical thrust force of the composite. Following this introduction, Sects. 2, 3, 4 and 5, some leading studies relevant to drilling thrust force performance will be discussed to extend the fundamental knowledge in assessing and optimizing the controlled drilling parameters for FRP composites. These parameters consist of mechanical performance of FRP composite and a wider range of drilling parameters, as well as the design of drill bit geometry. Then, the mechanics modelling and basics of the analytical models concerning the drilling thrust force of FRP composite will be discussed in the following sections of this chapter. Finally, the recent high performance mechanical drilling methods toward attaining delamination-free composite laminates and the conclusion will be presented.

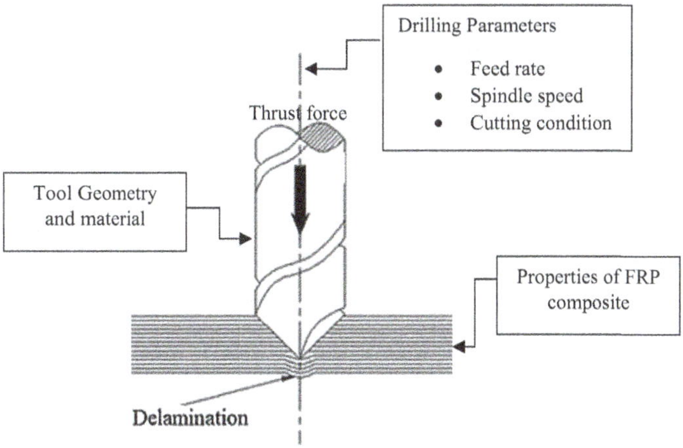

Fig. 4 Effect of input parameters to the thrust force

2 Overview Thrust Force in Drilling of FRP Composite

In the drilling process, the force generation perpendicular to the cutting force (Fc) represents the thrust force (F_A), as shown in Fig. 5. It is the key cutting characteristics activated in FRP drilling which signify the mechanical energy consumption of multi-tool work interactions governing the chip removal process. As mentioned before, the thrust force developed during the drilling of FRP composites has been cited as the key factor in drilling defects such as delamination damage, sub-surface damage, fiber breakage, matrix cracking, and fiber/matrix debonding.

According to Khashaba (2013), the resulting thrust force F_t, is generated along the drill bit axis due to contact with the chisel edge and cutting lips with the laminate composite, Fig. 6. The value of the resulting thrust force during drilling can be measured using a dynamometer. Further, previous studies show (Hocheng and

Fig. 5 Schematic depiction of forces acting on the chip in cutting process

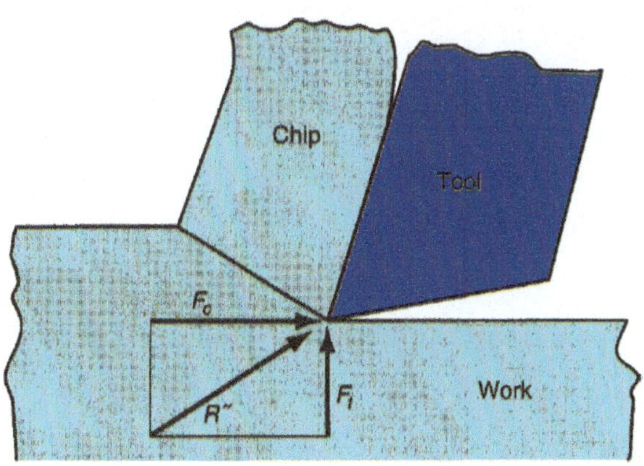

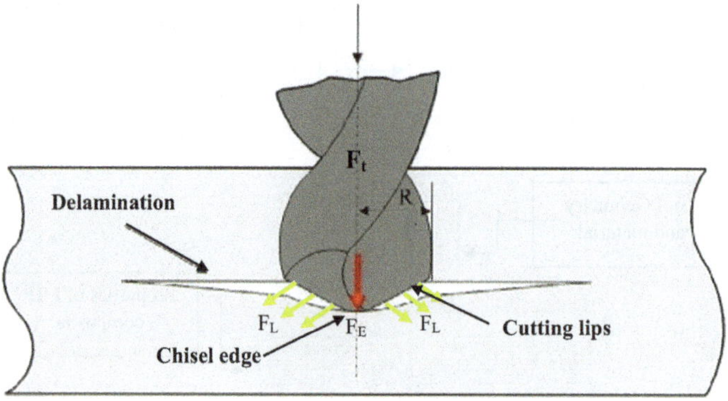

Fig. 6 Decomposition of the resulting thrust force Ft

Dharan 1990; Kim and Lee 2005; Zhenchao et al. 2014) that the concentrated thrust force from chisel edge makes a greater contribution than the distribution force from cutting lips. Therefore, it can be observed that a typical thrust force–time data for drilling carbon FRP composites, as shown in Fig. 7, involves six main stages. Initially (P1–P2), the thrust force increases sharply due to the initial chisel edge of the drill engagement with the laminate composite. This is followed by a maximum force as the chisel edge of the tool drills through the thickness of the laminate, P3: cutting process. Then, during phase 4, the deformation of the uncut material around the

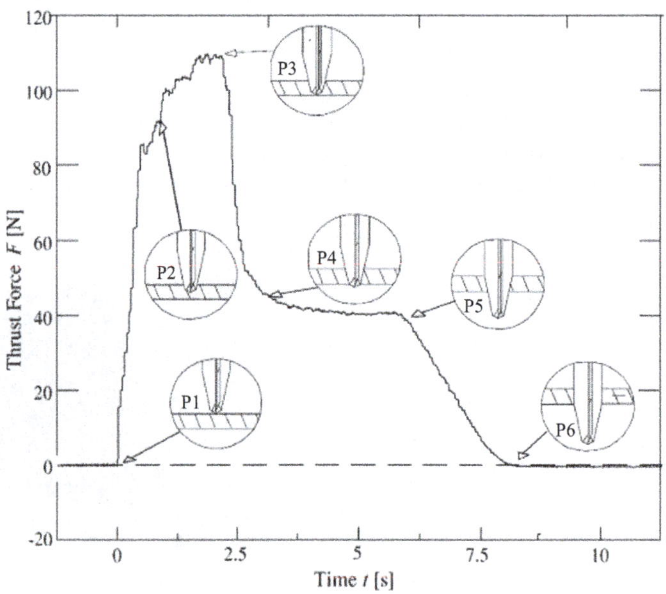

Fig. 7 Typical thrust forces profile during drilling process

drilled surface causes delamination, followed by a sharp reduction in the force as the tool's tip penetrates the laminate's last layer. When the whole chisel edge drills through the laminate, the reduction in force becomes more gradual as shown during phase 5 and 6. Finally, the drill exits the laminate and the force drops to zero (Murphy et al. 2002).

3 Experimental Study of Thrust Force in Drilling FRP Composite

3.1 Effect of Cutting Parameter on Drilling Thrust Force

Hole quality is one of the important criteria when assessing drilling performance because it influences the strength of composite parts post assembly. In this regard, a number of experimental studies have examined the relationship between thrust force and drilling strategy in order to reduce the machining defects and improve the drilling efficiency of FRP composites. The effects of input variables such as feed rate, spindle speed, and tool diameter on the thrust force during drilling FRP composite have been summarised in Table 2 and Fig. 8, a quick comparison of the published literature with respect to the drilling processes of FRP composites. It can be seen from Fig. 8. That all studies cited in this review showed that thrust force increased with feed rate and drill diameter with different settings of spindle speed and composite. This is similar to El-Sonbaty et al. (2004), whose experimental results also indicated that increasing drill diameter (8–13 mm) and feed rate (0.05–0.15 mm/rev) increased the maximum thrust force during the drilling of GFRP composites with HSS twist drill bit. This is likely caused by increased shearing and cross-sectional area of undeformed chip when drill diameter and feed rate are increased. These phenomena lead to the enhancement of the resistance of chip formation and consequently increase the cutting force and torque during drilling process.

Palanikumar and Muniaraj (2014) also investigated the effect of drill diameter and feed rate towards the drilling thrust force on hybrid metal matrix composites. The carbide-coated drill bits used in the dry experiment were of 4–12 mm diameter. In their study, the Taguchi and ANOVA methods were used to determine the desired setting for thrust force responses. The use of high cutting speed, lowest feed rate, and lowest level of drill diameter favoured minimum delamination and thrust force. Furthermore, they also developed an empirical model using response surface methodology (RSM), a combination of mathematical and statistical techniques to predict the thrust force in hybrid metal matrix composites, within the ranges of cutting parameters studied. Based on the empirical model, it can be concluded that the feed rate and drill diameters were found to make the largest contribution to the drilling thrust force and delamination factors.

Subsequently, Heidary et al. (2018) investigated the machining parameters (feed rate, spindle speed, and drill diameter) on drilling thrust force, delamination factor,

Table 2 Summaries of reported literature on effect of machining parameters on drilling of FRP composites

No.	References	Composite	Tool material and geometry	Input variables
Thrust force (F_t) decrease with increasing cutting speed and feed rate on drilling process				
1	El-Sonbaty et al. (2004)	GFRP	Material: HSS Geometry: twist drills Diameter: 8–13 mm	Spindle speed: 218–1850 RPM Feed rate: 0.05–0.23 mm/rev
2	Vimal Sam Singh et al. (2009)	GFRP	Material: Carbine Geometry: twist drill Diameter: 6, 8, 10 mm	Spindle speed: 500–2500 RPM Feed rate: 100–500 mm/min
3	Phadnis et al. (2013)	Cross-Ply CFRP	Material: HSS Geometry: twist drills Diameter: 6 mm	Spindle speed: 260–1700 RPM Feed rate: 50 mm/min
4	Nagaraja et al. (2013)	Woven CFRP	Material: HSS, USC and TiN-SC Geometry: twist drills Diameter: 4, 6, 8 mm	Spindle speed: 1200, 1500, 1800 RPM Feed rate: 10, 15, 20 mm/min
5	Palanikumar and Muniaraj (2014)	Hybrid metal matrix	Material: titanium nitride coated solid carbide Geometry: twist drills Diameter: 4, 8, 12 mm	Spindle speed: 1000–3000 RPM Feed rate: 0.05–0.15 mm/rev
6	Heidary et al. (2018)	GFRP/MWCNT	Material: HSS Geometry: twist drill Diameter: 4, 5 mm	Spindle speed: 315, 630 RPM Feed rate: 0.04, 0.06, 0.08, 0.1 mm/rev
7	Arputhabalan et al. (2019)	Natural hybrid fiber reinforced polymer composite	Material: HSS Geometry: twist drill Diameter: 8 mm	Spindle speed: 500–800 RPM Feed rate: 0.02, 0.04 mm/rev

(continued)

Table 2 (continued)

No.	References	Composite	Tool material and geometry	Input variables
8	Krishnaraj et al. (2012)	Woven CFRP	Material: carbine Geometry: twist drill Diameter: 6, 8, 10 mm	Spindle speed: 12,000–20,000 RPM Feed rate: 0.01, 0.05, 0.1, 0.3 mm/rev
9	Rawat and Attia (2009a)	Woven GFRP	Material: carbide (K10–K20) Geometry: twist drill Diameter: 5 mm	Spindle speed: 1500–15,000 RPM Feed rate: 0.02–0.8 mm/rev
Thrust force (F_t) increases with increasing cutting speed on drilling process				
10	Lin and Chen (1996)	UD CFRP	Material: carbine Geometry: twist drill Diameter: 7 mm	Speed: 9550–38,650 RPM Feed: 0.03, 0.05 0.07 mm/rev
11	Khashaba (2004)	Cross-winding, woven and chopped GFRP	Material: HSS Geometry: twist drills Diameter: 8 mm	Spindle speed: 445–1850 RPM Feed rate: 0.03–0.3 mm/rev
12	Feito et al. (2018)	Woven CFRP	Material: carbine Geometry: twist drills, step drill Diameter: 6 mm	Cutting speed: 25–100 m/min Feed rate: 0.05, 0.10, 0.15 mm/rev
13	Sorrentino et al. (2018a, b)	Unidirectional CFRP and GFRP	Material: carbine Geometry: twist drills Diameter: 10 mm	Cutting speed: 47, 251, 376 m/s Feed rate: 0.015, 0.05, 0.10, 0.15, 0.20, 0.30 mm/rev
Thrust force (F_t) no/slight effect with increasing cutting speed on drilling process				
14	Ramulu et al. (2001)	MD GFRP-titanium stracks	Material: HSS, HSS-Co, solid carbine Geometry: twist drill	Spindle speed: 325–2750 RPM Feed rate: 0.03–0.25 mm/rev

(continued)

Table 2 (continued)

No.	References	Composite	Tool material and geometry	Input variables
15	Rajamurugan et al. (2013)	GFR-polyester	Material: carbide (K10) Geometry: brad and spur Diameter: 4, 6, 8, 10, 12 mm	Spindle speed: 500–2000 RPM Feed rate: 50–300 mm/min
16	Eneyew and Ramulu (2014)	UD CFRP	Material: PCD Geometry: eight facet drill Diameter: 6.35 mm	Spindle speed: 1500–6000 RPM Feed rate: 0.064–0.32 mm/rev
17	Kumar and Sing (2017)	Woven GFRP	Material: HSS, carbide K20 and solid carbide Geometry: helical flute drill, straight shank drill, 4 facet drill	Spindle speed: 500–2000 RPM Feed rate: 0.1, 0.3, 0.5 mm/rev
18	Upputuri et al. (2019)	CFPR	Material: HSS Geometry: twist drill Diameter: 4, 6, 8 mm	Spindle speed: 1000–3000 RPM Feed rate: 50, 100, 150 mm/min

[a]GFRP = Glass Fiber-Reinforced Polymer, CFRP = Carbon Fiber-Reinforced Polymer

and residual flexural strength of GFRP composite. The experimental study (Taguchi method) was conducted using four different levels of feed rate and two levels of spindle speed and drill diameters. The optimum operating conditions for each response was developed using the grey relational analysis (GRA). The results indicated that the feed rate was the main contributing factor to the drilling thrust force, followed by drill diameter and spindle speed. By increasing the feed rate, the drilling process behaviour attributes high impact (punching effect) from the chisel edge on a composite plate which increases the drilling thrust force. This punching force produces greater bending and consequently results in more delamination damage growth. As expected, the lowest feed rate (0.04 mm/rev), drill diameter at 4 mm, and highest spindle speed (630 RPM) indicate lower thrust force and delamination factors than other parameter setting.

In contrast, the results of the effect of cutting speed on drilling are debated, as shown in Fig. 8. The researchers reported three different relationships between cutting speed and drilling thrust force under different drilling conditions. It is worth highlighting that a previous study (Lin and Chen 1996) on drilling carbon FRP composites found that the width of delamination damage is correlated to the ratio

Fig. 8 Summaries of relationship between thrust force and input parameter for drilling process of FRP composite

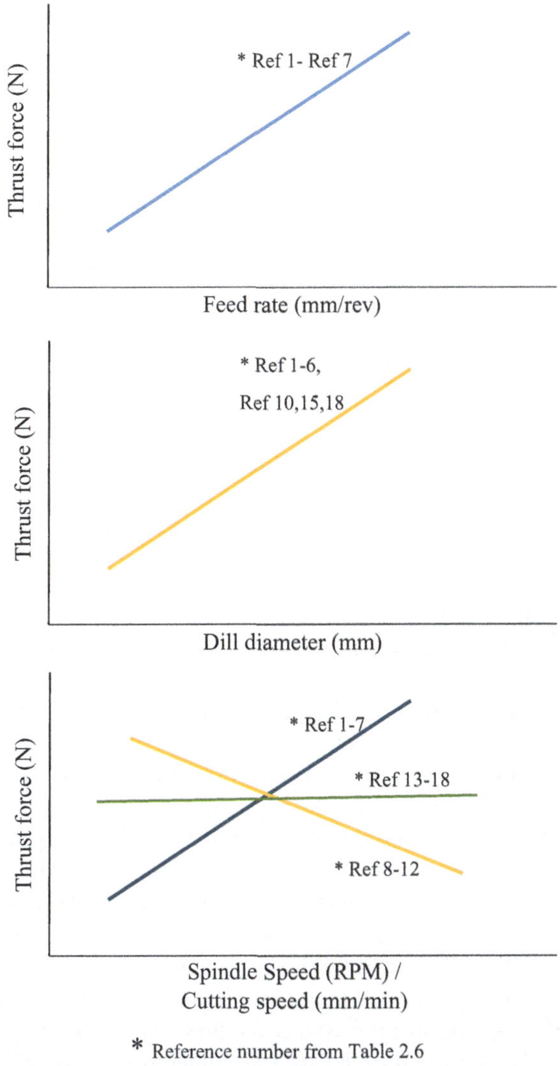

between drilling spindle speed and feed rate. The higher the ratio, the better the hole quality, and it will certainly increase production rate. However, they found that the thrust force increases drastically from 91 to 684 N as spindle speed increases. The finding is in line with previously reported studies on drilling carbon FRP and glass FRP composites (Feito et al. 2018; Khashaba 2004; Sorrentino et al. 2018b). This phenomenon occurs because the higher spindle speed accelerates tool wear and induces higher thrust force during drilling process. Surprisingly, some studies (Eneyew and Ramulu 2014; Kumar and Sing 2017; Rajamurugan et al. 2013; Ramulu et al. 2001) observed that the changing of spindle speed in drilling FRP composite does not significantly affect the thrust force performance in the drilling of FRP composite.

Owing to the benefits of high-speed drilling process, Rawat and Attia (2009a) performed an experimental investigation on the machinability of woven carbon FRP composite and tool-wear mechanism of tungsten carbide (WC) on dry high-speed drilling. It is known that high-speed machining is not only capable of reducing drilling thrust force but is also an advanced technology to improve productivity in industries. As expected, lower delamination factors were observed at 15,000 RPM as compared to 12,000 RPM, due to reduction in thrust and cutting forces. However, aggressive tool wear was the critical problem while drilling at such speeds using conventional twist drill bit. Their results showed that both the thrust force and cutting force were increased with increase in flank wear. This is likely due to the low thermal conduction of carbon FRP composite and inducted high temperature built up on the tool with continuous drilling at such high speeds. Beyond the critical flank wear of approximately 90 μm, there was a sudden rise in the matrix burnout and delamination in the cutting area.

On the contrary, several researchers (Arputhabalan et al. 2019; Krishnaraj et al. 2012; Palanikumar and Muniaraj 2014) found that at high spindle speed condition, the maximum drilling thrust force decreases progressively and the uncut laminate composite was sufficient to withstand it during the drilling process. This was due to the heat generated in the drilling zone assisted by the low coefficient of thermal conduction and glass transition temperature. As reported by Khashaba (2013), the thermal conductivity of the synthetic fiber and matrix are approximately 1.3 and 0.21 W/m °C respectively, considerably lower than metals (Steel = 53 W/m °C, Aluminium = 210 W/m °C). Thus, the accumulated drilling heat at the interface of tool and workpiece leads to thermal softening of the polymer matrix below the glass–rubber transition temperature (130–150 °C). Consequently, the cutting process required lower friction forces, thrust force, and cutting force to remove the softener material from the laminate composite surface. It is also highlighted that though the chopped GFRP composite required higher thrust force in cutting process than woven GFRP composite, it has lower delamination than the woven GFRP composite. This result referred to the presence of braids between the warp and filled tows that reduce the in-plane shear resistance as shown in Fig. 9. The delamination factors of woven GFR-epoxy composite are lower than that of woven GFR-polyester composite due to higher interface bond strength and shear strength of woven fiber and epoxy than polyester.

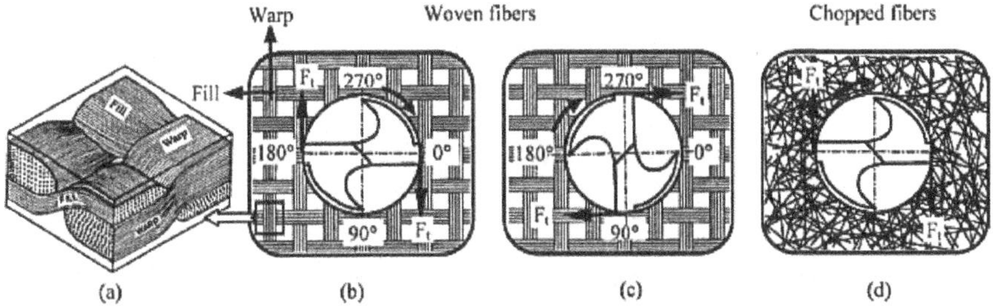

Fig. 9 Schematic diagrams illustrates **a–c** woven braid that made by the interlacing of warp and fill fiber and **d** chopped composites with random angles (Khashaba 2004)

3.2 *Effect of Tool Geometry and Material on Thrust Force Performance*

3.2.1 Tool Geometry

Machining advance FRP composites not only faces the challenge of insufficiency in work material which are of quality but also the high tool-wear issue. The rapid tool wear in machining FRP composites is due to the fiber abrasiveness, low thermal conductivity properties, and distinct mechanical properties of fiber and matrix in the composite. As aforementioned, the performance of cutting tool can also be an important factor for damage occurrence during the drilling process. Another consequence of tool-wear problem in the hole-making industry is frequent tool changes that may induce the productivity and machining cost. Similar to the thrust force damage, the tool-wear effect can also be diminished by selection of proper input variables such as tool geometry and material, workpiece and cutting condition. Since the properties of FRP composites and cutting parameters have been well documented in the previous section, this review is focused on variables of tool material and geometry on drilling of FRP composite. The tool employed to create the holes is called drill, which has a large length to diameter ratio. Among all the drill bit geometry, the twist drill is the most commonly employed in the secondary drilling process of FRP composites. The important nomenclature of twist drill is shown in Fig. 10. The cutting lips, chisel edge, and point angle are the main features that influence the thrust force behavior during drilling of FRP composites.

Geometrically, the ideal drilling process mostly takes place at cutting edge which includes the chisel edge, cutting lips, and the leading edges. The chisel edge and cutting lips play the main role in the material removal process, while, theoretically, no cutting action happens along the leading edge (Astakhov 2010). However, owing to the thrust force is generated during the drilling process, some material is removed by the leading edge as a finishing process for the hole surface. At high thrust force condition, an undesired conical hole shape will be created in the drilling process (rather than a perfect cylinder). Dharan and Won (2000) have claimed that primary cutting action mostly occurs at the outer parts of cutting lips while the chisel edge

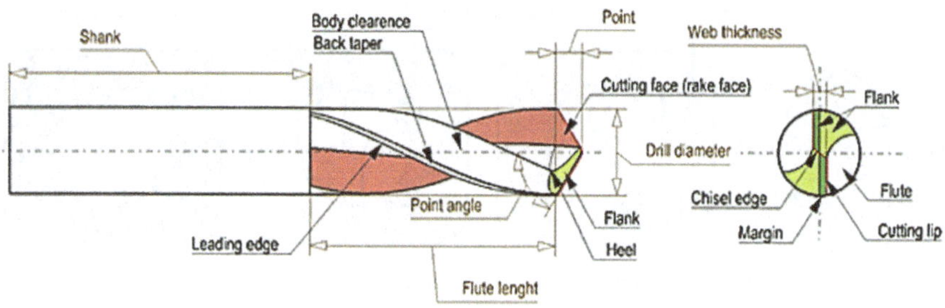

Fig. 10 Twist drill nomenclature (Astakhov 2010)

is used to position the drill before engagement with laminate and stabilize the drill throughout the cutting process. However, if improper parameters are set in the drilling operation, high thrust force from chisel edge area will push forward the uncut material (like extrusion action), thus, in turn, serious delamination happens around the hole, Fig. 11.

As discussed in the earlier section, it is believed that various drilling defects especially delamination occurs when the high thrust force is generated during drilling process. Tsao and Hocheng (2003) have conducted the drilling experiments to depict the effects of chisel length and the pilot hole on thrust force and delamination. Based on their study, the thrust force from chisel edge is the major factor for delamination. Besides that, they found that the drilling thrust force is reduced dramatically by cancelling the chisel edge effect with a pilot hole before actual drilling process. The delamination-free goal can be achieved by lowering the thrust force effect through optimizing the chisel length and feed rate in the drilling process.

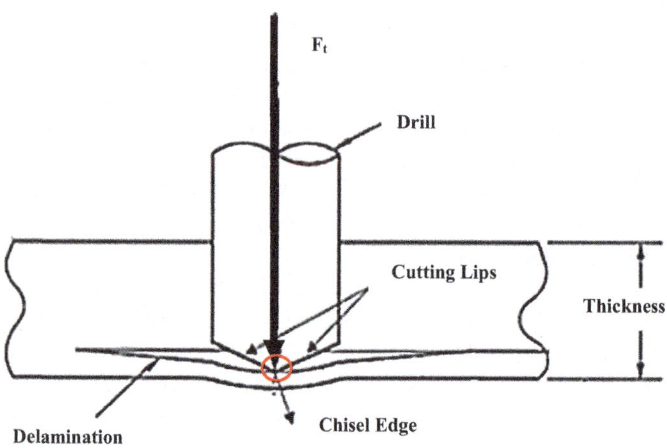

Fig. 11 Schematic depiction of delamination damage when drilling of FRP composite

Furthermore, Rawat and Attia (2009b) and Wang et al. (2013) have cited that the critical flank wear was normally observed at the primary cutting edge due to the friction and shearing action of the tool and abrasive of fiber in the drilling process. Therefore, the thrust force is largely increased with the rapid tool wear in the drilling process and thus resulted in a severe delamination damage around the drilled hole. In order to reduce the fracture or chipping problem on the drill bit, they suggested coating of a high-wear resistance material, such as a diamond material, on the tool to enhance the tool life and drilling performance.

Chen (1997) has established the linear relationship between the drilling thrust force and delamination factors of unidirectional carbon FRP by regression analysis under different drilling condition. The experimental results show that lower thrust force can be obtained at lower point angle and larger helix angle in order to reduce the delamination damage. This is mainly due to the tool orthogonal rake angle on primary cutting edge increases with point angle and helix angle. It is well known that the point angle is one of the features to determine the characteristics of the drill's cutting edges. An optimum point angle can enhance tool life and hole quality. Generally, the standard 118° is used to drill the general material such as metal. While, for the heterogenous and abrasive FRP composites, Campos Rubio et al. (2008) have suggested the small point angles (85°) of twist drill to improve the machinability of FRP composite in high-speed drilling. The high performance of the 85° point angle twist drill may be explained by the fact that, the shear plane area between tool and workpiece is smaller, thus probably required lower thrust forces and resulted in lower delamination factors of glass FRP composite.

From the review shown in Table 3, it is clear that a standard twist drill is the one that is most commonly applied in the drilling of FRP composites due to low cost and easy accessibility. Nevertheless, it is found that with the conventional standard twist drill it is difficult to achieve the damage-free hole in FRP composites and simultaneously maintain the thrust force below the critical value at high feed and speed condition. In order to overcome the limitation of the twist drill, in recent years, many researchers have reported the effect of a variety of special drill bit geometries on drilling-induced damage. Thus, many special design drill bits have been introduced for drilling of composite laminates, as shown in Fig. 12. It can be noticed that the drill bits used in drilling FRP composites could be divided into six categories based on the geometry: (1) Twist drill bit, (2) Step drill bit, (3) Brad point drill bit, (4) Slot drill bit, (5) Straight-flute drill bit, and (6) Core drill bit. Although, each type of drill bits has their unique functions, the final purpose mostly is to enhance the drilling efficiency and tool life.

As mentioned before, Won and Dharan (2002a, b) have conducted a series of drilling experiments on woven carbon FRP composite to investigate the effect standard twist drill with pre-drilled hole on drilling thrust force. They found that the pilot hole capable to reduce the required thrust force during drilling and allowing much higher feed rate in cutting process without the danger of delamination. Tsao and Hocheng (2007) have also investigated the performance of core drill on drilling thrust force at various parameters setting condition. The thrust force performance of conventional twist drill was compared with the special core drill bits. They found

Table 3 Mechanistic models of drilling FRP composite in literature

No.	References	Composite	Tool material and geometry	Predictor variables
1	Singh and Bhatnagar (2006)	UD-GFRP	Material: solid carbide Geometry: 4-facet, 8-facet, parabolic point, and Jodrill	Feed Speed Drill point geometry
2	Khashaba et al. (2010)	Woven GFRP	Material: carbine Geometry: twist drill Diameter: 8 and 10 mm	Feed Speed Drill diameter
3	Won and Dharan (2002a, b)	CFRP; AFRP	Material: HSS Geometry: twist drills	Feed Drill diameter
4	Singh et al. (2008)	UD-GFRP	Material: solid carbide Geometry: twist drills	Feed Drill point angle
5	Khashaba et al. (2010)	Woven GFRP	Material: cemented carbide Geometry: twist drills	Feed Speed Tool wear
6	Tsao and Hocheng (2007)	Woven CFRP	Geometry: core drill Diameter: 10 mm	Feed Speed Drill thickness Grit size
7	Mohan et al. (2005)	GFRP with chopped fiber mat	Material: carbide-coated drills Geometry: core drill Diameter: 4 different drill diameters	Feed Speed Drill Diameter Specimen thickness
8	El-Sonbaty et al. (2004)	GFRP	Material: HSS Geometry: core drill Diameter: 8, 9, 10, 11, 12 and 13 mm	Feed Speed Drill diameter Fiber volume fraction
9	Fernandes and Cook (2006)	CFRP	Material: HSS Geometry: one shot drill Diameter: 5 mm	Feed Drill diameter Number of drilled holes (tool wear)

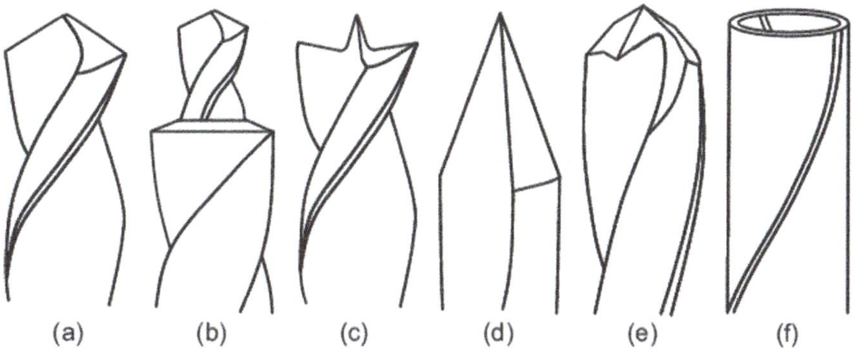

Fig. 12 Schematic of drill bit geometries used for drilling of FRP composites: **a** twist drill; **b** step drill; **c** brand point drill; **d** straight—flute drill; **e** slot drill; **f** core drill (Hocheng and Tsao 2008a)

that core drills offer the highest threshold values for thrust forces, Fig. 13. In general, the core drill is applied for drilling hard, brittle materials, as in civil engineering structure, jewels, and glass. Davim and Reis (2003) has reported that the "Brad & Spur" drill has less cutting pressure and thrust force than "Stub length" drill at the same cutting condition for glass FRP composite. In conjunction with tool geometry, the selection of cutting tool materials for a particular application is of extreme importance as in the case of drilling of FRP composites.

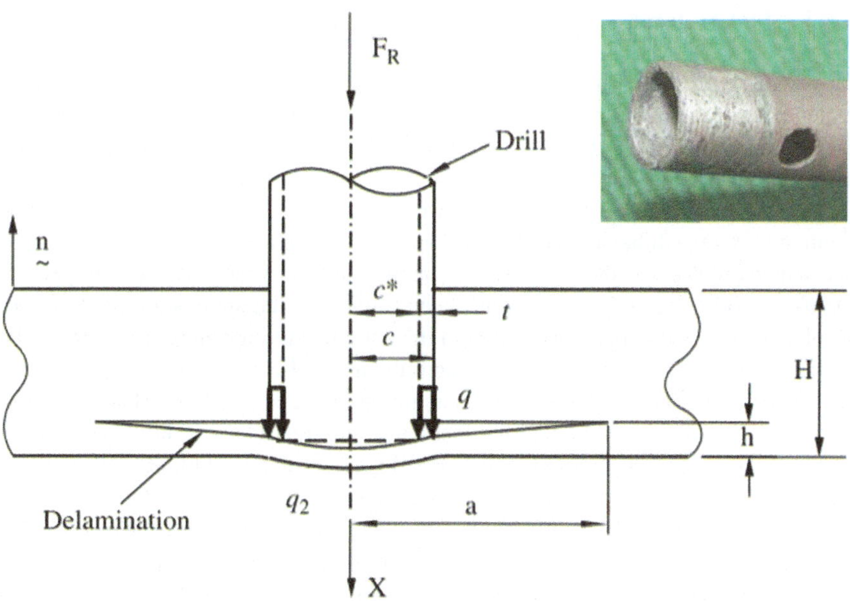

Fig. 13 Schematic depiction of delamination analysis of core drill (Tsao and Hocheng 2008b)

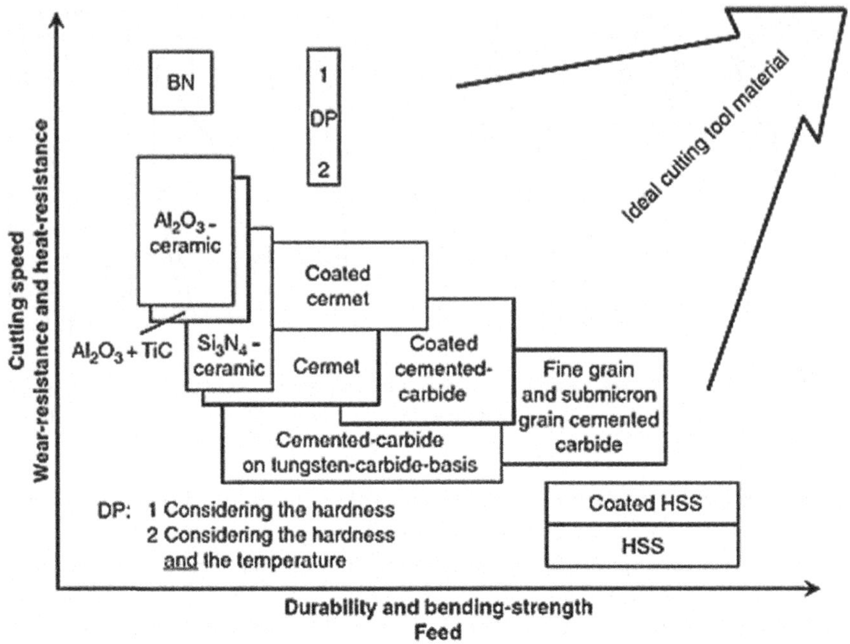

Fig. 14 Classification of various cutting tool materials (Klocke and Kuchle 2011)

3.2.2 Tool Material

In general, cutting tool material must consist of excellent hot hardness, toughness and impact strength, wear resistance, and chemical stability characteristics during the drilling process, Fig. 14 (Klocke and Kuchle 2011). However, it is difficult to obtain all the properties in a single material; each material has advantages and limitations to cutting process. According to the study of Sreejith et al. (1999) the polycrystalline diamond (PCD) exhibits high critical velocity and it is extremely beneficial for high-speed machining due to the high thermal conductivity and wear resistances properties. On the basis of high-cost issue, it has been found that the application of PCD tool is relatively low which is mostly used in the extreme cutting condition. Commonly, carbon tool steels were used as tool materials in metal cutting until the high-speed steels were developed for high-speed machining in the early 1900s. This is because the carbon tool steels pose lower hardness and wear resistance at moderate temperature (180 °C). The High-Speed Steel (HSS) is a highly alloyed tool steel which is capable used in higher speed drilling process (60 m/min) as compared to carbon tool steels. Subsequently, tungsten carbide (WC) and titanium carbide (TiC) were also introduced since the 1930s due to the rapid development of high-speed machining on FRP composites. Carbides are widely used as drilling tools because of their high hardness over the wide range of temperature, high elastic modulus, and low thermal expansion, as listed in Table 2. The comparison drilling thrust force performance of HSS and carbide drill bit for carbon FRP composites been conducted by Durão et al. (2013). Based on the delamination and circularity damage results, confirming the fact that the holes drilled with HSS drill show higher value in these responses as compared to the performance of carbine drill. Thus, it can be concluded that the HSS drill bit may not be suitable for drilling carbon FRP composites at high feed rate (0.2 mm/rev) condition due to low wear resistance and hot hardness properties present in HSS material will generate high thrust force.

As far as the high speed drilling with carbide tool is concerned studies by Murphy et al. (2002) is about the thrust force mechanism of uncoated, titanium nitride (TiN) and diamond-like carbon coated performance on drilling woven glass FRP composite. Surprisingly, they found almost similar results as that of the damage from the drilling process for all types of uncoated and coated drill bit. They have not highlighted on any benefits in using the coated drill bit for drilling FRP composites. In short, select suitable tool materials have a positive effect on most responses of the drilling operation especially for thrust force performance. In short, the uncoated tungsten carbide tools (K10–K20) consists the high thermal conductivity, low cost, and high toughness and hardness properties (Franke 2011; Rawat and Attia 2009b; Shyha et al. 2010). Therefore, it still is the desired tool material in most high-speed drilling studies for abrasive FRP composite machining.

4 Modelling of Cutting Forces in Drilling FRP Composite

It is well understood that drilling FRP composites is one of the most critical processes in component product manufacture. This is because the drilling process is carried out mainly in the last sequence of the manufacturing plan to create final features such as holes on composite components. As mentioned in the previous section, any inappropriate drilling parameters settings may lead to high thrust force, which, in turn, would cause rejection or unacceptable composite parts due to degradation such as delamination damage and poor surface quality. This is mainly attributed to the complexity associated with the drilling process and the geometry of the drill bit, as well as the anisotropic and highly abrasive nature of the fibers composite. Hence, exploring the relationship between the input parameters and the thrust force responses for improvement in the prediction accuracy of drilling performance is of research interest.

4.1 Mechanistic Models

In general, a statistical relationship between experimental variables can be characterized by a mathematical formula, which is widely known as empirical model or mechanistic model. Such model typically used statistical tools such as design of experiments (Taguchi or RSM), analysis of variance (ANOVA) and regression analysis to develop the relationship. The developed empirical relationship can be effectively used describe the trends and forecast the thrust force performance during drilling FRP composites. Therefore, as indicated in the earlier section, the correlation between predictor variables (spindle speed (v), feed (f) and tool geometry (t)) and the thrust force responses can be established using a liner/second order polynomial regression model. It is to be noted here that the polynomial regression models can be employed in two conditions (Michael Kutner et al. 2004):

1. When the experimental response function is true curvilinear to the predictor variable.
2. When the response function is unknown or complex, the polynomial function is adequate for presented the actual function.

The polynomial regression model can, therefore, be adequately implemented in this study as an approximation to obtain the information about the relationship of the response function. In this case, the polynomial model is expressed in a quadratic equation shown as follows:

$$\begin{aligned} \mathrm{E}\{Y\} = \beta_0 + \beta_1 X_1 + \beta_2 X_2 + \beta_3 X_3 + \beta_{11} X_1^2 + \beta_{22} X_2^2 \\ + \beta_{33} X_3^2 + \beta_{12} X_1 X_2 + \beta_{13} X_1 X_3 + \beta_{23} X_2 X_3 \end{aligned} \quad (4.1)$$

The β_0 is the constant coefficient, β_i is the coefficient of three main factors, β_{ii} is the square or interaction coefficients between the pairs of the main factors. In general, the constant coefficient β_0 is the y-axis intercept. While, the β_i coefficient is the slope of the response function line, as depicted in Fig. 15. The interaction coefficients in model can directly estimate through experimental data to establish relationship between maximum or average cutting force and the influential factors, i.e. cutting parameters and drill bit parameters. The empirical model can be effectively used to predict the thrust force of drilled holes at the 95% confidence level under selected drilling parameter setting. Researchers (Wan et al. 2019) have summarised the empirical models created through different combinations of drilling parameters (i.e. spindle speed and feed rate), drill bit parameters (i.e. drill diameter and drill point angle), and composite parameters (i.e. volume fraction material and the laminate thickness), as shown in Table 3. It can be noticed that many researchers are devoted

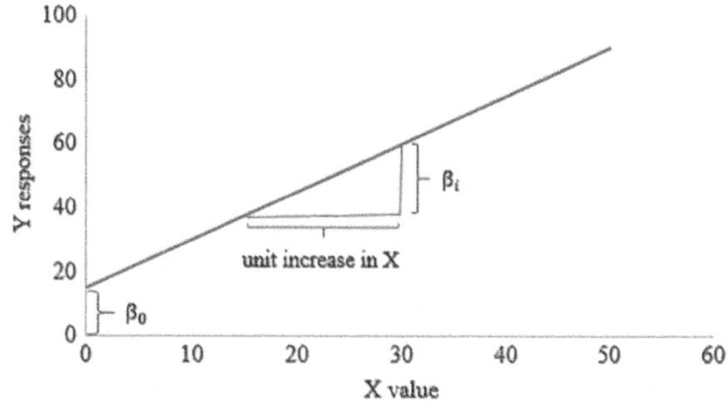

Fig. 15 Illustration of coefficients of regression model

to the analysis of the relationship between thrust force and its predicted factors, such as different feed rate, cutting speed, drill geometry, and properties of FRP composites. They found that the mechanistic models under different drilling conditions are easy to understand and capable of predicting thrust force in actual drilling process effectively.

Won and Dharan (2002a, b) modified Shaw's simplified metal cutting equations (1957) to predict the thrust force during drilling of carbon FRP composite without considering the tool wear. The predicted coefficients in their empirical model were feed rate and drill diameter. However, previous studies (Rawat and Attia 2009b; Wang et al. 2013) cited that the critical flank wear was normally observed at the primary cutting edge due to the friction and shearing action of the tool and abrasion of fiber in the drilling process. Therefore, the thrust force is greatly largely increased with rapid tool wear in the drilling process and resulted in severe delamination damage around the drilled hole. Therefore, by considering the tool wear issue, Fernandes and Cook (2006) established an empirical model of the maximum thrust force during the drilling of carbon FRP composite with a "one shot" drill bit and three different laminate thickness: 2 mm, 4 mm, and 5 mm give the final equation for thrust force in

$$\begin{aligned} F_{tmax,2mm} &= (0.003n + 1.0467) \times (76.56(fd)^{0.39} + 1.047d^2) \\ F_{tmax,4mm} &= (0.0036n + 1.2128) \times (76.56(fd)^{0.39} + 1.047d^2) \\ F_{tmax,5mm} &= (0.0035n + 1.5159) \times (76.56(fd)^{0.39} + 1.047d^2) \end{aligned} \tag{4.2}$$

where n is the number of drilled holes, f is the feed rate (mm/rev), while d is drill diameter (mm). As expected, the validation results indicate that, as shown in Fig. 16, the experimental thrust forces are close to the predicted maximum thrust force during the drilling of carbon fiber using a one-shot drill bit.

Drilling experiments of glass FRP composite were conducted by Mohan et al. (2005) and Singh et al. (2008) who used different tool geometries and fiber architecture. Thrust force results were verified with the analysis of variance (ANOVA) at 95% confidence level to define significant factors according to the variation of each factor. Thus, point angle and feed rate were used to develop the drilling thrust force

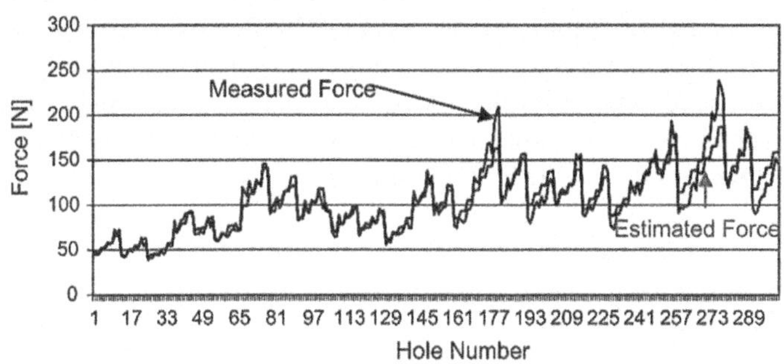

Fig. 16 Experimental and estimated maximum thrust force for drilling carbon FRP composite

model by adopting regression analysis.

$$F_t = -98.0319 + 1.4365p + 402.8315f \quad (4.3)$$

where F_t is the thrust force, ρ is the point angle, and f is the feed rate. Furthermore, Khashaba et al. (2010) also developed the thrust force empirical model through multiple linear regression models to directly consider the drill wear for drilling woven GFRP composites. The relationship between drilling force and significant factors such as drill pre-wear (w), feed (f), and speed (v) were created.

$$F_t = -1.61 + 3.17v + 977.781f + 12.793w \quad (4.4)$$

In addition to that, the relationship between thrust force and predictor variables can be defined by the sign and value of the coefficient in the quadratic equation. The positive coefficient sign represents that the changing trend of the predictor variables is directly proportional to the thrust force, whereas the negative sign means there is an inverse relationship between the predictor variables and drilling thrust force. In Eqs. 4.2 and 4.3, the coefficients for the feed rate presented with a positive sign for all models. This implies that for every 1 mm/rev, the feed rate increases, the thrust force will increase by 402.8315 N and 977.781 N for the tow equations respectively. The estimated relationship of the quadratic responses is in agreement with previously reported studies of drilling of FRP composite and findings presented in previous sections.

Although the ANOVA results proved that the empirical models can effectively predict the maximum thrust force in the drilling process, surprisingly, the deviation errors can be observed in the aforementioned studies. The empirical or regression model may not clearly predict and describe the thrust force during the drilling process. Conversely, previous studies (Hocheng and Tsao 2005; Pyo Jung et al. 2005; Zhenchao et al. 2014) have pointed out that the thrust force and delamination damage on the drilled FRP composite was not only affected by drilling parameters but also the heterogeneous properties of FRP composite material. Nevertheless, it can be concluded that the regression models can be used with reasonable or acceptable accuracy to predict the maximum thrust force of the drilling FRP composites process over the entire range of drilling parameters only. Care has to be taken in cases where different drilling parameter settings and cutting conditions are employed. This will be further investigated in the following section for accurate monitoring and prediction of the critical thrust force in order to monitor or control the delamination damage in the FRP composites during the drilling process.

4.2 Critical Thrust Force Models

Among drilling-induced damages, delamination has been recognised as one of the crucial failure mechanisms in drilling operation, which is a highly undesired problem in assembly process (Capello 2004). A number of studies have performed the notched behaviour study on drilled FRP composites through the open-hole and multi-bolted single-lap joints test in order to monitor and forecast the structure strength. Their results showed that delamination has a high potential to reduce the joining strength and may affect the long-term performance of drilled FRP composites (Gamdani et al. 2015; Wisnom and Hallett 2009). A number of studies have been published with respect to critical thrust force models to monitor or control the delamination damage in the FRP composites during the drilling process. It is evident that the delamination factor, especially the push-out delamination, is significantly affected by thrust force (which is influenced by machining parameters setting) and the stiffness of workpiece (Akmal et al. 2018; Durão et al. 2015; Pyo Jung et al. 2005; Saoudi et al. 2016; Tan et al. 2017; Tsao 2012; Zhenchao et al. 2014).

It is believed that the path towards delamination-free (push-out) drilling process can only be achieved when the thrust force does not exceed the inter-laminar bonding strength of the composite structure, namely "critical thrust force" (Zhenchao et al. 2014). All the analytical models discussed in this section have been presented in Table 4. Owing to the high correlation observed between thrust force and delamination, Hocheng and Dharan (1990) were the first to establish the analytical model to determine the critical thrust force for delamination onset and propagation of FRP composites. They used linear elastic fracture mechanics (LEFM) method to obtain the critical thrust force model to prevent or monitor the delamination onset. In order to satisfy the applicability of LEFM theory to the propagation of laminate composites, the crack or delamination must be coplanar and symmetrical in a plane with the material, and lastly, no plastic zone must be present at the crack tip. The delamination onset mechanism has been shown in Fig. 17. The laminate workpiece is clamped with a uniformly distributed load along the inner edges. The "D" in the figure represents the diameter of the drill bit, "FA" is the thrust force from chisel edge of drill, "x" is the tool displacement, "a" is existing crack size, and "h" and "H" are the uncut thickness under the tool and thickness of the workpiece respectively. As the drill bit moves forward, the thrust force from the tool pushes the uncut laminae and deforms elastically at that zone. At this point, if the external energy (drilling thrust force) exceeds the internal change of strain energy in the material, crack extension into the new surface occurs. In order to prevent the delamination onset in the drilling process, the critical thrust force for peel-up at entry and push-out at exit can be expressed as follows:

The critical thrust force at the onset of delamination during tool entry (peel-up) can be calculated:

$$F_c^* = k_p \pi \left[\frac{8G_{IC}E(H-h)^3}{3\left(1-v^2\right)} \right]^{1/2} \quad (4.5)$$

Table 4 Analytical model for drilling process of FRP composites

No.	References	Model	Workpiece	Assumptions
1	Hocheng and Dharan (1990)	$F_C^* = k_p\pi\left[\frac{8G_{IC}E(H-h)^3}{3(1-v^2)}\right]^{1/2}$	Unidirectional FRP composite	– Isotropic – Delamination in circular shape – Concentrated thrust force
2	Jain and Yang (1991)	$F_{\text{crit}} = 3\pi\left[\frac{D_{22}}{D_{11}}\sqrt{2G_{\text{Ic}}D_{\text{C}}^*}\right]^{1/4}$	Unidirectional FRP composite	– Orthotropic – Elliptical shape of delamination area – Concentrated thrust force
3	Lachaud et al. (2001)	$F_Z = 8\pi\left(\frac{G_{IC}D}{(1/3)-(D'/8D)}\right)^{1/2}$	Multidirectional FRP composite	– Orthotropic – Elliptical shape of delamination area – Distributed thrust force
4	Woo et al. (2005)	$P_c = \sqrt{\frac{G_{\text{IC}}\cdot b}{\Omega-\Xi}},$	FRP-metallic strips	– Energy balance theory and energy release rate concept – Concentrated thrust force
5	Zhenchao (2014)	$P = \frac{2\sqrt{3}\pi}{\xi}\sqrt{G_{IC}D}$	Metal-FRP stacks	– Orthotropic – Concentrated force at chisel edge and distributed force at cutting lips
6	Tan et al. (2017)	F* = $4\sqrt{2}\pi\sqrt{(\text{G}_{\text{IC}}\ D)_c V_C + (\text{G}_{\text{IC}}\ D)_g(1-V_C)}$	Hybrid Carbon/Glass FRP composite	– Orthotropic – Uniformly distributed and concentration load at the cutting lips and edge

whereas, critical load at the onset of crack propagation during tool exit (push-out) can be calculated as:

$$F_A^* = \pi\left[\frac{8G_{IC}E(h)^3}{3(1-v^2)}\right]^{1/2} \tag{4.6}$$

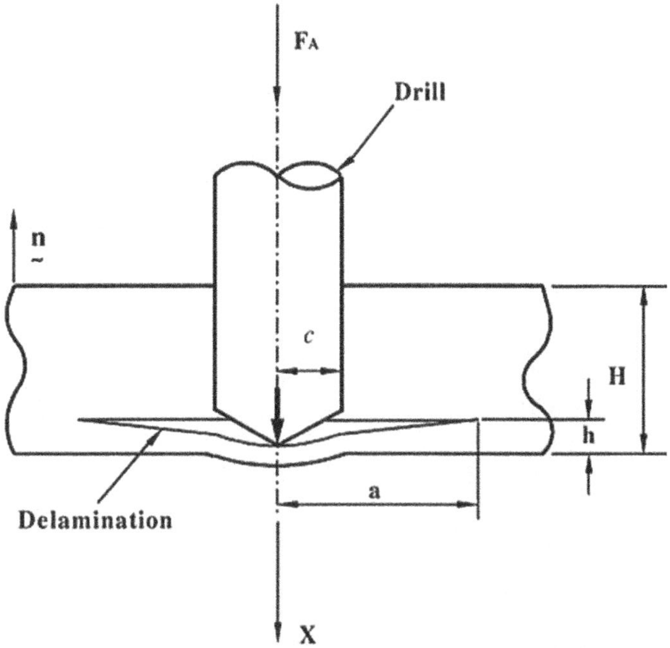

Fig. 17 Circular plate model for delamination analysis (Hocheng and Dharan 1990)

where "GIC" is the critical energy release rate for delamination Mode-I per unit area, "kp" the peeling factor is the defined ratio of critical peeling force to the critical cutting force, "E" is Young's modulus of the material, "v" is the Poisson ratio, and "h" is the uncut thickness under the cutting tool. Most researchers have assumed that the delamination onset tends to propagate in an opening mode, Mode I, by thrust force in the drilling process, as shown in Fig. 18. In fact, the "GIC" for the thin plane case is much lower than other fracture modes, which makes the critical thrust force relatively lower than the experimental value. Hocheng and Dharan's (1990) study is a good start for analytical modelling for predicting the critical thrust force in order to improve the delamination damage, but its applicability is limited attributing to the assumption of isotropy of each layer. Again, this has induced a conservative prediction for the critical thrust force for FRP composites.

Subsequently, Jain and Yang (1991) extended HoCheng-Dharan's model, taking into consideration the orthotropic or anisotropic properties of the composites and hypothesising an elliptical shape of delamination area (Fig. 19). The fracture mechanics laminated plate theory and cutting mechanics were employed to establish analytical models for carbon FRP composites in the drilling process. The critical thrust force model is given by:

$$F_{\mathrm{crit}} = 3\pi \left[\frac{D_{22}}{D_{11}} \sqrt{2G_{\mathrm{Ic}} D_{\mathrm{C}}^{*}} \right]^{1/4}, \tag{4.7}$$

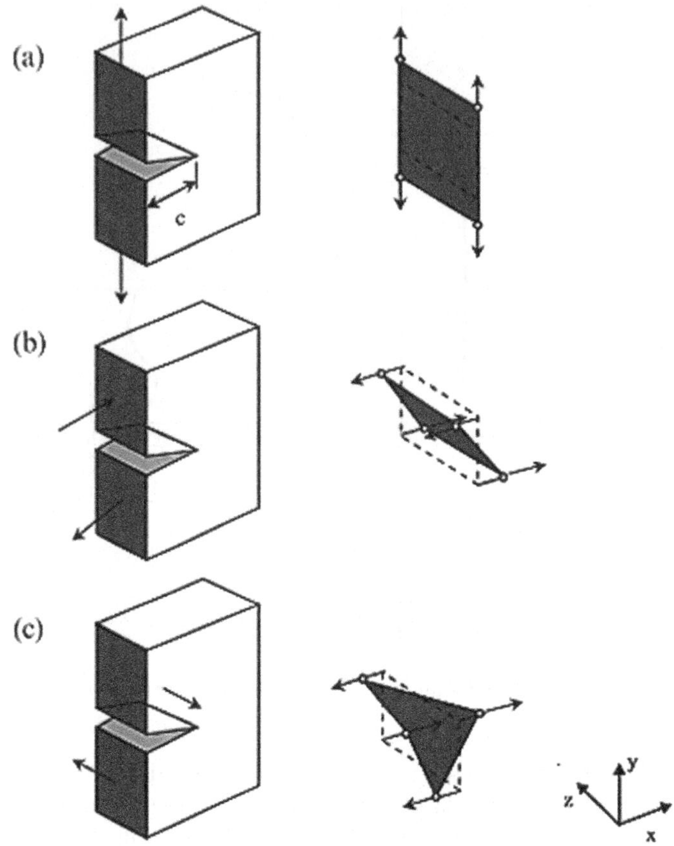

Fig. 18 Crack deformation modes, **a** Mode I, **b** Mode II and **c** Mode III (Irwin and de Wit 1983)

Fig. 19 Elliptical shape of delamination area (Jain and Yang 1991)

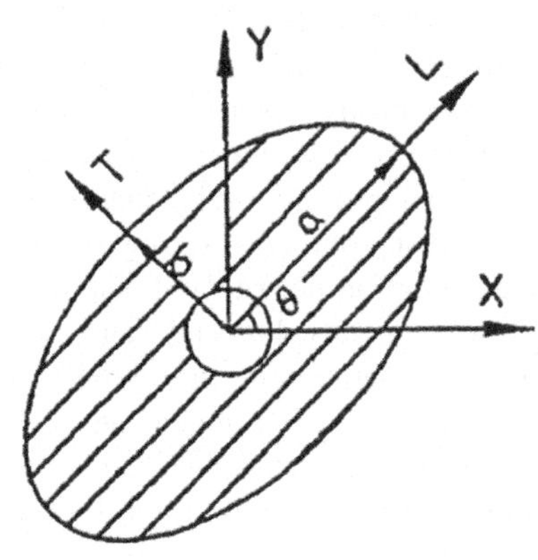

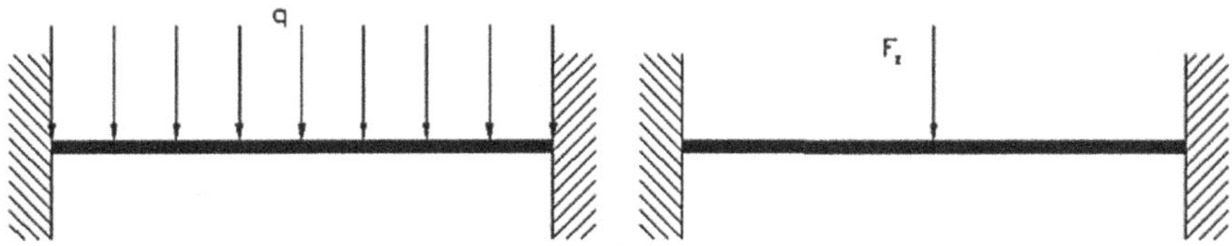

Fig. 20 Drill/plate contact with **a** uniform distributed force and **b** concentration force (Lachaud et al. 2001)

where

$$D_C^* = 2D_{11} + \frac{2(D_{12} + 2D_{66})}{3}\sqrt{\frac{D_{11}}{D_{22}}}, \tag{4.8}$$

where "D" is the flexural rigidity of the carbon FRP composite. In the model, the critical energy release rate "G_{IC}" was a constant, and the thrust force from the chisel edge was a concentrated force, with good agreement with predicting values and the validation of results.

Lachaud et al. (2001) have established a critical thrust force model for multi-directional tracking sequence carbon FRP composite based on the classical plate theory. They assumed that the thrust force is uniformly distributed on the chisel edge and cutting lips of the drill, which is contrary to previously mentioned models (concentration force) (Fig. 20). In addition, the unilateral properties of carbon fibers lead to the delamination area deformed in elliptical shape when subjected to bending/compression stress from the drill bit. Therefore, the critical model for uniformly distributed force is:

$$F_Z = 8\pi\left(\frac{G_{IC}D}{(1/3) - (D'/8D)}\right)^{1/2}, \tag{4.9}$$

However, based on the experimental measurements of static punching, it can be concluded that the distributed analytical model can accurately predict the thrust force at the location of uncut thickness as compared to the concentration critical thrust force model. Therefore, Zhenchao et al. (2014) presented an analytical model for predicting the critical thrust force during drilling metal-FRP stacks composite. The strategies of drilling metal-FRP composites are different and more complex than monolithic FRP composites. It is clear that as per the understanding of the cracking mechanism, the edge condition, and the deformation behaviour of the metal plate during the metal-FRP stacks drilling, the previous analytical models have to be modified for predicting the critical thrust force for delamination onset. They analysed the critical thrust force model based on two cases: (1) drilling from metal to FRP and (2) drilling FRP to metal (backend support concept), Fig. 21. Besides, they claimed that the thrust force at the chisel edge is a concentrated form and a uniform distributed load at the cutting

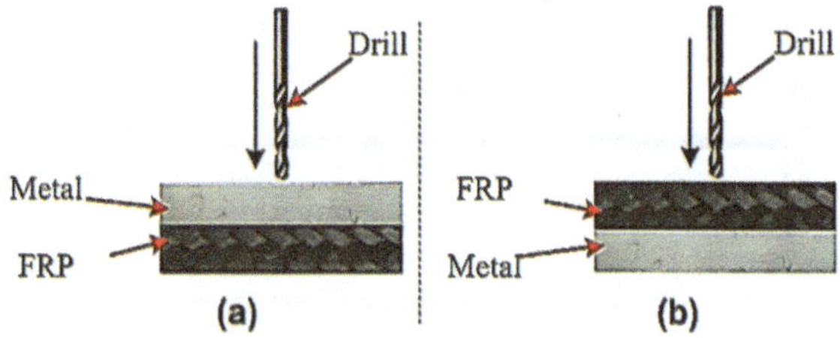

Fig. 21 Drilling direction of metal-FRP stacks and FRP-metal stacks (Zhenchao et al. 2014)

edge. The critical thrust force model for drilling metal to FRP was developed based on the theorem of virtual work and LEFM as follows:

$$P = \frac{2\sqrt{3}\pi}{\xi}\sqrt{G_{IC}D} \tag{4.10}$$

where "ξ" denotes the proportional coefficient is varied in relation to the resultant force to the drilling thrust force, mostly in the range of 50–70% depending on the process parameter and tool geometry. The non-rotation test results are in line with the solutions of the analytical model. This model makes the parameter optimised and more convenient since the critical thrust force was defined.

Additionally, an analytical study and rule of hybrid mixture (ROM) approach have been attempted in Tan et al. (2017) to establish the critical thrust force and control delamination during drilling hybrid FRP composite. Some important underlying assumptions are needed: (1) fibers are arranged in hexagonal close packing throughout the matrix in the hybrid composite; (2) the adhesive bonding between fiber and matrix is strong and free of a void in the hybrid composites; (3) the fiber, matrix, and composite behave as iso-strain ($\varepsilon_{\text{hybrid}} = \varepsilon_{\text{fiber}} = \varepsilon_{\text{matrix}}$) during loading. Thus, the coefficient of bending stiffness in the model have to be modified by substituting the ROHM equation to predict the hybrid properties, as shown below:

$$\mathrm{F}^* = 4\sqrt{2}\pi\sqrt{(\mathrm{G_{IC}}\,D)_c V_C + (\mathrm{G_{IC}}\,D)_g (1 - V_C)} \tag{4.11}$$

where Vc is the volume fraction of carbon fiber, D is the bending stiffness, G_{IC} is the critical energy release rate, c is the carbon FRP composite, and g is the glass FRP composite.

As indicated earlier, the changing thrust force is known to play a vital role in influencing the size of delamination zone. When the thrust force was further increased after a critical value, the delamination damage showed a similar rate of growth. Subsequently, in this section, the critical thrust force results from Eq. 4.11 were compared with critical models from previous studies, in Eqs. 4.6 and 4.10. Their models were utilised to predict the critical thrust force during drilling of hybrid FRP composite material used in Tan et al. (2017), based on the relevant material properties for critical thrust force model, Table 5. The critical thrust force values

Table 5 Material properties for monolithic and hybrid FRP composite

Property	Glass FRP	Carbon FRP	[a]Hybrid FRP
E_1 (GPa)	12.50	39.02	[a]25.95
E_2 (GPa)	8.97	36.98	[a]23.26
G_{12} (Jm^{-2})	207.00	176.00	[a]191.50
V_{12}	0.26	0.28	0.27

[a]*Which is calculated by* $\sigma_h = \sigma_C V_C + \sigma_G(1 - V_C)$, $V_C = 0.52$

Table 6 Comparison thrust force with previous models

Ply No. (uncut layers)	Tan et al. (2017)	(Zhenchao et al. 2014)	Hocheng and Tsao (2003)
	Critical thrust force (N)		
1	47.64	41.67	15.20
2	134.75	117.86	43.00
3	247.55	216.56	78.99
4	381.13	333.42	121.63
5	532.65	465.97	169.98

from these three models have been summarised in Table 6 and Fig. 22. The critical thrust force values from Zhenchao et al. (2014) are close to the critical value from Tan et al.'s (2017) model. The small variation may be due to the positive hybrid effect in hybrid FRP composite which was not considered in Zhenchao's model. Again, as expected, the results from Hocheng and Tsao's model showed a large

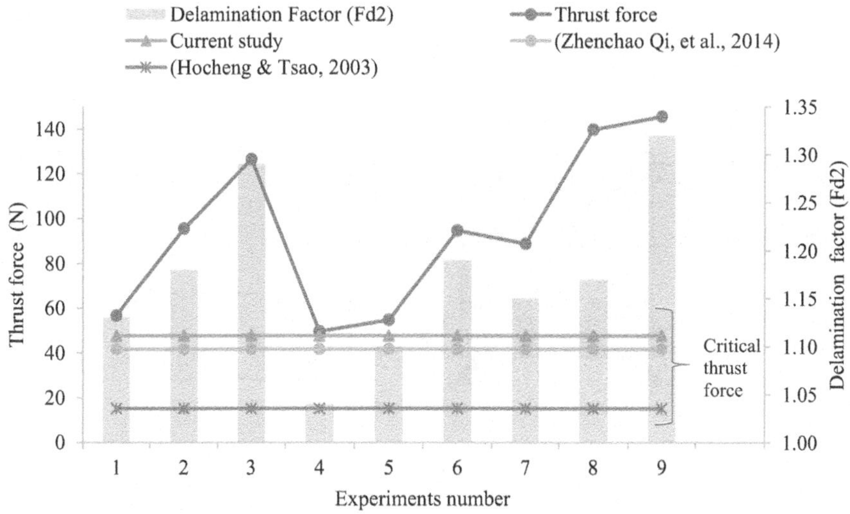

Fig. 22 Comparison critical thrust force from different models toward drilling thrust force and delamination factor

deviation as compared to the other two models. This is likely due to the assumption that each layer of the FRP laminate is homogenous, isotropic, and the delamination area is circular. This indicates that Hocheng and Tsao's is inadequate for predicting the onset delamination of hybrid FRP composite. Therefore, they concluded that previous models may not accurately predict the critical thrust force for hybrid FRP composites due to the bending stiffness properties which may be influenced by the positive hybrid effect. It is important to note that the critical thrust force can be an attractive benchmark or reference for industrial practice to control the drilling thrust force so that delamination damage can be alleviated for better assembly performance of the drilled FRP composites.

5 Approaches to Reduce Thrust Force/Delamination in Drilling of Composite Laminates

Based on the review, it is suggested that regardless of the finding in critical thrust force section, the laminate can be drilled as fast as permissible at entry as the uncut thickness is sufficient to withstand the thrust force. Then, the value of feed rate can be progressively decreases as the tool approaches the exit plane. This method may require a numerical control programming on the drilling machine to control the feed during the drilling process when the thrust force nears the critical thrust force value, i.e., the feed can be slowed down just before reaching the bottom-most ply. This strategy is one of the paths towards a "delamination-free" during drilling FRP composites. Therefore, Khashaba (2004) applied a variable feed technique on drilling cross-winding glass FRP composites using CNC milling/drilling machine. Figure 23 shows the feed cycle along the hole depth in the drilling of cross-winding

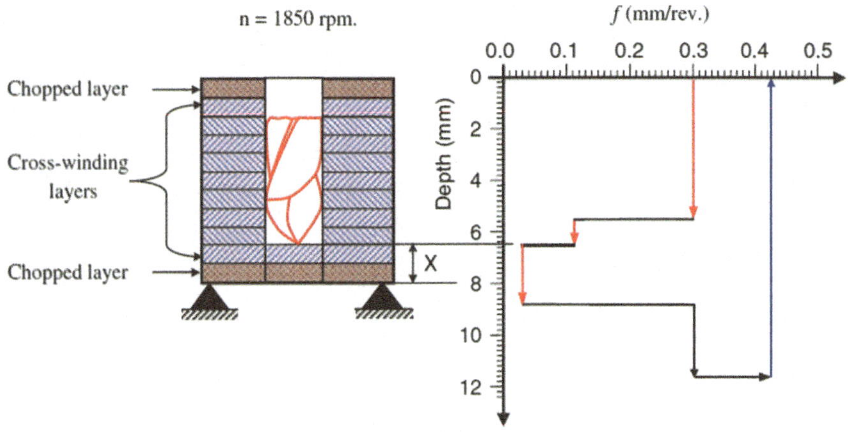

Fig. 23 Feed variation during drilling cross-winding composite (Khashaba 2004)

glass FRP composites. The delamination-free results can be achieved through the variable feed technique because of the gradual decreases in drilling thrust force.

With regard to this, Akmal et al. (2018) performed analytical as well as experimental investigations on the critical feed rate model for woven (90°/0°) flax FRP composite based on the HoCheng-Dharan model. They found that the lowest feed rate of 0.16 mm/rev and thrust force 124.70 N were the critical values for high tensile strength of flax FRP composite in the drilling process. However, the delamination damage can be observed for the parameters setting below 124.70 N. This could be attributed to other uncontrolled factors such as vibration due to drilling, dissimilar failure mechanisms of natural flax fibers compared to synthetic fibers, premature damage of the epoxy matrix, and other factors.

Karnik et al. (2008) and Rawat and Attia (2009a) performed an experimental investigation on the machinability of woven carbon FRP composite and tool-wear mechanism of tungsten carbide (WC) on dry high-speed drilling (10,000–15,000 RPM). It is well known that high-speed machining is not only capable of reducing the drilling thrust force and delamination damage but is also an advanced technology to improve the productivity and reduce production cost in industries. However, their results showed that both the thrust force and cutting force increased with increase in flank wear. This is likely caused by the low thermal conduction of carbon FRP composites, inducting the high temperature build up on the tool with continuous drilling at such high speeds. Beyond the critical flank wear of approximately 90 μm, there was a sudden rise in the matrix burnout and delamination in the cutting area. In short, aggressive tool wear was the critical problem while drilling at high speed using conventional twist drill bit.

On the basis of aforesaid limitations, the vibration-assisted twist drilling was introduced to enhance the drilling quality of FRP composites. Sadek et al. (2013) and Arul et al. (2006) performed a series of experiments using vibration drilling on the FRP composite to assess thrust force, flank wear, and delamination factors. The results revealed that the thrust force of vibration drilling is smaller than that of conventional drilling. It can also reduce cutting temperature by 50% and delamination damage associated with the drilling of FRP composites. Therefore, it can be stated that the vibration drilling method is appropriate for producing a hole in FRP composites.

Nonetheless, non-conventional drilling technology has been quite ineffective in the rapid removal of materials due to the wear issue in mass production. Additionally, the lack of experience and knowledge in handling these non-conventional drilling processes and their high investment cost may deter the application of non-conventional drilling processes in composite industries. Therefore, further work on optimising the performance of non-conventional drilling processes through robust approaches is needed.

6 Summary

The FRP composites are well recognized for their unique properties and widely used in various engineering application has caused a need for an understanding of the machinability of FRP composites. Unlike the well-established principle of metal cutting, the drilling of FRP composites possesses peculiar material properties and complex cutting edges of the drill bit that adds to the complexity in the cutting process. Delamination damage is one of the crucial failure mechanisms in drilling operation; which is a highly undesired problem in the assembly process. As mentioned before, the changing of drilling thrust force has shown significant influence in the size of delamination damage. Judging from the experimental thrust force responses, the feed rate and drill diameters were found to make the largest contribution to the drilling thrust force and delamination factors as compared to spindle speed. This is likely caused by increased shearing and cross-sectional area of the undeformed chip when drill diameter and feed rate are increased. These phenomena lead to the enhancement of the resistance of chip formation and consequently increase the cutting force and torque during the drilling process. If improper parameters are set in the drilling operation, high thrust force from chisel edge area will push forward the uncut material (like extrusion action), thus, in turn, serious delamination happens around the hole. Generally, for achieving high-quality holes without compromising the tool life and production rate, the overall desired and suggested parameters were combination low feed rate, high cutting speed and low point angle. An analytical model was developed to identify the critical thrust force at the onset of delamination during the drilling process. Validation results of the thrust force in the analytical and experimental study have indicated that the delamination could be alleviated once the critical thrust force is not exceeded. The developed critical thrust force can be an attractive benchmark or reference for industrial practice to control the drilling process so that delamination damage can be prevented for better assembly performance of the drilled FRP composites.

References

Abrão AM, Faria PE, Rubio JCC, Reis P, Davim JP (2007) Drilling of fiber reinforced plastics: a review. J Mater Process Technol 186(1–3):1–7. https://doi.org/10.1016/j.jmatprotec.2006.11.146

Akmal A, Nasir A, Iskandar A, Chye T, Aizat N (2018) Experimental study towards determination of critical feed for minimization of delamination damage in drilling flax natural fibre composites. Procedia CIRP 77:191–194. https://doi.org/10.1016/j.procir.2018.08.283

Altin Karataş M, Gökkaya H (2018) A review on machinability of carbon fiber reinforced polymer (CFRP) and glass fiber reinforced polymer (GFRP) composite materials. Defence Technol 14(4):318–326. https://doi.org/10.1016/j.dt.2018.02.001

Arputhabalan J, Prabhu S, Palanikumar K, Venkatesh S, Vijay K (2019) Assay of machining attributes in drilling of natural hybrid fiber reinforced polymer composite. Mater Today Proc 16:1097–1105. https://doi.org/10.1016/j.matpr.2019.05.201

Arul S, Vijayaraghavan L, Malhotra SK, Krishnamurthy R (2006) The effect of vibratory drilling on hole quality in polymeric composites. Int J Mach Tools Manuf 46(3–4):252–259. https://doi.org/10.1016/j.ijmachtools.2005.05.023

Astakhov VP (2010) Straight flute and twist drills. In: Geometry of single-point turning tools and drills: fundamentals and practical applications, pp 205–339. https://doi.org/10.1007/978-1-84996-053-3_4

Azrin Hani A, Hashim M, Lim T, Mariatti M, Ahmad R (2015) Impact behaviour of woven coir-epoxy composite: effects of woven density and woven fabric treatment. J Mater: design Appl 1(0):1–12. https://doi.org/10.1177/1464420714567744

Campos Rubio J, Abrao AM, Faria PE, Correia AE, Davim JP (2008) Effects of high speed in the drilling of glass fibre reinforced plastic: evaluation of the delamination factor. Int J Mach Tools Manuf 48(6):715–720. https://doi.org/10.1016/j.ijmachtools.2007.10.015

Capello E (2004) Workpiece damping and its effect on delamination damage in drilling thin composite laminates. J Mater Process Technol 148(2):186–195. https://doi.org/10.1016/S0924-0136(03)00812-4

Chen W (1997) Some experimental investigations in the drilling of carbon fibre-reinforced plastic (CFRP) composite laminates. Int J Mach Tools Manuf 37(8):1097–1108

Davim J, Reis P (2003) Study of delamination in drilling carbon fiber reinforced plastics (CFRP) using design experiments. Compos Struct 59(4):481–487. https://doi.org/10.1016/S0263-8223(02)00257-X

Dharan CK, Won M (2000) Machining parameters for an intelligent machining system for composite laminates. Int J Mach Tools Manuf 40(3):415–426. https://doi.org/10.1016/S0890-6955(99)00065-6

Durão LMP, Manuel J, Tavares RS, Hugo V, Albuquerque C De, Gonçalves DJS (2013) Damage evaluation of drilled carbon / epoxy laminates based on area assessment methods. Compos Struct 96: 576–583. https://doi.org/10.1016/j.compstruct.2012.08.003

Durão LMP, Panzera TH, Scarpa F, Filho SLMR, Oliveira PR (2015) Damage assessment of fibre reinforced laminates. Compos Struct 133:939–946. https://doi.org/10.1016/j.compstruct.2015.08.020

El-Sonbaty I, Khashaba UA, Machaly T (2004) Factors affecting the machinability of GFR/epoxy composites. Compos Struct 63(3–4):329–338. https://doi.org/10.1016/S0263-8223(03)00181-8

Eneyew ED, Ramulu M (2014) Experimental study of surface quality and damage when drilling unidirectional CFRP composites. J Mater Res Technol 3(4):354–362. https://doi.org/10.1016/j.jmrt.2014.10.003

Feito N, Díaz-Álvarez J, López-Puente J, Miguelez MH (2018a) Experimental and numerical analysis of step drill bit performance when drilling woven CFRPs. Compos Struct 184(August 2017):1147–1155. https://doi.org/10.1016/j.compstruct.2017.10.061

Feito N, Díaz-Álvarez J, López-Puente J, Miguelez MH (2018b) Experimental and numerical analysis of step drill bit performance when drilling woven CFRPs. Compos Struct 184(October 2017):1147–1155. https://doi.org/10.1016/j.compstruct.2017.10.061

Fernandes M, Cook C (2006) Drilling of carbon composites using a one shot drill bit. Part II: empirical modeling of maximum thrust force. Int J Mach Tools Manuf 46(1):76–79. https://doi.org/10.1016/j.ijmachtools.2005.03.016

Franke V (2011) Drilling of long fiber reinforced thermoplastics—influence of the cutting edge on the machining results. CIRP Ann Manuf Technol 60(1):65–68. https://doi.org/10.1016/j.cirp.2011.03.078

Gamdani F, Boukhili R, Vadean A (2015) Tensile strength of open-hole, pin-loaded and multi-bolted single-lap joints in woven composite plates. Mater Des 88:702–712. https://doi.org/10.1016/j.matdes.2015.09.008

Geng D, Liu Y, Shao Z, Lu Z, Cai J, Li X, Zhang D et al (2019) Delamination formation, evaluation and suppression during drilling of composite laminates: a review. Compos Struct 216(February):168–186. https://doi.org/10.1016/j.compstruct.2019.02.099

Harris B (1999) Engineering composite materials. Composites 18. https://doi.org/10.1016/0010-4361(87)90420-4
Heidary H, Karimi NZ, Minak G (2018) Investigation on delamination and flexural properties in drilling of carbon nanotube/polymer composites. Compos Struct 201(November 2017):112–120. https://doi.org/10.1016/j.compstruct.2018.06.041
Hocheng H, Dharan CKH (1990) Delamination during drilling in composite laminates. Trans ASME J Eng Ind 112:236–239
Hocheng H, Tsao CC (2003) Comprehensive analysis of delamination in drilling of composite materials with various drill bits. J Mater Process Technol 140(1–3):335–339. https://doi.org/10.1016/S0924-0136(03)00749-0
Hocheng H, Tsao CC (2005) The path towards delamination-free drilling of composite materials. J Mater Process Technol 167(2–3):251–264. https://doi.org/10.1016/j.jmatprotec.2005.06.039
Hocheng H, Tsao CC (2006) Effects of special drill bits on drilling-induced delamination of composite materials. Int J Mach Tool Manuf 46(12–13):1403–1416. https://doi.org/10.1016/j.ijmachtools.2005.10.004
Irwin GR, de Wit R (1983) Fracture Mechanics. J. Test Evaluation 11:56–65. https://doi.org/10.1007/1-4020-3153-X
Jain S, Yang DCH (1991) Effects of feed rate and chisel edge on delamination in composite drilling. Process Manuf Compos Mater 27(49):37–51
Jawaid M, Abdul Khalil HPS (2011) Cellulosic/synthetic fibre reinforced polymer hybrid composites: a review. Carbohydrate Polymers 86(1):1–18. https://doi.org/10.1016/j.carbpol.2011.04.043
Karnik SR, Gaitonde VN, Rubio JC, Correia AE, Abrão AM, Davim JP (2008) Delamination analysis in high speed drilling of carbon fiber reinforced plastics (CFRP) using artificial neural network model. Mater Des 29(9):1768–1776. https://doi.org/10.1016/j.matdes.2008.03.014
Khashaba UA (2004) Delamination in drilling GFR-thermoset composites. Compos Struct 63(3–4):313–327. https://doi.org/10.1016/S0263-8223(03)00180-6
Khashaba UA (2013) Drilling of polymer matrix composites: a review. J Compos Mater 47(15):1817–1832. https://doi.org/10.1177/0021998312451609
Khashaba UA, El-Sonbaty IA, Selmy AI, Megahed AA (2010a) Machinability analysis in drilling woven GFR/epoxy composites: part I – effect of machining parameters. Compos Part A: Appl Sci Manuf 41(3):391–400. https://doi.org/10.1016/j.compositesa.2009.11.006
Khashaba UA, El-Sonbaty IA, Selmy AI, Megahed AA (2010b) Machinability analysis in drilling woven GFR/epoxy composites: part II—effect of drill wear. Compos Part A Appl Sci Manuf 41(9):1130–1137. https://doi.org/10.1016/j.compositesa.2010.04.011
Kim GW, Lee KY (2005) Critical thrust force at propagation of delamination zone due to drilling of FRP/metallic strips. Compos Struct 69(2):137–141. https://doi.org/10.1016/j.compstruct.2004.06.013
Klocke F, Kuchle A (2011) Cutting tool materials and tools. Manufacturing processes 1: cutting, pp 95–196. https://doi.org/10.1007/978-3-642-11979-8_4
Krishnaraj V, Prabukarthi A, Ramanathan A, Elanghovan N, Kumar MS, Zitoune R, Davim JP (2012) Optimization of machining parameters at high speed drilling of carbon fiber reinforced plastic (CFRP) laminates. Compos Part B Eng 43(4):1791–1799. https://doi.org/10.1016/j.compositesb.2012.01.007
Kumar D, Sing KK (2017) Experimental analysis of delamination, thrust force and surface roughness on drilling of glass fibre reinforced polymer composites material using different drills. Mater Today Proc 4(8):7618–7627. https://doi.org/10.1016/j.matpr.2017.07.095
Lachaud F, Piquet R, Collombet F, Surcin L (2001) Drilling of composite structures. Compos Struct 52(5):511–516. https://doi.org/10.1016/S0263-8223(01)00040-X
Lin SC, Chen IK (1996) Drilling carbon fiber-reinforced composite material at high speed. Wear 194(1–2):156–162. https://doi.org/10.1016/0043-1648(95)06831-7
Michael Kutner H, Nachtsheim Christopher J, Neter J (2004) In: Gordon B (ed) Applied linear regression models, 4th edn. McGraw Hill/Irwin, Americas, New York

Mohan NS, Ramachandra A, Kulkarni SM (2005) Influence of process parameters on cutting force and torque during drilling of glass-fiber polyester reinforced composites. Compos Struct 71(3–4):407–413

Murphy C, Byrne G, Gilchrist MD (2002) The performance of coated tungsten carbide drills when machining carbon fibre-reinforced epoxy composite materials. Proc Inst Mech Eng Part B, J Eng Manuf 216(2):143–152. https://doi.org/10.1243/0954405021519735

Nagaraja, Herbert MA, Shetty D, Shetty R, Shivamurthy B (2013) Effect of process parameters on delamination, thrust force and torque in drilling of carbon fiber epoxy composite. Res J Recent Sci 2(8):47–51

Palanikumar K, Muniaraj A (2014a) Experimental investigation and analysis of thrust force in drilling cast hybrid metal matrix (Al–15%SiC–4%graphite) composites. Measurement 53:240–250. https://doi.org/10.1016/j.measurement.2014.03.027

Palanikumar K, Muniaraj A (2014b) Experimental investigation and analysis of thrust force in drilling cast hybrid metal matrix (Al-15%SiC-4%graphite) composites. Measur: J Int Measur Confederation 53:240–250. https://doi.org/10.1016/j.measurement.2014.03.027

Phadnis VA, Makhdum F, Roy A, Silberschmidt VV (2013) Drilling in carbon/epoxy composites: experimental investigations and finite element implementation. Compos Part A: Appl Sci Manuf 47:41–51. https://doi.org/10.1016/j.compositesa.2012.11.020

Pyo Jung J, Woo Kim G, Yong Lee K (2005) Critical thrust force at delamination propagation during drilling of angle-ply laminates. Compos Struct 68(4):391–397. https://doi.org/10.1016/j.compstruct.2004.04.004

Rahme P, Landon Y, Lachaud F, Piquet R, Lagarrigue P (2015) Composites: part A delamination-free drilling of thick composite materials. Compos Part A 72:148–159. https://doi.org/10.1016/j.compositesa.2015.02.008

Rajamurugan TV, Shanmugam K, Palanikumar K (2013) Analysis of delamination in drilling glass fiber reinforced polyester composites. Mater Des 45:80–87. https://doi.org/10.1016/j.matdes.2012.08.047

Ramulu M, Branson T, Kim D (2001) A study on the drilling of composite and titanium stacks. Compos Struct 54(1):67–77. https://doi.org/10.1016/S0263-8223(01)00071-X

Rawat S, Attia H (2009a) Characterization of the dry high speed drilling process of woven composites using Machinability Maps approach. CIRP Ann Manuf Technol 58(1):105–108. https://doi.org/10.1016/j.cirp.2009.03.100

Rawat S, Attia H (2009b) Wear mechanisms and tool life management of WC–Co drills during dry high speed drilling of woven carbon fibre composites. Wear 267(5–8):1022–1030. https://doi.org/10.1016/j.wear.2009.01.031

Sadek A, Attia MH, Meshreki M, Shi A (2013) Characterization and optimization of vibration-assisted drilling of fibre reinforced epoxy laminates. CIRP Ann Manuf Technol 62:91–94

Saoudi J, Zitoune R, Mezlini S, Gururaja S, Seitier P (2016) Critical thrust force predictions during drilling: analytical modeling and X-ray tomography quantification. Compos Struct 153:886–894. https://doi.org/10.1016/j.compstruct.2016.07.015

Shaw M (1957) On the drilling of metals ii-the torque and thrust of drilling. J Manuf Sci Eng Trans ASME 79:139–148

Shyha I, Soo SL, Aspinwall D, Bradley S (2010) Effect of laminate configuration and feed rate on cutting performance when drilling holes in carbon fibre reinforced plastic composites. J Mater Process Technol 210(8):1023–1034. https://doi.org/10.1016/j.jmatprotec.2010.02.011

Singh I, Bhatnagar N (2006) Drilling of uni-directional glass fiber reinforced plastic (UD-GFRP) composite laminates. Int J Adv Manuf Technol 27(9): 870–876. https://doi.org/10.1007/s00170-004-2280-7

Singh I, Bhatnagar N, Viswanath P (2008) Drilling of uni-directional glass fiber reinforced plastics: experimental and finite element study. Mater Des 29(2):546–553

Singh RVS, Latha B, Senthilkumar VS (2009) Modeling and analysis of thrust force and torque in drilling GFRP composites by multi-facet drill using fuzzy logic. Int J Rec Trend Eng 1(5):66–70

Sorrentino L, Turchetta S, Bellini C (2018a) A new method to reduce delaminations during drilling of FRP laminates by feed rate control. Compos Struct 186(November 2017):154–164. https://doi.org/10.1016/j.compstruct.2017.12.005

Sorrentino L, Turchetta S, Bellini C (2018b) A new method to reduce delaminations during drilling of FRP laminates by feed rate control. Compos Struct 186(September 2017):154–164. https://doi.org/10.1016/j.compstruct.2017.12.005

Sreejith PS, Krishnamurthy R, Narayanasamy K, Malhotra SK (1999) Studies on the machining of carbon/phenolic ablative composites. J Mater Process Technol 88(1):43–50. https://doi.org/10.1016/S0924-0136(98)00377-X

Suppakul P, Bandyopadhyay S (2002) The effect of weave pattern on the mode-I interlaminar fracture energy of E-glass/vinyl ester composites. Compos Sci Technol 62(5), 709–717. https://doi.org/10.1016/S0266-3538(01)00220-2

Tan CL, Azmi AI, Muhammad N (2017) Critical thrust force for on-set delamination of hybrid FRP composite during drilling process. Key Eng Mater 740:111–117. https://doi.org/10.4028/www.scientific.net/kem.740.111

Tsao CC (2012) Effect of induced bending moment (IBM) on critical thrust force for delamination in step drilling of composites. Int J Mach Tools Manuf 59:1–5. https://doi.org/10.1016/j.ijmachtools.2012.03.001

Tsao CC, Hocheng H (2003) The effect of chisel length and associated pilot hole on delamination when drilling composite materials. Int J Mach Tools Manuf 43(11):1087–1092. https://doi.org/10.1016/S0890-6955(03)00127-5

Tsao CC, Hocheng H (2007) Parametric study on thrust force of core drill. J Mater Process Technol 192–193:37–40. https://doi.org/10.1016/j.jmatprotec.2007.04.062

Tsao CC, Hocheng H (2008a) Effects of peripheral drilling moment on delamination using special drill bits. J Mater Process Technol 201(1–3):471–476. https://doi.org/10.1016/j.jmatprotec.2007.11.225

Tsao CC, Hocheng H (2008b) Evaluation of thrust force and surface roughness in drilling composite material using Taguchi analysis and neural network. J Mater Process Technol 203(1–3):342–348. https://doi.org/10.1016/j.jmatprotec.2006.04.126

Upputuri HB, Nimmagadda VS, Duraisamy E (2019) Optimization of drilling parameters on carbon fiber reinforced polymer composites using fuzzy logic. Mater Today: Proc. https://doi.org/10.1016/j.matpr.2019.05.400

Wan M, Li SE, Yuan H, Zhang WH (2019) Cutting force modelling in machining of fiber-reinforced polymer matrix composites (PMCs): a review. Compos Part A Appl Sci Manuf 117(October 2018):34–55. https://doi.org/10.1016/j.compositesa.2018.11.003

Wang X, Kwon PY, Sturtevant C, Kim DD, Lantrip J (2013) Tool wear of coated drills in drilling CFRP. J Manuf Process 15(1):127–135. https://doi.org/10.1016/j.jmapro.2012.09.019

Wisnom MR, Hallett SR (2009) Composites: part A the role of delamination in strength, failure mechanism and hole size effect in open hole tensile tests on quasi-isotropic laminates. Compos Part A 40(4):335–342. https://doi.org/10.1016/j.compositesa.2008.12.013

Won MS, Dharan CKH (2002a) Chisel edge and pilot hole effects in drilling composite laminates. J Manuf Sci Eng 124(2):242. https://doi.org/10.1115/1.1448317

Won M, Dharan C (2002b) Drilling of aramid and carbon fiber polymer composites. J Manuf Sci Eng Trans ASME 124:778–783

Wong T, Wu S, Croy G (1982) Material, an analysis of delamination in drilling composite. In: 14th national SAMPE technology conference, Atlanta, pp 471–483

Zhenchao Q, Zhang K, Li Y, Liu S, Cheng H (2014) Critical thrust force predicting modeling for delamination-free drilling of metal-FRP stacks. Compos Struct 107:604–609. https://doi.org/10.1016/j.compstruct.2013.07.036

Numerical and Analytical Approaches in Machining of FRP Composites

Ozden Isbilir

Abstract The superior mechanical properties of the fiber reinforced polymer laminated composites attracts design engineers, so these materials are used widely in various industries including aerospace, defence and military. Fiber reinforced polymer composites are generally produced near net shape, however machining operations are oftenly employed to provide geometric and dimensional accuracy. Nevertheless, their unique properties affect their machinability that can hinder the size of their application and the costs. This chapter presents a brief overview on the analytical and numerical modeling approaches in machining processes of fiber reinforced polymer composites based on the reported studies. Initially, the modelling of the mechanical response of a fiber reinforced polymer composite is explained in lamina level. Damage mechanisms are discussed and a number of damage initiation criteria are reported. The analytical and numerical models of different machining processes are introduced and discussed. The current drawbacks are explained and trends summarised for the future studies.

1 Introduction

Manufacturing is an indicator of a country's production power, hence lies at the heart of any industrialized society. The importance of manufacturing can be observed from the economical reports of any country. Manufacturing activities can cover between 15 and 20% of the total monetary value of all goods and services produced. Due to this fact, the size of manufacturing activities can be straightly linked to the size of a country's economy and may reflect the prosperity and living standards of a society.

Manufacturing includes different types of processes such as casting, moulding, forming, machining, joining, heat treatments, surface treatments, and modern times' popular method: additive manufacturing. These processes have been gradually developed over the years due to the increased needs of the society. Demands for higher

O. Isbilir (✉)
Mechanical Engineering Department, Engineering Faculty, Karabuk University, Karabuk, Turkey
e-mail: ozdenisbilir@karabuk.edu.tr

M. T. Hameed Sultan et al. (eds.), *Machining and Machinability of Fiber Reinforced Polymer Composites*, Composites Science and Technology,
https://doi.org/10.1007/978-981-33-4153-1_3

quality, minimum cost and higher productivity have brought the technological developments in the last couple of decades. Hence, the modern manufacturing operations have extreme complexity and technological sophistication. Manufacturing processes need to be performed in the way to provide these technical demands.

Machining has been the core of the manufacturing processes since the industrial revolution. Machining can be explained simply the removal of extra material from a workpiece by using a cutting tool and a machine tool to reach the required dimensions with surface quality. In modern manufacturing environments, a high amount of automation and computer numeric controlled (CNC) machines are preferred for competition and rapid adaptation where it brings a high amount of costs. Therefore, the economical success of a company depends on the level of ability that the company operates these systems efficiently. Nevertheless, this is not a straightforward task due to the high number of variables and complexity of the nature of any machining operation. Typical inputs of machining processes include the followings:

- Machine tool features
- Cutting tool feature
- Cutting parameters
- Work material feature including composition, mechanical properties, thermal properties, chemical properties, microstructure, geometry, etc.
- Cutting environment feature including the use of coolant and its properties, removal of chips, etc.

Machining process outputs are dependent on the process inputs and include the followings:

- Surface integrity including surface defects, microstructural alterations, surface roughness, residual stress, etc.
- Geometric and dimensional accuracy
- Tool life
- Induced cutting forces
- Induced cutting temperature
- Chip feature
- Noise and vibrations.

Composite materials offer great properties such as high strength-to-weight ratio, high modulus-to-weight ratio, good damage tolerance, good fatigue and corrosion resistance, which make them highly preferably against conventional materials for many applications. Composite materials are broadly used in a variety of industries including aerospace, defence, military, construction and so on. Nevertheless, the wider applications of these materials can be hindered owing to the costs and difficulties associated with their manufacturing. Fiber reinforced composites are greatly produced near net shape, but machining operations are often needed to provide geometric and dimensional accuracy. Machining of FRPs may be performed before or after the fabrication. The machining processes before the lay-up and curing include the preparation of the reinforcement material and prepregs. However, machining of FRPs after fabrication will be discussed in this chapter.

FRPs may require machining processes due to the high technical demands after curing. Conventional machining methods such as drilling, edge trimming, milling, and turning are commonly applied methods for FRPs. Some of the non-traditional machining methods such as abrasive waterjet cutting, laser beam cutting and electrical discharge machining are also preferred. Most of these methods were developed for the metal machining industries. These processes can be used for the machining of FRPs with extra care since FRPs are different from metals in many ways. In brief, FRPs have inhomogeneous structure that contain very different constituents. The reinforcement has a stiff and brittle feature whereas the matrix is weak and ductile what means the constituents exhibit very different mechanical behaviour. The machining of metals is mostly described with shearing and high plastic deformation. The cutting mostly forms continuous chip. The cutting process reaches a steady-state under a constant cutting condition where the machining outputs variables can be predicted with an acceptable tolerance. Whereas, the machining of FRPs is identified by mostly brittle behaviour and dusty chip formation. Due to the inhomogeneous and anisotropic structure of the FRPs, mechanical behaviours of the workpieces are completely different from metals. The cutting forces are affected by the angle of the fibers and dissimilarity of the constituents, thus they typically oscillate. The induced cutting temperatures can be influenced by the thermal properties of the constituents and orientation of the reinforcements. The polymeric matrix can be affected from cutting temperatures since they are not durable at high temperatures like metals. Different thermal expansion coefficients of the constituents may cause different deformations and stresses. For all these reasons, preventative actions should be taken carefully from the negative effects of excessive heat.

In general terms, the success of any machining operation depends on the choice of appropriate process input parameters since these parameters can play an important role on the process output variables, the machining cost and the machining time. An operator may select some of the input parameters using his own experience or from the various sources. However, a small change in a single parameter or an improper selection of any parameter can influence the process performance significantly since the process depends on many factors. Hence, achieving the optimum process performance is rarely possible even for a skilled operator. If the material is a type of composite, optimisation even becomes much more difficult. An effective method to deal with this problem is to seek the relationship between the input and output variables of the process by modelling it using suitable mathematical approaches. In the following of this chapter, the focus will be on the analytical and numerical modelling of the machining processes of the fiber reinforced composites.

2 Modelling of Fiber Reinforced Composite Material

Modelling the behaviour of any material is the backbone particularly in finite element analysis. Although having four decades of extensive research experience, a complete and approved procedure for the estimation of the behaviour of composite materials has not yet been fully accomplished because of their complex nature. The behaviour and the performance of the polymer composite structures depend on a variety of parameters including different behaviour of constituent materials, forms and proportions of the constituents, heterogeneous microstructure, the existence of interfaces, anisotropy, manufacturing processes, lay-up, geometry, loading conditions, and failure mechanisms.

2.1 Constitutive Model

The constitutive model of a material or commonly named constitutive law is the mathematical formulation used to estimate mechanical response of a material under loading condition before damage initiation. The mechanical behaviour of solid materials is generally defined as stress-strain relation, however, the stress can be a function of strain, strain rate, temperature, grain size, hardness and material properties.

When fiber reinforced polymer composites are considered, high-strength fibers are embedded in a polymer-based matrix, such as epoxies. Glass, carbon or kevlar fiber reinforced polymer composites are among the most commonly used examples of these type of polymer matrix composites. Laminated composite structures are made through the overlapping of several layers (laminate) with various fiber orientations as required by the stacking sequence. One of the main differences between fiber reinforced polymer composites and traditional engineering materials is that a composite's mechanical behaviour depends on direction of loading due to reinforcements. Engineers must be able to estimate the behaviour of each lamina individually to analyse the response of an entire laminating composite structure. Therefore, material constitutive law in this part is summarised for the lamina level.

A lamina in a unidirectional composite is characterised by having all fibers oriented in the same direction whereas fibers aligned in different directions in multidirectional composites. The former model allows treating the lamina as an orthotropic material. But in practice, fibers are not completely parallel to each other and uniformly oriented within the lamina. According to the latter model, lamina can be treated as an anisotropic material which is the most general form of material. In these macro-scale approaches, composite material is assumed to have linear elasticity with a homogeneous microstructure where material properties are considered the same at every point within the material. The stress-strain relationship is expressed in the following compact form:

$$[\sigma] = [C][\varepsilon]$$

where $[\sigma]$, $[C]$, $[\varepsilon]$ are known as the stress matrix, the stiffness matrix and strain matrix, respectively.

Similar to that, the strain–stress relation can be expressed by inverting the stiffness matrix in the stress-strain relation in the following form.

$$[\varepsilon] = [C]^{-1}[\sigma] = [S][\sigma]$$

where $[S]$ is the elastic compliance matrix, and is symmetric.

The heterogeneous microstructure, dissimilar properties of constituents, directionality of reinforcements cause to anisotropy in the composite structure. The generalised form of the stress-strain relationship for an anisotropic material is given by:

$$\begin{bmatrix} \sigma_{11} \\ \sigma_{22} \\ \sigma_{33} \\ \sigma_{23} \\ \sigma_{31} \\ \sigma_{12} \end{bmatrix} = \begin{bmatrix} C_{11} & C_{12} & C_{13} & C_{14} & C_{15} & C_{16} \\ C_{12} & C_{22} & C_{23} & C_{24} & C_{25} & C_{26} \\ C_{13} & C_{23} & C_{33} & C_{34} & C_{35} & C_{36} \\ C_{14} & C_{24} & C_{34} & C_{44} & C_{45} & C_{46} \\ C_{15} & C_{25} & C_{35} & C_{45} & C_{55} & C_{56} \\ C_{16} & C_{26} & C_{36} & C_{46} & C_{56} & C_{66} \end{bmatrix} \begin{bmatrix} \varepsilon_{11} \\ \varepsilon_{22} \\ \varepsilon_{33} \\ 2\varepsilon_{23} \\ 2\varepsilon_{31} \\ 2\varepsilon_{12} \end{bmatrix}$$

where, the 1, 2, and 3 axes are reinforcement direction, in-plane and out-of-plane directions, respectively (shown in Fig. 1). As can be noticed from the stiffness matrix, 21 independent elastic constants must be determined by experiments.

If there is one plane of symmetry such as a parallel plane to the 1–2 plane, then the material is called monoclinic. The stress-strain relationship for a monoclinic material is given by:

$$\begin{bmatrix} \sigma_{11} \\ \sigma_{22} \\ \sigma_{33} \\ \sigma_{23} \\ \sigma_{31} \\ \sigma_{12} \end{bmatrix} = \begin{bmatrix} C_{11} & C_{12} & C_{13} & 0 & 0 & C_{16} \\ C_{12} & C_{22} & C_{23} & 0 & 0 & C_{26} \\ C_{13} & C_{23} & C_{33} & 0 & 0 & C_{36} \\ 0 & 0 & 0 & C_{44} & C_{45} & 0 \\ 0 & 0 & 0 & C_{45} & C_{55} & 0 \\ C_{16} & C_{26} & C_{36} & 0 & 0 & C_{66} \end{bmatrix} \begin{bmatrix} \varepsilon_{11} \\ \varepsilon_{22} \\ \varepsilon_{33} \\ 2\varepsilon_{23} \\ 2\varepsilon_{31} \\ 2\varepsilon_{12} \end{bmatrix}$$

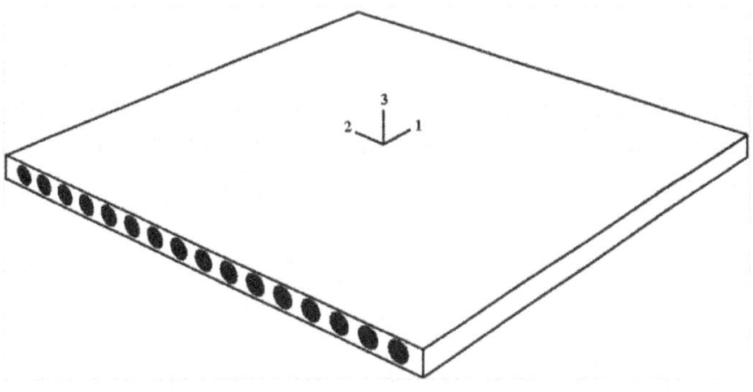

Fig. 1 Typical fiber reinforced composite lamina

If the material has two mutually orthogonal planes of symmetry, the plane which is orthogonal to these planes is also a plane of symmetry. This material is named as orthotropic. If the out-of-plane stress components are not ignorable and the lamina is unidirectional, orthotropic material is commonly adopted for the constitutive law. There are nine independent constants and twelve nonzero terms associated with the stiffness matrix of an orthotropic material. It should be noted that there is no interaction between normal and shear behaviour. The stress-strain relations takes the following form.

$$\begin{bmatrix} \sigma_{11} \\ \sigma_{22} \\ \sigma_{33} \\ \sigma_{23} \\ \sigma_{31} \\ \sigma_{12} \end{bmatrix} = \begin{bmatrix} C_{11} & C_{12} & C_{13} & 0 & 0 & 0 \\ C_{12} & C_{22} & C_{23} & 0 & 0 & 0 \\ C_{13} & C_{23} & C_{33} & 0 & 0 & 0 \\ 0 & 0 & 0 & C_{44} & 0 & 0 \\ 0 & 0 & 0 & 0 & C_{55} & 0 \\ 0 & 0 & 0 & 0 & 0 & C_{66} \end{bmatrix} \begin{bmatrix} \varepsilon_{11} \\ \varepsilon_{22} \\ \varepsilon_{33} \\ 2\varepsilon_{23} \\ 2\varepsilon_{31} \\ 2\varepsilon_{12} \end{bmatrix}$$

If the has the same elastic properties in all directions in a plane, it is called transversely isotropic. In such case, there is an axis of material symmetry in addition to three planes of symmetry. If the 3-axis is considered to coincide with the axis of symmetry, other two axes can be in any direction where 1–2 plane is called an isotropic plane. A transversely isotropic material has five independent elastic constants and twelve nonzero terms. The constitutive law of transversely isotropic material is expressed as:

$$\begin{bmatrix} \sigma_{11} \\ \sigma_{22} \\ \sigma_{33} \\ \sigma_{23} \\ \sigma_{31} \\ \sigma_{12} \end{bmatrix} = \begin{bmatrix} C_{11} & C_{12} & C_{13} & 0 & 0 & 0 \\ C_{12} & C_{11} & C_{13} & 0 & 0 & 0 \\ C_{13} & C_{13} & C_{33} & 0 & 0 & 0 \\ 0 & 0 & 0 & C_{44} & 0 & 0 \\ 0 & 0 & 0 & 0 & C_{44} & 0 \\ 0 & 0 & 0 & 0 & 0 & C_{66} \end{bmatrix} \begin{bmatrix} \varepsilon_{11} \\ \varepsilon_{22} \\ \varepsilon_{33} \\ 2\varepsilon_{23} \\ 2\varepsilon_{31} \\ 2\varepsilon_{12} \end{bmatrix}$$

where $C_{66} = \frac{C_{11}-C_{12}}{2}$.

For an isotropic material, all planes are planes of material symmetry and are isotropic. There are only two independent constants and 12 nonzero terms in the stiffness matrix associated with isotropic material. The stress-strain relations takes the following form.

$$\begin{bmatrix} \sigma_{11} \\ \sigma_{22} \\ \sigma_{33} \\ \sigma_{23} \\ \sigma_{31} \\ \sigma_{12} \end{bmatrix} = \begin{bmatrix} C_{11} & C_{12} & C_{12} & 0 & 0 & 0 \\ C_{12} & C_{11} & C_{12} & 0 & 0 & 0 \\ C_{12} & C_{12} & C_{11} & 0 & 0 & 0 \\ 0 & 0 & 0 & C_{44} & 0 & 0 \\ 0 & 0 & 0 & 0 & C_{44} & 0 \\ 0 & 0 & 0 & 0 & 0 & C_{44} \end{bmatrix} \begin{bmatrix} \varepsilon_{11} \\ \varepsilon_{22} \\ \varepsilon_{33} \\ 2\varepsilon_{23} \\ 2\varepsilon_{31} \\ 2\varepsilon_{12} \end{bmatrix}$$

where $C_{44} = \frac{C_{11} - C_{12}}{2}$.

A lamina or a ply is thin compared to other dimensions of the whole laminate. Hence, a lamina can be considered as in-plane stress condition, where all the off-plane stress components ($\sigma_3, \sigma_4, \sigma_5$) are assumed zero. In such a case, the constitutive law of an individual lamina can be simplified (into 2D) as below.

$$\begin{bmatrix} \sigma_{11} \\ \sigma_{22} \\ \sigma_{12} \end{bmatrix} = \begin{bmatrix} Q_{11} & Q_{12} & 0 \\ Q_{12} & Q_{22} & 0 \\ 0 & 0 & C_{66} \end{bmatrix} \begin{bmatrix} \varepsilon_{11} \\ \varepsilon_{22} \\ 2\varepsilon_{12} \end{bmatrix}$$

2.2 *Damage Mechanisms*

When engineers design structures, they consider some function for them. If the objective of a structure is to carry loads, then it must have adequate load-bearing capacity over the service life without losing its integrity. This is a common concern in design stages regardless of the type of material. However, there can be various issues concerning the type of material used in the application, particularly for composite materials. Having heterogeneous microstructure, significant property differences of constituents, the presence of interfaces, directionality of fibers and anisotropy cause to complex material behaviour compared to monolithic materials. Due to different phases of composite structures, deformation, transfer of the stresses and damage mechanisms are much more sophisticated without any doubt. Composite materials present a broad range of damage mechanisms. When we consider intrinsic tailoring, anisotropic behaviour and lay-out of the laminated composite structures, these damage mechanisms can be classified generally as fiber phase damages, matrix phase damages, and interface-based damages.

Fiber failure is one of the most serious damage mechanisms in a fiber reinforced polymer composite structure. When applied loads cause to fracture in the fibers, it is identified as fiber failure. Fiber failure is an intralaminar form and major type damage, hence the breakage of fibers causes to the entire damage of a fiber reinforced polymer composite structure. Fiber failure can be observed in tension, compression and shear. If a unidirectional composite is imposed in tension in the fiber direction, individual fibers fail where they are the weakest. After the first local fiber failures, stress redistributed between fibers and matrix that can affect other fibers in the vicinity of the failed ones and possibly ends up with new fiber failure. When a unidirectional composite is loaded in compression or shear, buckling and kinking of fibers can occur. Under compressive loads, out-of-phase or in-phase deformations can provoke buckling in fibers. Moreover, as Budiansky (1983), Budiansky and Fleck (1993), Niu and Talreja (2000) indicated initial misalignment of fibers due to manufacturing processes or any misalignment in the loading system can create local shear and produce kinking. This both mechanisms produce instability in the strength, can promote other types of damages in the composite structure and leads to structural failure.

Fiber reinforced polymer laminated composites provide high mechanical properties in the reinforcement direction. However, these properties are generally weak in the transverse directions. Therefore, damage mechanisms can occur and develop easily in the matrix phase. As Nairn (2000) expressed, matrix damages are an intralaminar form of damages such as cracks or voids and usually the first observed form of damage in fiber reinforced polymer composites. These damages can occur due to tensile and compressive loading, fatigue loading, and thermal loading. They can be associated with manufacturing-induced defects or occurred due to the development of interface damages such as fiber/matrix debonds (delamination). Although matrix damages do not cause an entire failure by itself, they can result in a significant degradation in the stiffness of the composite structure. They can promote other types of damage, such as delamination and fiber breakage and consequently lead to a structural failure.

As mentioned above, the presence of interfaces makes composites different from monolithic materials. The behaviour and the performance of a fiber reinforced polymer composite structure are remarkably affected by the properties of the interface between constituents. The interface bonds the fibers and matrix adhesively, hence its service is critical in deformation and transfer of stresses between fiber and matrix. Thus controlling interfacial properties can provide management of the performance of a composite structure. In unidirectional laminated composites, interface damages can be observed as intralaminar and interlaminar form. Debonding is the intralaminar type damage and induces at the interface between fiber and matrix when the interface is weak. In such damage, fiber and matrix do not adhere to each other, hence they act separately. This leads to the formation of matrix damages.

In laminated composites, fracture of the interfaces between adjacent laminae in a laminate lead to separation of them. This is an interlaminar form of damage, or commonly called delamination. Delamination can induce at free edges and unprotected surfaces through the thickness (out-of-plane direction). If the laminated structure is loaded, out-of-plane normal and shear stress components develop and can cause cracks between the adjacent laminae. As Choi et al. (1999) and Khashaba (2004) stated the development of these cracks produces delamination that leads to an accelerated deterioration of the structural integrity and even may lead to ultimate damage of the composite structure.

2.3 Damage Initiation Criteria

Once any damage mechanisms occur in the structure, the material behaviour is considered to have been irreversibly affected. In this case, the constitutive model needs to be updated according to the onset and progression of the damage. This typically involves appropriate damage criteria for the onset of the damage and degradation rules for the progress of the damage. The damage criteria used to predict damage mechanisms in composite materials can be categorised in several ways. Some damage criteria are based on strength whereas some of them are based on strain and some of them are based on fracture mechanics theories. Another classification can be made according to the region, intralaminar and interlaminar damages. Some damage criteria focus only on intralaminar damages, whereas some of them also include interlaminar damages. The damage criteria can be also grouped as the ones neglect the interactions between different stress components or the ones consider interactions between different stress components. The latter ones aim to distinguish different damage mechanisms. As mentioned above, researchers have developed different damage criteria. The most common damage criteria, particularly for laminated fiber reinforced polymer composite materials, are summarised in Table 1 with the references. In the table, X, Y, Z are the normal strengths in the fiber (longitudinal), normal to fiber (transverse), out-of-plane (through-thickness) directions; S, S_L, S_T are the in-plane shear strength in 2D, the longitudinal and transverse shear strength components in 3D; subscripts 1, 2 and 3 denote the longitudinal, transverse and through-thickness directions; subscripts for strengths and damage index T and C refer to tension and compression; subscripts for damage index F and M denote fiber and matrix; subscript "is" refers to "in situ". Interested readers for all other symbols and abbreviations used for the damage criteria are encouraged to consult the referenced papers given in the table.

It should be mentioned that the analysis at the scale of an individual lamina is macro-level modelling. The equations used to define stiffness and damage models are based on experimental results at macro-scale. However, for more detailed analysis micro-level modelling can be utilised. In such a case, more complex models may be required for individual behaviour of the constituents and interfaces. This means that stress-strain relationships and damage models must be based on experimental studies at micro-scale. In this level model, the presence of voids or other imperfections may be included as well.

2.4 Damage Modelling

Modelling of damage is usually complex due to the nature of laminated structüre composite. There are two steps of modelling of damage. The first step is the onset of damage, which is predicted by the damage criteria mentioned above. Although these criteria do not usually cause an entire failure of the structure, it is necessary to account the degradation in the structural performance due to any damage to estimate the overall mechanical response of the composite accurately. The second step covers the evolution of the damage. By the application of damage mechanics approach, material stiffness is degraded when a damage criterion is met. This common approach can be applied to both intralaminar and interlaminar damage through the material constitutive behaviour. There are two degradation strategies. Instantaneous damage approach assumes a sudden degradation at the onset of damage initiation. Thus, the common practice in this approach is the reduction of the related stiffness values instantly to

Table 1 Damage initiation criteria

Name of the damage model	Condition	Damage initiation formulation	Remarks
Maximum stress	if $\sigma_{11} \geq 0$	$F_{FT} = \frac{11}{X_T}$	
	if $\sigma_{11} < 0$	$F_{FC} = \frac{11}{X_C}$	
	if $\sigma_{22} \geq 0$	$F_{MT} = \frac{22}{Y_T}$	
	if $\sigma_{22} < 0$	$F_{MC} = \frac{22}{Y_C}$	
		$F_S = \frac{12}{S}$	
Maximum strain	if $\sigma_{11} \geq 0$	$F_{FT} = \frac{\varepsilon_{11}}{\varepsilon_{XT}}$	
	if $\sigma_{11} < 0$	$F_{FC} = \frac{\varepsilon_{11}}{\varepsilon_{XC}}$	
	if $\sigma_{22} \geq 0$	$F_{MT} = \frac{\varepsilon_{22}}{\varepsilon_{YT}}$	
	if $\sigma_{22} < 0$	$F_{MC} = \frac{\varepsilon_{22}}{\varepsilon_{YC}}$	
		$f_s = \frac{\varepsilon_{12}}{\varepsilon_S}$	
Azzi and Tsai (1965)	if $\sigma_{11} \geq 0, X = X_T,$ otherwise$X = X_C$ if $\sigma_{22} \geq 0, Y = Y_T,$ otherwise$Y = Y_C$	$F = \left(\frac{\sigma_{11}}{X}\right)^2 + \left(\frac{\sigma_{22}}{Y}\right)^2 + \left(\frac{\sigma_{12}}{S}\right)^2 - \frac{\sigma_{11}\sigma_{22}}{X^2}$	2-D
Tsai and Wu (1971)	$F = A_{11}\sigma_{11}^2 + 2A_{12}\sigma_{11}\sigma_{22} + A_{22}\sigma_{22}^2 + A_{66}\sigma_{12}^2 + B_1\sigma_{11} + B_2\sigma_{22}$ where; $A_{11} = \frac{1}{X_T X_C}, A_{22} = \frac{1}{Y_T Y_C}, A_{12} = -\frac{1}{2\sqrt{X_T X_C Y_T Y_C}}, A_{66} = \frac{1}{S_A S_T}, B_1 = \frac{1}{X_T} - \frac{1}{X_C}, B_2 = \frac{1}{Y_T} - \frac{1}{Y_C}$		2-D
Hoffman (1967)	$F = \left(\frac{1}{X_T} - \frac{1}{X_C}\right)\sigma_{11} + \left(\frac{1}{Y_T} - \frac{1}{Y_C}\right)\sigma_{22} + \left(\frac{1}{X_T X_C}\right)\sigma_{11}^2 + \left(\frac{1}{Y_T Y_C}\right)\sigma_{22}^2 + \left(\frac{1}{S^2}\right)\sigma_{12}^2 - \left(\frac{1}{X_T X_C}\right)\sigma_{11}\sigma_{22}$		2-D

(continued)

Table 1 (continued)

Name of the damage model	Condition	Damage initiation formulation	Remarks
	$F = C_1(\sigma_{22} - \sigma_{33})^2 + C_2(\sigma_{33} - \sigma_{11})^2 + C_3(\sigma_{11} - \sigma_{22})^2 + C_4\sigma_{11} + C_5\sigma_{22} + C_6\sigma_{33} + C_7\sigma_{23}^2 + C_8\sigma_{13}^2 + C_9\sigma_{12}^2$ where; $C_1 = \frac{1}{2}\left(\frac{1}{Z_T Z_C} + \frac{1}{Y_T Y_C} - \frac{1}{X_T X_C}\right)$, $C_2 = \frac{1}{2}\left(\frac{1}{X_T X_C} + \frac{1}{Z_T Z_C} - \frac{1}{Y_T Y_C}\right)$, $C_3 = \frac{1}{2}\left(\frac{1}{X_T X_C} + \frac{1}{Y_T Y_C} - \frac{1}{Z_T Z_C}\right)$, $C_4 = \left(\frac{1}{X_T} - \frac{1}{X_C}\right)$, $C_5 = \left(\frac{1}{Y_T} - \frac{1}{Y_C}\right)$, $C_6 = \left(\frac{1}{Z_T} - \frac{1}{Z_C}\right)$, $C_7 = \frac{1}{\sigma_{23}^2}$, $C_8 = \frac{1}{\sigma_{13}^2}$, $C_9 = \frac{1}{\sigma_{12}^2}$		3-D
Hashin and Rotem (1973)	if $\sigma_{11} \geq 0$	$F_{FT} = \frac{11}{X_T}$	2-D
	if $\sigma_{11} < 0$	$F_{FC} = \left\lvert \frac{11}{X_C} \right\rvert$	
	if $\sigma_{22} \geq 0$	$F_{MT} = \left(\frac{22}{Y_T}\right)^2 + \left(\frac{12}{S}\right)^2$	
	if $\sigma_{22} < 0$	$F_{MC} = \left(\frac{22}{Y_C}\right)^2 + \left(\frac{12}{S}\right)^2$	
Hashin (1980)	if $\sigma_{11} \geq 0$	$F_{FT} = \left(\frac{11}{X_T}\right)^2 + \left(\frac{12}{S_L}\right)^2$	2-D
	if $\sigma_{11} < 0$	$F_{FC} = \left\lvert \frac{11}{X_C} \right\rvert$	
	if $\sigma_{22} \geq 0$	$F_{MT} = \left(\frac{\sigma_{22}}{Y_T}\right)^2 + \left(\frac{\sigma_{12}}{S_L}\right)^2$	
	if $\sigma_{22} < 0$	$F_{MC} = \left(\frac{\sigma_{22}}{2S_T}\right)^2 + \left[\left(\frac{Y_C}{2S_T}\right)^2 - 1\right]\frac{\sigma_{22}}{Y_C} + \left(\frac{\sigma_{12}}{S_L}\right)^2$	
Hashin (1980)	if $\sigma_{11} \geq 0$	$F_{FT} = \left(\frac{\sigma_{11}}{X_T}\right)^2 + \left(\frac{\sigma_{12}}{S_L}\right)^2 + \left(\frac{\sigma_{13}}{S_L}\right)^2$	3-D

(continued)

Table 1 (continued)

Name of the damage model	Condition	Damage initiation formulation	Remarks
	if $\sigma_{11} < 0$	$F_{FC} = \left\lvert \frac{\sigma_{11}}{X_C} \right\rvert$	
	if $\sigma_{22} + \sigma_{33} \geq 0$	$F_{MT} = \left(\frac{\sigma_{22}+\sigma_{33}}{Y_T}\right)^2 + \frac{1}{S_T^2}(\sigma_{23}^2 - \sigma_{22}\sigma_{33}) + \frac{1}{S_L^2}(\sigma_{12}^2 + \sigma_{13}^2)$	
	if $\sigma_{22} + \sigma_{33} < 0$	$F_{MC} = \frac{1}{Y_C}\left[\left(\frac{Y_C}{2S_T}\right)^2 - 1\right](\sigma_{22} + \sigma_{33}) + \frac{1}{4S_T^2}(\sigma_{22} + \sigma_{33})^2 + \frac{1}{S_T^2}\left(\sigma_{23}^2 - \sigma_{22}\sigma_{33}\right) + \frac{1}{S_L^2}\left(\sigma_{12}^2 + \sigma_{13}^2\right)$	
Puck and Schürmann (1998)	If $\left(\varepsilon_{11} + \frac{\nu_{f12}}{E_{f11}} m_{\sigma f}\sigma_{22}\right) > 0$	$F_{FT} = \frac{1}{\varepsilon_{11T}}\left(\varepsilon_{11} + \frac{\nu_{f12}}{E_{f11}} m_{\sigma f}\sigma_{22}\right)$	2-D
	If $\left(\varepsilon_{11} + \frac{\nu_{f12}}{E_{f11}} m_{\sigma f}\sigma_{22}\right) < 0$	$F_{FC} = \frac{1}{\varepsilon_{11C}}\left\lvert\left(\varepsilon_{11} + \frac{\nu_{f12}}{E_{f11}} m_{\sigma f}\sigma_{22}\right)\right\rvert + (10\gamma_{12})^2$	
	If $\sigma_{22} \geq 0$	$F_{M-A} = \sqrt{\left(\frac{\tau_{21}}{S_{21}}\right)^2 + \left(1 - p_{\perp\parallel}^{(+)}\frac{Y_T}{S_{21}}\right)^2\left(\frac{\sigma_{22}}{Y_T}\right)^2} + p_{\perp\parallel}^{(+)}\frac{\sigma_{22}}{S_{21}} + \left\lvert\frac{\sigma_{11}}{\sigma_{1D}}\right\rvert$	
	If $\sigma_{22} < 0$ and $0 \leq \left\lvert\frac{\sigma_{22}}{\tau_{21}}\right\rvert \leq \frac{R_{\perp\perp}^A}{\lvert\tau_{21C}\rvert}$	$F_{M-B} = \frac{1}{S_{21}}\sqrt{\tau_{21}^2 + \left(p_{\perp\parallel}^{(-)}\sigma_{22}\right)^2} + p_{\perp\parallel}^{(-)}\sigma_{22} + \left\lvert\frac{\sigma_{11}}{\sigma_{1D}}\right\rvert$	
	If $\sigma_{22} < 0$ and $0 \leq \left\lvert\frac{\tau_{21}}{\sigma_{22}}\right\rvert \leq \frac{\lvert\tau_{21C}\rvert}{R_{\perp\perp}^A}$	$F_{M-C} = \left[\left(\frac{\tau_{21}}{2\left(1+p_{\perp\perp}^{(-)}\right)S_{21}}\right)^2 + \left(\frac{\sigma_{22}}{Y_C}\right)^2\right]\frac{Y_C}{(-\sigma_{22})} + \left\lvert\frac{\sigma_{11}}{\sigma_{1D}}\right\rvert$	
LaRC03 by Davila and Camanho (2003)	$\sigma_{11} \geq 0$	$F_{FT} = \frac{\varepsilon_{11}}{\varepsilon_{11}^T}$	2-D

(continued)

Table 1 (continued)

Name of the damage model	Condition	Damage initiation formulation	Remarks
	$\sigma_{11} < 0$ and $\sigma_{22}^{m} < 0$ kinking	$F_{FC} = \frac{\lvert\tau_{12}^{m}\rvert + \eta^{L}\sigma_{22}^{m}}{S_{is}^{L}}$	
	$\sigma_{11} < 0$ and $\sigma_{22} \geq 0$	$F_{FC} = (1-g)\left(\frac{\sigma_{22}^{m}}{Y_{is}^{T}}\right) + g\left(\frac{\sigma_{22}^{m}}{Y_{is}^{T}}\right)^{2} + \left(\frac{\tau_{12}^{m}}{S_{is}^{L}}\right)^{2}$	
	$\sigma_{22} \geq 0$	$f_{MT} = (1-g)\left(\frac{\sigma_{22}}{Y_{is}^{T}}\right) + g\left(\frac{\sigma_{22}}{Y_{is}^{T}}\right)^{2} + \left(\frac{\tau_{12}}{S_{is}^{L}}\right)^{2}$	
	$\sigma_{22} < 0$	$F_{MC} = \left(\frac{\tau_{eff}^{T}}{S^{T}}\right)^{2} + \left(\frac{\tau_{eff}^{L}}{S_{is}^{L}}\right)^{2}$	
	$\sigma_{22} < 0$ and $0 < \sigma_{11} < Y^{C}$	$F_{MC} = \left(\frac{\tau_{eff}^{mT}}{S^{T}}\right)^{2} + \left(\frac{\tau_{eff}^{mL}}{S_{is}^{L}}\right)^{2}$	
LaRC04 by Pinho et al. (2005)	$\sigma_{11} \geq 0$	$F_{F} = \frac{\sigma_{11}}{X^{T}}$	3-D
	$\sigma_{11} < 0$ and $\sigma_{2^m2^m} < 0$ kinking	$F_{F} = \frac{\lvert\tau_{1^m2^m}\rvert}{S_{is}^{L} - \eta^{L}\sigma_{2^m2^m}}$	
	$\sigma_{22} \geq 0$	$F_{Mtens} = (1-g)\frac{\sigma_{22}}{Y_{is}^{T}} + g\left(\frac{\sigma_{22}}{Y_{is}^{T}}\right)^{2} + \frac{\Lambda_{23}^{0}\tau_{23}^{2} + \chi(\gamma_{12})}{\chi\left(\gamma_{12\lvert is}^{u}\right)}$	
	$\sigma_{22} < 0$ and $\sigma_{11} \geq -Y^{C}$	$F_{M} = \left(\frac{\tau^{T}}{S^{T} - \eta^{T}\sigma_{n}}\right)^{2} + \left(\frac{\tau^{L}}{S_{is}^{L} - \eta^{L}\sigma_{n}}\right)^{2}$	
	$\sigma_{22} < 0$ and $\sigma_{11} < -Y^{C}$	$F_{M} = \left(\frac{\tau^{Tm}}{S^{T} - \eta^{T}\sigma_{n}^{m}}\right)^{2} + \left(\frac{\tau^{Lm}}{S^{L} - \eta^{L}\sigma_{n}^{m}}\right)^{2}$	

(continued)

Table 1 (continued)

Name of the damage model	Condition	Damage initiation formulation	Remarks
	$\sigma_{11} < 0$ and $\sigma_{2^m 2^m} \geq 0$	$F_{M/F} = (1-g)\frac{\sigma_{2^m 2^m}}{Y_{is}^T} + g\left(\frac{\sigma_{2^m 2^m}}{Y_{is}^T}\right)^2 + \frac{\wedge_{23}^0 \tau_{2^m 3^\psi}^2 + \chi(\gamma_{1^m 2^m})}{\chi\left(\gamma_{12\|is}^u\right)}$	
LaRC05 by Pinho et al. (2012)	If $\sigma_{11} \geq 0$	$F_F = \frac{\sigma_{11}}{X^T}$	3-D
	If $\sigma_{11} \leq -\frac{X_c}{2}$ kinking else splitting	$F_{\text{KINK}} = F_{\text{SPLIT}} = \left(\frac{\tau_{23}^m}{S_T^{is} - \eta_T \sigma_{22}^m}\right)^2 + \left(\frac{\tau_{12}^m}{S_L^{is} - \eta_L \sigma_{22}^m}\right)^2 + \left(\frac{\sigma_{22+}^m}{Y_T^{is}}\right)^2$	
		$F_M = \left(\frac{\tau_T}{S_T^{is} - \eta_T \sigma_N}\right)^2 + \left(\frac{\tau_L}{S_L^{is} - \eta_L \sigma_N}\right)^2 + \left(\frac{\sigma_{N+}}{Y_T^{is}}\right)^2$	

zero regarding to the damage mode. Whereas progressive damage approach considers a gradual degradation rather than a sudden degradation after the onset of any damage initiation. A damage variable is defined for each of the damage modes to indicate the evolution of the damage. Damage variables can be based on an equivalent displacement or energy dissipation. In progressive damage models, the effective stress is expressed as follow:

$$[\overline{\sigma}] = [D][\sigma]$$

where $[\overline{\sigma}]$, $[D]$ and $[\sigma]$ denote the effective stresses, damage variables, and nominal stresses, respectively.

3 Analytical Approaches in Machining of Fiber Reinforced Polymer Composites

3.1 Analytical Modelling of Orthogonal Cutting

Everstine and Rogers (1971) are the pioneers for the analytical of the orthogonal cutting of fiber reinforced polymer composite structures. They used the general theory of large plane deformations for idealised fiber reinforced materials, which was formulated by Pipkin and Rogers (1971). The authors restricted the analysis by plane strain condition with the assumption of incompressible workpiece with strong reinforcement. They established an analytical model to predict cutting force when the fiber orientation is 0° as shown in Fig. 2. The first analytical cutting force model for fiber reinforced polymer composite based on continuum mechanics approach is given below.

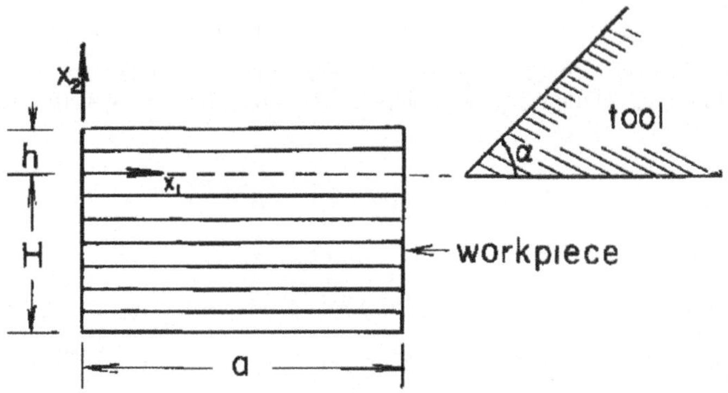

Fig. 2 Analytical model of orthogonal cutting of composite materials with 0° fiber orientation. Reprinted, with kind permission, from Everstine and Rogers (1971)

$$F_C = \sqrt{\frac{2(1-\upsilon)\mu}{E}}\pi h T_U(\tan\alpha + \tan\delta)$$

where F_C is the cutting force, ν is Poisson's ratio of contraction to extension in the plane of transverse isotropy, μ is the shear modulus along the fibers, E is the elasticity modulus in the transverse direction, T_U is the ultimate tensile strength normal to the fiber orientation, h is the depth of cut, α is the rake angle of the cutting tool. This model is appropriate to estimate the cutting force when fiber orientation is only 0°. This is one of the main drawbacks of the study and limits its wider application in the machining of fiber reinforced polymer composite laminates. Besides, the influence of the interaction between the cutting tool and composite workpiece at the flank face on the cutting force was ignored. Hence, it can cause inaccuracy in the prediction of the cutting force.

Merchant (1945a, b) explained the mechanics of the metal cutting process by orthogonal cutting with well-known shear plane theory. Based on his theory, Takeyama and Iijima (1988) developed a force model in orthogonal cutting of unidirectional glass fiber reinforced polymer. The model estimates the cutting force and thrust force regarding to the fiber orientation, as follows:

$$F_C = \frac{bt_1\tau(\theta')\cos(\beta-\gamma)}{\cos(\phi+\beta-\gamma)\sin\phi}$$

$$F_T = \frac{bt_1\tau(\theta')\sin(\beta-\gamma)}{\cos(\phi+\beta-\gamma)\sin\phi}$$

where F_C is the cutting force, F_T is the thrust force, b is the width of cut, t_1 is the depth of cut, $\tau(\theta')$ is the yield stress of GFRP, θ' is the angle between shear plane and fiber direction, β is the angle of friction on the rake face, γ is the rake angle, ϕ is the shear angle.

Another investigation for the analytical modelling of the orthogonal cutting of fiber reinforced polymer composite was conducted by Bhatnagar et al. (1995). Authors modified Merchant's classical metal cutting model for unidirectional carbon fiber reinforced polymer composites where the fiber orientation angle is less than 90°. Based on the fiber orientation angle, the cutting and thrust forces were given as:

$$F_C = \tau_0 A_0 \frac{\cos(\beta-\gamma)}{\sin\theta\cos(\theta+\beta-\gamma)}$$

$$F_T = \tau_0 A_0 \frac{\sin(\beta-\gamma)}{\sin\theta\cos(\theta+\beta-\gamma)}$$

where F_C is the cutting force, F_T is the thrust force, A_0 is the area of the undeformed chip, τ_0, the shear strength, θ is the angle of fibers, β is the mean friction angle, γ is the rake angle. The model predicts force components in orthogonal cutting accurately when the fiber orientation angle is less than 90°. However, the predictions become less accurate due to the change in the chip mechanisms. When the fiber orientation angle is larger than 90°, the shear plane and fiber orientation angle coincide and influences the chip formation mechanism.

Pwu and Hocheng (1998) investigated chip formation in orthogonal cutting of unidirectional fiber reinforced polymer composite structure. Authors were interested in the effect of bending failure on chip formation. The analytical model based on beam theory, composite mechanics and linear elastic fracture mechanics. The cutting force is expressed depending on the chip formation as:

$$F_C = \frac{\sigma_{\max} b H^2}{6(L - t_c)}$$

where F_C is the cutting force, $T_{\max}$ is the maximum stress of each lamina in the fiber direction, b is the width of the chip in the out-of-plane direction, H is the chip thickness in the feed direction, L is the length of chip, and t_c is the depth of cut. The predicted cutting force matched with the experiments. However, it should be underlined that in this model the fibers are forced to be perpendicular to the cutting direction and this restricts its possible emploments in different cutting processes of fiber reinforced composites.

Zhang et al. (2001) stated that the mechanics modelling of cutting needs to be performed differently when the angle of fiber lower than 90° and larger than 90°. They established an analytical model to estimate force components in orthogonal cutting of unidirectional fiber reinforced polymer composites (see Fig. 3). They segregated the cutting zone into three regions based on the deformation regions of the composite, and obtained the force components individually in each region, then superposed them. Authors applied the Merchant's thin shear plane theory in the first region, and expressed the force components as,

$$F_{y1} = \tau_1 \mathrm{h} a_c \frac{\cos\phi \tan(\phi + \beta - \gamma) - \sin\phi}{\frac{\tau_1}{\tau_2}\cos(\theta - \phi)\sin\theta - \sin(\theta - \phi)\cos\theta}$$

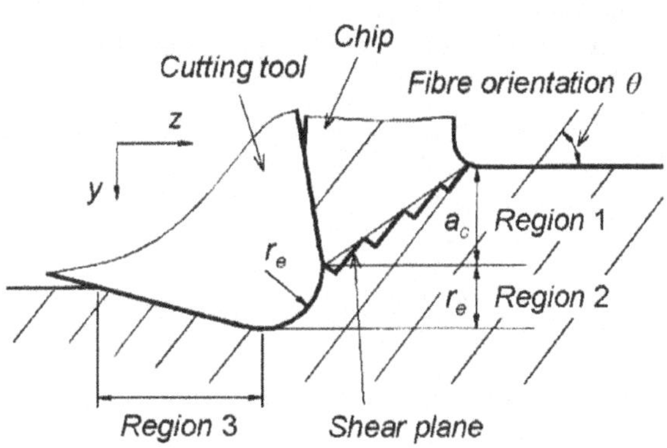

Fig. 3 Analytical model of orthogonal cutting of composite materials with fiber orientation smaller than 90°. Reprinted, with kind permission, from Zhang et al. (2001)

$$F_{z1} = \tau_1 \mathrm{ha}_c \frac{\sin\phi \tan(\phi + \beta - \gamma) + \cos\phi}{\frac{\tau_1}{\tau_2}\cos(\theta - \phi)\sin\theta - \sin(\theta - \phi)\cos\theta}$$

where τ_1 and τ_2 are the shear strengths of the composite, θ is the angle of fibers, ϕ is the angle of the shear plane, β is the angle of the friction on the rake face, h is the thickness of the workpiece perpendicular to the y–z plane, γ is the rake angle, and a_c is the depth of cut. In the second region, the authors used indentation mechanics and calculated the force components, as follows:

$$F_{y2} = P_{\text{real}}(\cos\theta - \mu\sin\theta)$$

$$F_{z2} = P_{\text{real}}(\sin\theta + \mu\cos\theta)$$

where P_{real} is the real resultant force, μ is the friction coefficient. In the third region, the force components estimated by using the contact mechanics.

$$F_{y3} = \frac{1}{2} r_e E_3 h(1 - \mu\cos\alpha\sin\alpha)$$

$$F_{z3} = \frac{1}{2} r_e E_3 h \cos^2\alpha$$

where r_e is the edge radius, α is the clearance angle of the cutting tool, and E_3 is the effective elasticity modulus in region 3. As these force components superposed, the thrust force and cutting force were expressed, as follows:

$$F_y = F_{y1} + F_{y2} + F_{y3}$$

$$F_z = F_{z1} + F_{z2} + F_{z3}$$

Sahraie Jahromi and Bahr (2010) developed the model suggested by Zhang et al. (2001) to a further stage. Authors proposed a micro-scale model considered the representative volume element (RVE) of the fiber reinforced polumer laminated composite to predict cutting force components in the orthogonal cutting when the fiber orientation between 90° and 180° (as shown in Fig. 4) They investigated the deflection of the RVE by using the principle of minimum potential energy and expressed analytical force components as follows:

When $\theta < \gamma + 90°$

$$F_C = -G_{LT}\frac{\pi d_f^2}{4}\frac{t}{2c + d_f}\cos\theta + \max\left(\left|S^{\text{shear}}\right|, \left|S^{\text{bend}}\right|\right) n_{\text{RVE}}$$

$$F_T = -G_{LT}\frac{\pi d_f^2}{4}\frac{t}{2c + d_f}\sin\theta$$

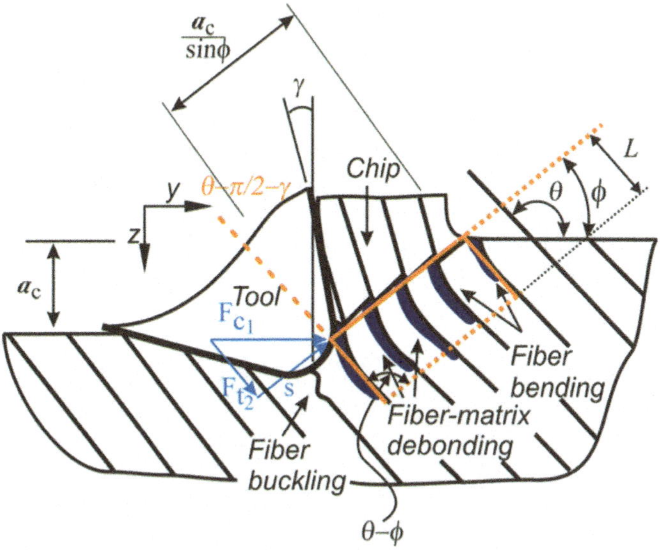

Fig. 4 Analytical model of orthogonal cutting of composite materials with fiber orientation larger than 90°. Reprinted, with kind permission, from Sahraie Jahromi and Bahr (2010)

$$-\mu\left(-G_{LT}\frac{\pi d_f^2}{4}\frac{t}{2c+d_f}\cos\theta + \max\left(\left|S^{\text{shear}}\right|, \left|S^{\text{bend}}\right|\right)n_{\text{RVE}}\upsilon_f\right)$$

When $\theta \geq \gamma + 90°$

$$F_C = -G_{LT}\frac{\pi d_f^2}{4}\frac{t}{2c+d_f}\cos\theta + \frac{\max\left(\left|S^{\text{shear}}\right|, \left|S^{\text{bend}}\right|\right)}{\cos\gamma}n_{\text{RVE}}(1+\mu\sin\gamma)\nu_f$$

$$F_T = -G_{LT}\frac{\pi d_f^2}{4}\frac{t}{2c+d_f}\sin\theta - \max\left(\left|S^{\text{shear}}\right|, \left|S^{\text{bend}}\right|\right)\tan\gamma\, n_{\text{RVE}}\nu_f$$

where F_C is total cutting force, F_T is total thrust force, G_{LT} is the shear modulus, θ is the fiber orientation, S^{shear} is the critical lateral force associated with matrix shearing, S^{bend} is the critical lateral force associated with fiber bending, γ is the rake angle of the tool, μ is the friction coefficient, ν_f is the fiber volume fraction. The major significance of the model is the representation of the fiber reinforced polymer laminated composite as multiphase material rather than homogeneous material. Thus, the model promising a wider application in the machining of fiber reinforced polymer laminated composites.

3.2 Analytical Modelling of Drilling

Guo et al. (2011) established an analytical model to estimate thrust force and torque in the drilling of fiber reinforced polymer laminated composites. Authors divided the cutting region into two zones according to the cutting region of a twist drill, as the chisel edge region and the cutting lips region. They considered contact theory for the chisel edge zone and oblique cutting for the cutting lip zone which was divided into three deformation regions as in the model suggested by Zhang et al. (2001). The total thrust force (F_{chith}) and torque (F_{chitq}) on the chisel edge are determined by;

$$F_{\text{chith}} = \int_{-t/\sin\pi-\psi}^{t/\sin\pi-\psi} \frac{E_3}{1-\nu^2}\left(k_{\text{chi}}t + \frac{a_c}{2}\right)\tan\gamma_w \cos\gamma_f dr$$

$$F_{\text{chitq}} = \int_{-t/\sin\pi-\psi}^{t/\sin\pi-\psi} \frac{E_3}{1-\nu^2}\left(k_{\text{chi}}t + \frac{a_c}{2}\right)\tan\gamma_w \sin\gamma_f r dr$$

where ψ is the rake angle of the chisel edge, γ_f is the angle of feed rate, γ_w is half of the chisel edge angle, k_{chi} is the chisel edge coefficient, E_3 is Young's modulus in the third orientation, ν is the Poisson's ratio, r is the radius of the drill, t is half of the lip spacing, a_c is the feed. For the force and torque analysis in the cutting lip region, initially, the cutting lips were divided into smaller pieces to transform of the angles of the tool into orthogonal cutting condition. Then cutting forces and torque in each section were determined employing the model proposed by Zhang et al. (2001). After these force components were superposed to calculate the total thrust force and torque for cutting lips region as follows:

$$F_{\text{lipth}} = 2\int_{d_1/2}^{d/2} \left\{dF'_T \cos\xi \sin p - \left[\left(dF'_C 2 + dF'_T 2\right)^{1/2} \sin i \tan\beta \cos i + dF'_C \sin i\right] (\cos i \cos p + \sin i \sin p \sin\gamma_n)\right\}$$
$$\frac{1}{dl \sin p}\frac{r}{\left(r^2-t^2\right)^{1/2}}dr$$

$$F_{\text{liptq}} = 2\int_{d_1/2}^{d/2} \left[dF'_C \cos i + \left(dF'_C 2 + dF'_T 2\right)^{1/2} \sin^2 i \tan\beta\right]\frac{1}{dl \sin p}\frac{r}{\left(r^2-t^2\right)^{1/2}}dr$$

Finally, thrust force and torque values for the chisel edge region and cutting lips region can be summed.

Delamination is one of the most important factors in the assessment of the quality of composite structure since it can significantly affect the structural integrity. For this reason, delamination-free manufacturing is one of the favourite topics for scientists in this research field. Since thrust force is assumed as the major reason for delamination in a drilling process, researchers seek the relationship between the level of thrust force without delamination onset and other process parameters including drill geometry, material properties, the use of support plate and so on. In the literature analytical models for the critical thrust force is divided into two groups based on the constitutive model of composite; isotropic models and orthotropic models.

Ho-Cheng and Dharan (1990) studied the critical thrust force in the drilling of fiber reinforced polymer laminated composite by a conventional twist drill. Authors proposed an analytical model for the peel-up and push-out at the entry and exit sides of hole by using linear elastic fracture mechanics and classical plate bending theory. In this first model, the material is considered as isotropic and the axial force is at the centre of the plate. The critical thrust force (F_A) for the onset of push out (exit) delamination, and the horizontal critical cutting force (F_C) for the onset of peel up (entrance) delamination are expressed as follows:

$$F_A = \pi \left[\frac{8G_{IC}Eh^3}{3(1-\nu^2)} \right]^{1/2}$$

$$F_C = k_p\pi \left[\frac{8G_{IC}E(H-h)^3}{3(1-\nu^2)} \right]^{1/2}$$

where E is elastic modulus, ν is the Poisson's ratio, G_{IC} is the critical strain energy release rate, h is uncut-plies thickness under drill bit at the exit side, k_P is the peeling factor which is assumed to be a function of tool geometry and friction between the tool and the workpiece, H is the total thickness of the circular plate. The main weakness of the model was the exclusion of the anisotropic characteristic of the reinforced composite.

Hocheng and Tsao (2003, 2005, 2006) introduced the analytical models of critical thrust force in the drilling of fiber reinforced polymer laminates for special drill bits including saw drill, candlestick drill, core drill and step drill since tool geometry alters the application of the force (see Fig. 5). Authors proposed these models for the exit side delamination by using linear elastic fracture mechanics and classical plate bending theory with the assumption of heterogeneous isotropic material behaviour and circular delamination shape. Critical thrust forces for a saw drill bit (F_S), a candlestick drill bit (F_C), a core drill bit (F_R) and a step drill bit (F_T) were expressed as follows:

$$F_S = \pi\sqrt{\frac{32G_{IC}M}{1-2s^2+s^4}}$$

$$F_C = \pi(1+\alpha)\sqrt{\frac{32G_{IC}M}{1+\alpha^2(1-2s^2+s^4)}}$$

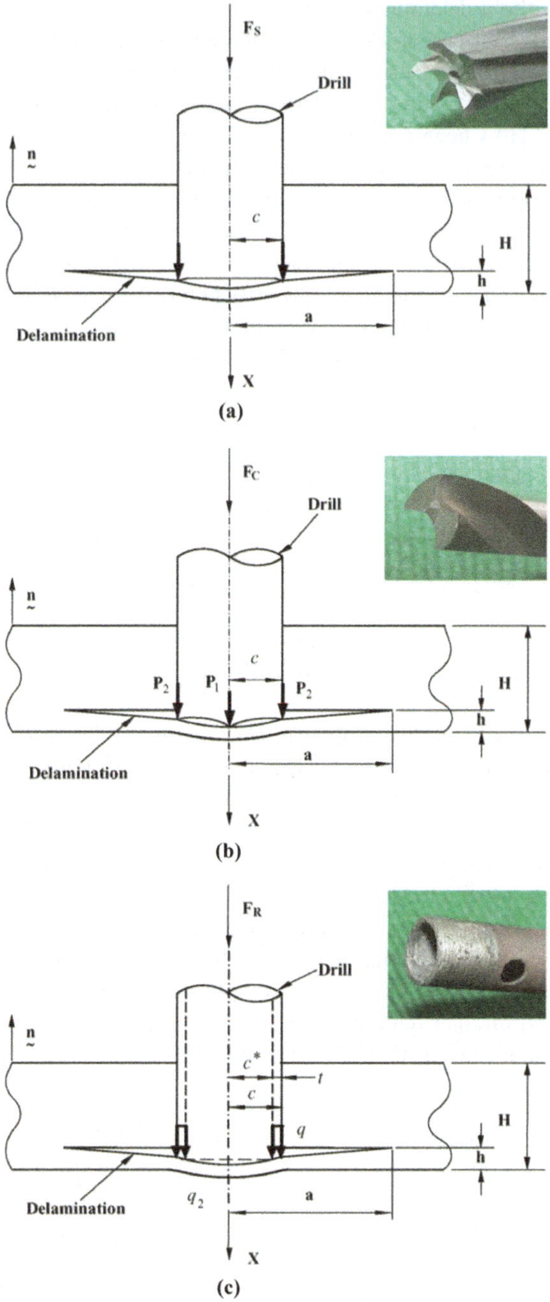

Fig. 5 Analytical models of critical thrust force in drilling of composite materials: **a** circular plate with saw drill, **b** circular plate with candle stick drill, **c** circular plate with core drill. Reprinted, with kind permission, from Hocheng and Tsao (2005)

$$F_R = \pi \left\{ \frac{32 G_{IC} M}{1 - \left[\left(2 - 2\beta + \frac{3\beta^2}{2}\right) + \frac{4(1-\beta)^2}{\beta(2-\beta)} \ln(1-\beta)\right] s^2 + \left[\frac{(2 - 4\beta + 5\beta^2 - 3\beta^3 + \beta^4)}{2} + \frac{2(1-\beta)^2(2 - 2\beta + \beta^2)}{\beta(2-\beta)} \ln(1-\beta)\right] s^4} \right\}^{1/2}$$

$$F_T = \frac{\sqrt{2}\pi}{1-\nu}\left[\frac{32 G_{IC} M \left\{(1-\nu) + 2(1+\nu)\xi^2\right\}^2}{(1+\nu)\left\{2(1-\nu)(1+2\nu^2) - (12 - 4\nu + 3\nu^2 + 3\nu^3)\xi^2 - 8(1+3\nu)\xi^2 \ln \xi\right\}}\right]^{1/2}$$

Tsao and Hocheng (2005a, b, 2008), Tsao (2006, 2007a, b, 2012) and Tsao et al. (2012) investigated the effects of back-up plate, eccentricity, pilot hole deviation, peripheral moment, induced bending moment and finally back-up force on critical thrust force and delamination extensively and proposed analytical models in their studies.

Isotropic and quasi-isotropic analytical models consider circular crack for the delamination in the drilling process. However, for unidirectional fiber reinforced polymer laminated composites delamination crack can be likely in a different shape, such as elliptic. Jain and Yang (1993) considered the anisotropy of the unidirectional fiber reinforced polymer laminated material and considered that the axial force is at the centre of the plate. Authors developed an analytical model based on orthotropic material by utilizing linear elastic fracture mechanics and classical plate bending theory. The proposed critical thrust force (P^*) and critical feed (f^*) at the onset of exit delamination were expressed as follows:

$$P^* = 3\pi \sqrt[4]{\frac{D_{22}}{D_{11}}} \sqrt{2 G_{IC} D_c^*}$$

with

$$D_c^* = 2D_{11} + \frac{2(D_{12} + 2D_{66})}{3} \sqrt{\frac{D_{22}}{D_{11}}}$$

and

$$f^* = \left\{ \frac{d^{1.2}}{1.91} \left[\frac{3\pi}{d^2 H_B} \left(\sqrt[4]{\frac{D_{22}}{D_{11}}} \right) \sqrt{2 G_{IC} D_c^*} - \frac{0.101}{d} \right] \right\}^{2.5}$$

where D_{ij}, are elements of stiffness matrix, G_{IC} is the critical strain energy release rate, d is the diameter of drill in mm, H_B is the Brinell hardness number in Kg/mm^2 of the workpiece material, f is the feed in mm/rev. This approach was mainly made for UD laminated composites. Due to this limitation, Jain and Yang (1994) broadened their previous approach for multidirectional laminated composites. They formulated the critical thrust force in the drilling of a cross-ply laminate as below.

$$P^* = \pi\sqrt{6G_{IC}[3D_{11} + 2(D_{12} + 2D_{66}) + 3D_{22}]}$$

Piquet et al. (2000) investigated the critical thrust force in the drilling of composites. Authors developed an analytical model based on orthotropic material by implementing linear elastic fracture mechanics and classical plate bending theory. They considered orthotropic behaviour for composite workpiece and a uniformly constant distributed pressure q over the section of the hole with a resultant thrust force F_z. They also assumed the shape of the delamination crack circular. The critical thrust force was expressed as

$$F_z = 8\pi\sqrt{\frac{G_{IC}D}{\frac{1}{3} - \frac{D'}{8D}}}$$

With

$$D = \frac{1}{8}(3D_{11} + 2D_{12} + 4D_{66} + 3D_{22})$$

$$D' = \frac{D_{11} + D_{22}}{2} + \frac{D_{12} + D_{66}}{3}$$

where D_{ij}, are elements of the stiffness matrix, G_{IC} is the critical strain energy release rate.

Rahmé et al. (2011) established the critical thrust force in the drilling of fiber reinforced laminates in different loading cases as shown in Fig. 6. The model was based on classical plate theory and virtual work theory. They assumed a circular delamination shape with orthotropic material behaviour. In the case of uniformly distributed load, the critical thrust force was expressed as:

$$F_Z = 8\pi\sqrt{2G_{IC}D}$$

With

$$D = \frac{1}{8}(3D_{11} + 2D_{12} + 4D_{66} + 3D_{22})$$

If the load is concentrated in the middle, the critical thrust force is formulated as:

$$F_Z = 4\pi\sqrt{2G_{IC}D}$$

In the case of a triangular type distributed load, the critical thrust force is given as:

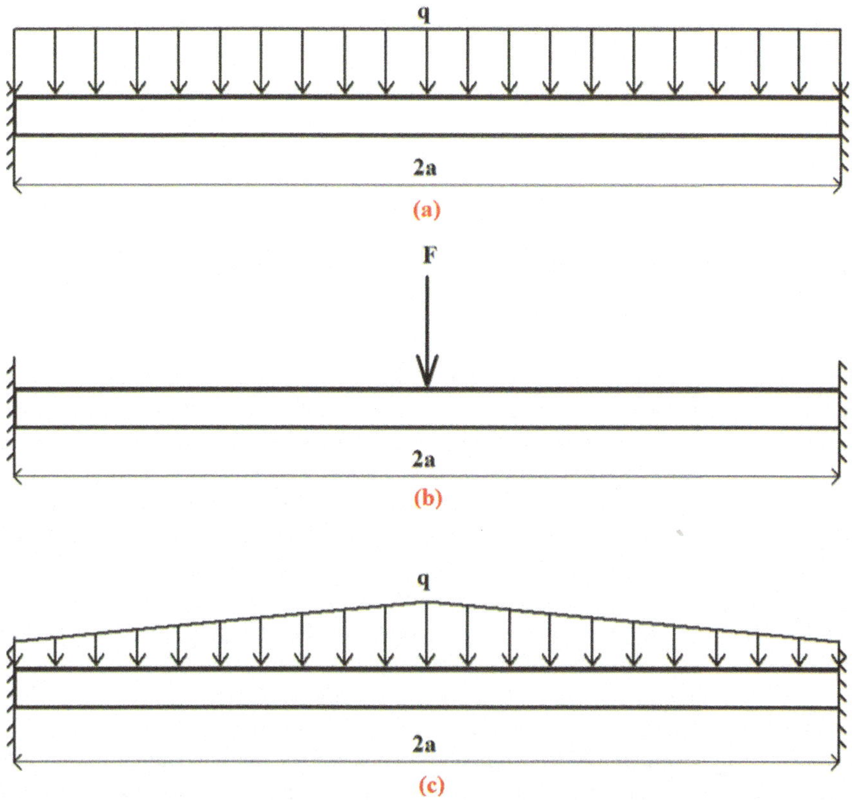

Fig. 6 Analytical models of critical thrust force in drilling of composite materials: **a** uniformly distributed load, **b** concentrated loading, **c** triangular loading

$$F_Z = \frac{\pi}{840}\left(440Aa^3 + \sqrt{270950400G_{IC}D + 57408A^2a^6}\right)$$

When the load is uniformly distributed over a disc with a radius of c, the critical thrust force is expressed as:

$$F_Z = \frac{8\pi a^2}{\left(2a^2 - c^2\right)}\sqrt{2G_{IC}D}$$

In the case of a drill bit, the geometry of the tool can be segregated into two regions; the web and the cutting edges. When there is a concentrated load in the middle (the force from the web) and distributed load over (the force from the cutting edges), the critical thrust force at the onset of delamination is given below.

$$F_Z = 16\pi\sqrt{\frac{6G_{IC}D}{13}}$$

In this last loading case, stresses on the drill web and cutting edges were assumed the same. Then the forces summed to obtain the critical thrust force. However, in a real drilling condition, the stress distribution will likely to be different. The advantage of these analytical models to represent the different force penetration of different types of cutting tools and their effects on the delamination of the laminated structure.

4 Numerical Approaches in Machining of Fiber Reinforced Polymer Composites

Finite element modelling has become one of the favourite fields of numerical methods for researchers in machining. Regarding to the complexity and development in the technology, there are four levels of finite element model of machining of composites, namely macroscale, mesoscale, microscale, and nanoscale models. Early investigators developed macroscale models to predict outputs of the machining of composites. Macroscale models are easier to implement and the computational cost is the least among all levels of models, however, macroscale models are not adequate to simulate the different damage modes such as debonding of fibers, breakage of fibers, cracks of matrix, etc. This means the accuracy of the chip formation is typically poor. Therefore, the outcomes of the machining process by using macroscale models are generally unreliable. With the advances in computer technology and numerical methods, researchers have improved their models. Mesoscale and microscale models are more refined models compared to macroscale ones and able to overcome some drawbacks of earlier models. However, the computational time requirement is generally much longer in microscale models.

4.1 Numerical Modelling of Orthogonal Cutting

Arola and Ramulu (1997) developed a two-dimensional finite element model of the orthogonal machining of unidirectional graphite/epoxy composite by using commercial software, ABAQUS (see Fig. 7). Authors created the workpiece with equivalent homogeneous material (EHM) and the damage was checked with the maximum stress and Tsai-Hill criteria. The predicted thrust force results showed significant differences with the experiments since the damage model was macroscale and did not meet fully with the mechanical behaviour of the fiber reinforced polymer composite.

Nayak et al. (2005) established two-dimensional finite element model of the orthogonal cutting of unidirectional glass fiber reinforced polymer composite to investigate the effects of the fiber angle, tool geometry, depth of cut on cutting forces and sub-surface damage (see Fig. 8). Authors initially modelled the workpiece with equivalent homogeneous material but the model presented an inadequate estimation of thrust force and damage mechanism. Then, they implemented a microscale

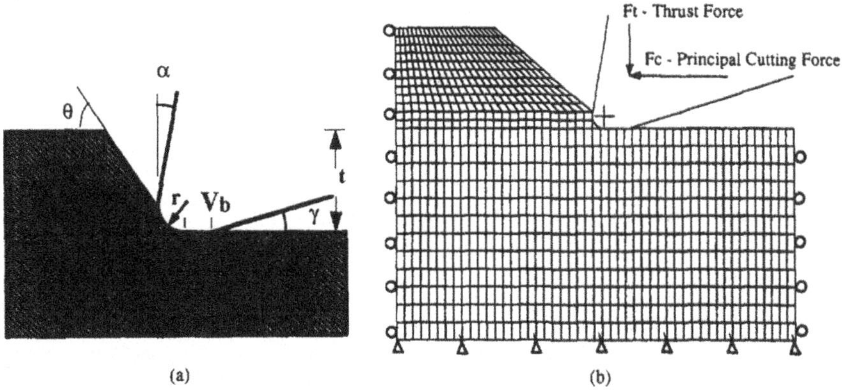

Fig. 7 Finite element model of orthogonal cutting of composite materials. Reprinted, with kind permission, from Arola and Ramulu (1997)

approach where the fiber and matrix were modelled separately with isotropic elastic behaviour. The study showed that the estimated cutting force matched with the experiments, but the thrust force did not in the macroscale model due to the use of EHM approach. In microscale model, the tendency of both horizontal cutting force and thrust force matched with the experiments.

Rao et al. (2007a, b) investigated orthogonal machining of unidirectional carbon and glass fiber reinforced polymer composites by using a two-dimensional finite element model. Authors created used macroscale and microscale model in ABAQUS to get advantages of both techniques to predict outputs of the orthogonal cutting as shown in Fig. 9a. The mechanical behaviour of fiber and matrix were considered elastic and elasto-plastic, respectively. Debonding behaviour of the fiber matrix interface was modelled with a cohesive zone approach. The damage of fiber was ensured as the maximum principal stress reached the tensile strength of the fiber whereas the damage of the matrix was assured with the degradation of the elastic modulus of matrix when the ultimate strength was reached (see Fig. 9b). The results provided a good visual understanding of different damage indexes, namely fiber, matrix and the interface. However, implementation of the approach for a full chip formation is computationally costly since authors focused on a quasi-static model.

Calzada et al. (2012) established a finite element model and experimental investigation on orthogonal machining of unidirectional carbon fiber reinforced polymer laminated composites. Authors developed a two-dimensional finite element analysis based on the combination of two approaches. A macroscale approach by using equivalent homogenous material (EHM) was applied in the regions away from machining zone whereas microstructure based approach with two individual phases (carbon fiber and epoxy) were implemented at the vicinity of the cutting zone. The carbon fiber was characterized by fully elastic and anisotropic material. The carbon fibers were considered to fail once the damage started. The epoxy matrix was modelled as elasto-plastic and isotropic material. The interface between fibers and matrix was modelled

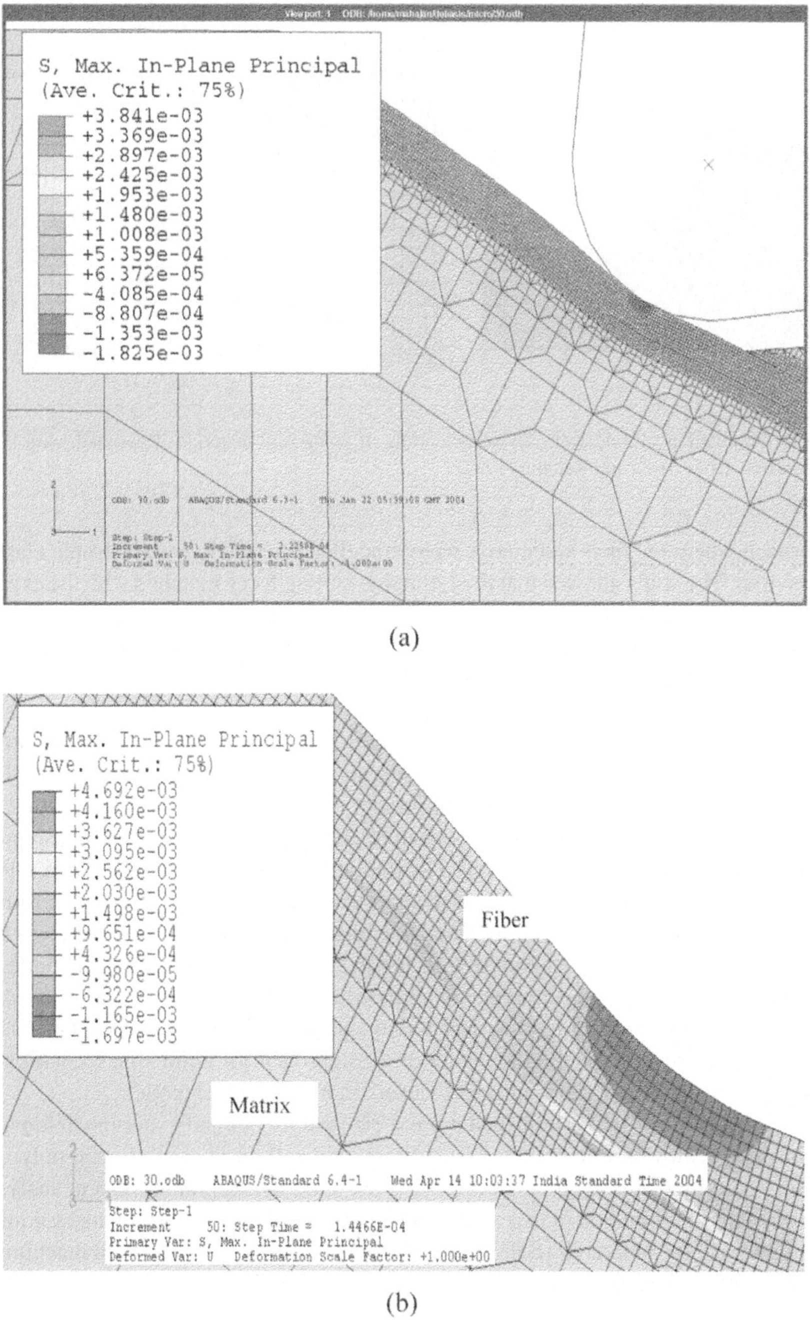

Fig. 8 Distribution of maximum stress in orthogonal cutting of UD-GFRP with 45° fiber orientation using FEM **a** during process, **b** end of process. Reprinted, with kind permission, from Nayak et al. (2005)

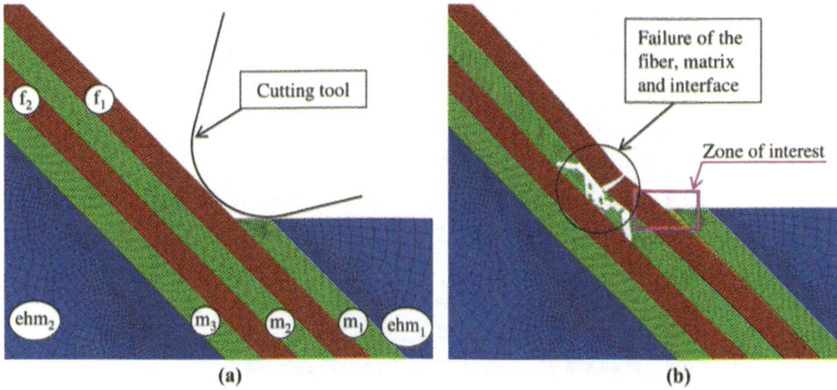

Fig. 9 Orthogonal machining of UD-GFRP with 45° fiber orientation using FEM **a** finite element model **b** damage distributions when 0.15 mm DOC and 10° rake angle. Reprinted, with kind permission, from Rao et al. (2007a)

with continuum elements which allowed failure in either tension or compression modes. The EHM region was modelled as elastic and anisotropic without having damage models. The FEM results showed failure of the fibers, chip sizes with regard to fiber length and cutting forces could be accurately predicted in any fiber orientation (see Fig. 10). The model was also promising to design tool geometry in the machining of CFRP. However, the implementation of the model in larger sizes could be challenging due to the extreme computational demands.

Abena et al. (2015) established a two-dimensional finite element analysis of orthogonal machining of unidirecitional carbon fiber reinforced polymer composite using the commercial software ABAQUS. The behaviour of the composite workpiece represented by a mesoscale material model, a combination of micro and macroscale models. The micromechanical model was established at the vicinity of the cutting region, while the EHM approach was applied for the rest of the workpiece to provide stiffness while reducing computational time of the analysis. At the machining region, the fiber, matrix, and interface were modelled individually with different constitutive responses and failure models. The mechanical response of the epoxy matrix was described with an elasto-plastic behaviour with isotropic hardening. Von Mises yield criterion was employed for the damage. Transversely isotropic and perfectly elastic material behaviour was implemented for the behaviour of the carbon fibers. Maximum principal stress criterion was applied for the damage model of the fibers. A traction-separation law which included the mechanical response and damage under tension, compression and shear was employed for the cohesive character of the matrix-fiber interface by using small thickness of cohesive elements. Predicted cutting forces of the FE model showed close match with experiments, however, the model underestimated the thrust forces.

Abena et al. (2017) improved their previous model into a three-dimensional case. In this mesoscale model, the authors proposed three approaches for the cohesive zone;

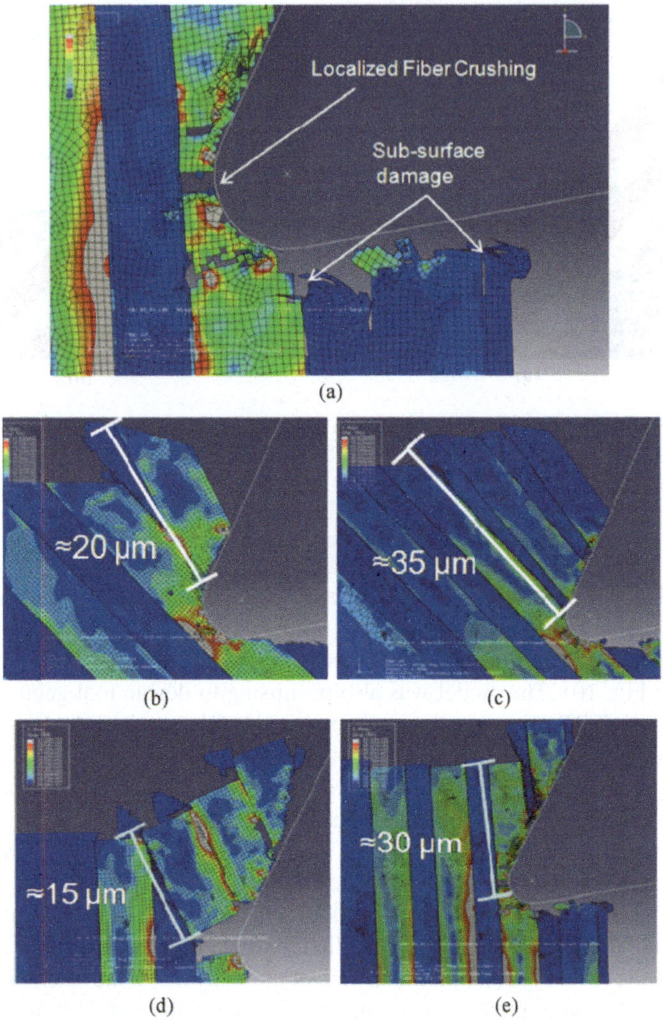

Fig. 10 Damage distributions in orthogonal machining of UD-CFRP using FEM. **a** In 45° and 90° fiber orientation, **b** 45° fiber orientation with 15 μm DOC, **c** 45° fiber orientation with 30 μm DOC, **d** 90° fiber orientation with 15 μm DOC, **e** 90° fiber orientation with 30 μm DOC. Reprinted, with kind permission, from Calzada et al. (2012)

(i) zero thickness cohesive elements, (ii) small thickness cohesive elements, (iii) surface-based cohesive behaviour (see Fig. 11). All approaches employed traction-separation law for the mechanical behaviour of the fiber matrix interface. The first approach was able to simulate fiber matrix debonding under tensile and shear behaviour, but failed under the compression behaviour due to excessive deformation of the elements. The second approach was not compatible with the chip formation. On the other hand, the third approach overcame some of the weaknesses of the

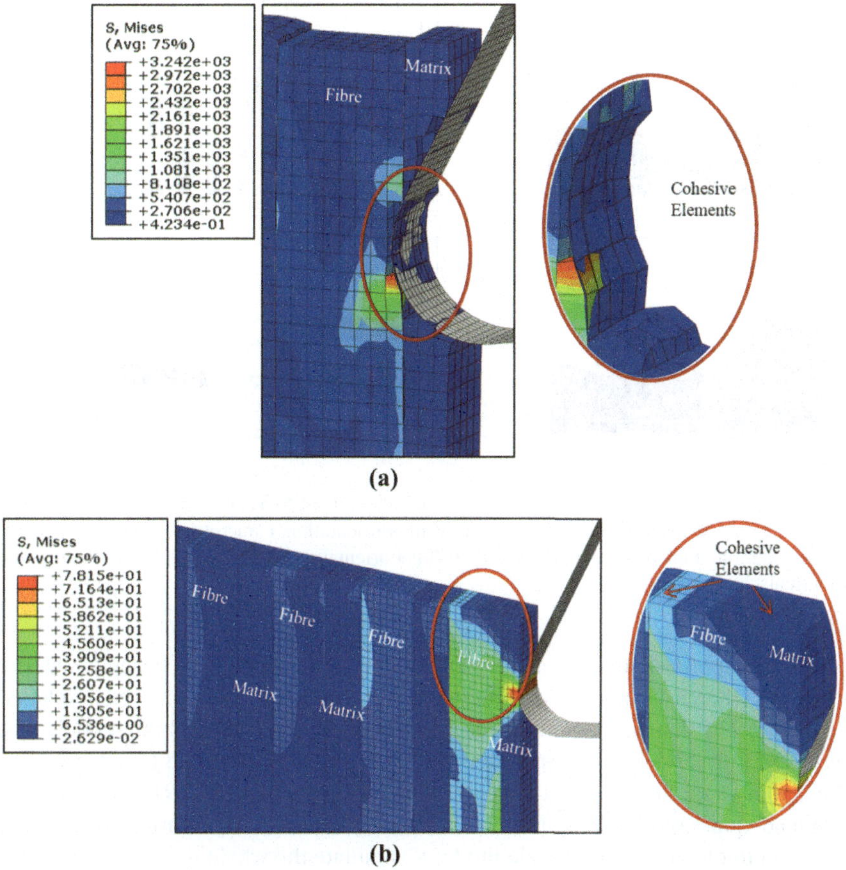

Fig. 11 Finite element models of three-dimensional orthogonal cutting of unidirectional CFRP composites **a** Approach (i): zero thickness cohesive elements **b** Approach (ii): small thickness cohesive elements. Reprinted, with kind permission, from Abena et al. (2017)

first and second approaches, but this approach showed less damage on the interface. Nevertheless, predicted thrust forces were inaccurate in both the second and third approaches.

4.2 Numerical Modelling of Milling

Rentsch et al. (2011) proposed two-dimensional finite element models so-called orthogonal cutting to represent milling process of unidirectional carbon fiber reinforced laminates by using the commercial software, ABAQUS. The behaviour of the composite workpiece represented by both macro and microscale models. In

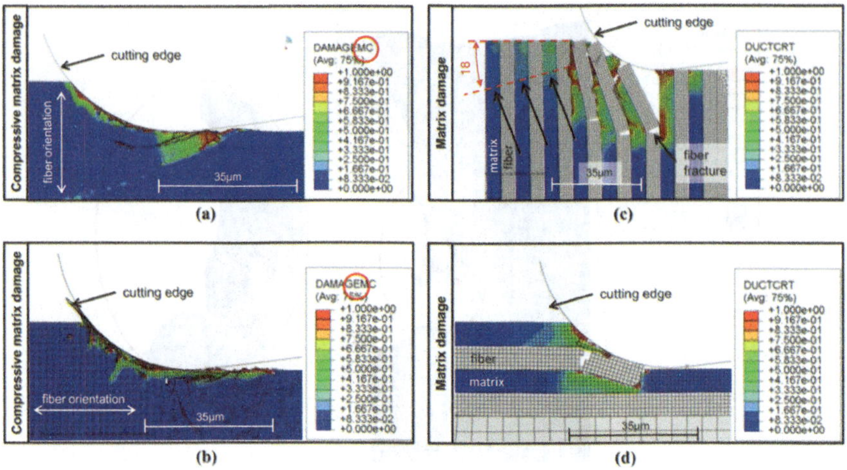

Fig. 12 Matrix damage distributions in milling of CFRP using FEM **a** macroscale approach with 90° fiber orientation, **b** macroscale approach with 0° fiber orientation, **c** microscale approach with 90° fiber orientation, **d** microscale approach with 0° fiber orientation. Reprinted, with kind permission, from Rentsch et al. (2011)

macroscale approach, the material was modelled as elastic with anisotropic material properties whereas, in microscale approach fiber, matrix and interface were modelled separately. The damage was modelled with the Hashin criteria in both models (see Fig. 12). Both models underestimated the cutting forces compared to experiments. The author suggested more accurate modelling of the milling of FRP composites to obtain adequate results. With the current modelling and computational facilities, it is hard to implement a microscale model to simulate the whole milling process due to extreme computational costs.

4.3 Numerical Modelling of Drilling

Zitoune et al. (2005) utilized finite element method to investigate drilling of unidirectional carbon fiber reinforced composite based on a orthogonal cutting model. In general, orthogonal models are good starting points for understanding mechanisms in 2-D, however, they have a lack to represent the exact cutting mechanisms for a complex process. With the developments of computer technology and numerical approaches, more realistic 3-D models have started to be applied. Researchers have proposed more accurate finite element model including more detailed constitutive law and damage models.

Zitoune and Collombet (2007) developed a 3-D finite element model to estimate the critical thrust force in the drilling of unidirectional carbon fiber reinforced laminates. They decomposed the total thrust force into two forces; the one on the cutting

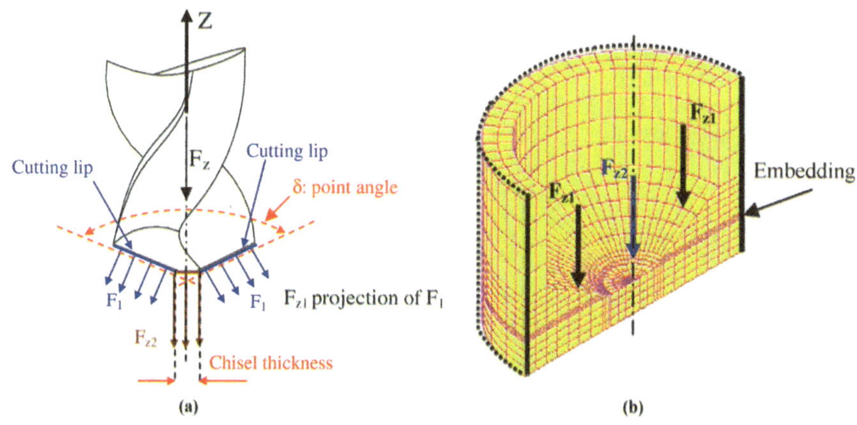

Fig. 13 Numerical prediction of the critical thrust force in drilling of fiber reinforced composite structures: **a** Decomposition of thrust force Fz, **b** Half cross view of the finite element model. Reprinted, with kind permission, from Zitoune and Collombet (2007)

lips and the one on the chisel edge (see Fig. 13). Through the experimental study, the authors observed that exit delamination starts at the surface contact with the chisel edge, so they organized their analytical and numerical models. They used a quasi-isotropic material and ignored the effect of the intra-laminar damage modes, hence the accuracy of the model for anisotropic behaviour is unknown.

Durão et al. (2006, 2008) established a non-linear 3-D finite element model with cohesive interface elements to estimate the onset and progression of delamination in the drilling of fiber reinforced composite material (see Fig. 14). Thin cohesive elements used between the laminae and incorporated mixed-mode cohesive damage based on fracture mechanics. Application of the cohesive zone method does not need predefined crack front which is an advantage over the virtual crack extension method. Authors considered delamination defect but ignored the intra-laminar damage mode such as fiber and matrix-based damages in the finite element model. Besides, they concentrated to drill only the last two plies of the laminate in the simulation. They employed simplified drill geometries. All these issues can lead to an underestimation in the outputs of the numerical model.

Isbilir and Ghassemieh (2012, 2013, 2014) introduced mesoscale 3-D finite element models to study the influences of cutting parameters and real drill geometry on drilling outputs in the drilling of carbon fiber reinforced composite laminates as shown in Fig. 15. Authors firstly used 3-D continuum shell elements for the composite workpiece. Since continuum shell elements are not completely reflecting the out-of-plane mechanical behaviour of composite structure, they utilized 3-D solid elements to obtain an accurate model of the 3-D drilling process. They implemented orthotropic material behaviour for each UD lamina, the Hashin's damage criteria for the initiation of the intra-laminar damage modes. Delamination was modelled by the integration of the cohesive zone method. They used zero thickness, surface-based cohesive contact instead of cohesive elements with a bilinear traction–separation law

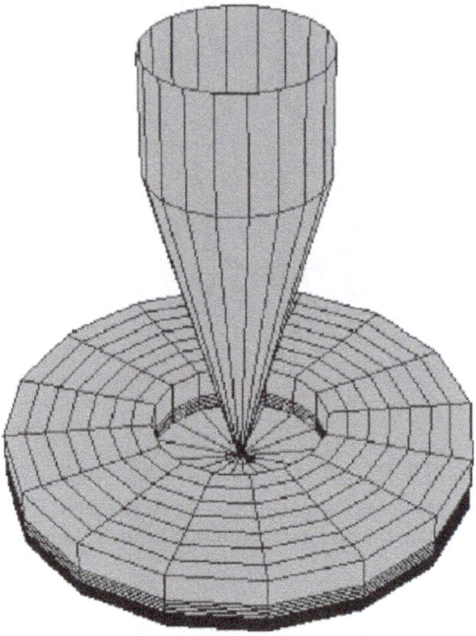

Fig. 14 Finite element model of delamination onset in carbon/epoxy composites drilling. Reprinted, with kind permission, from Durão et al. (2008)

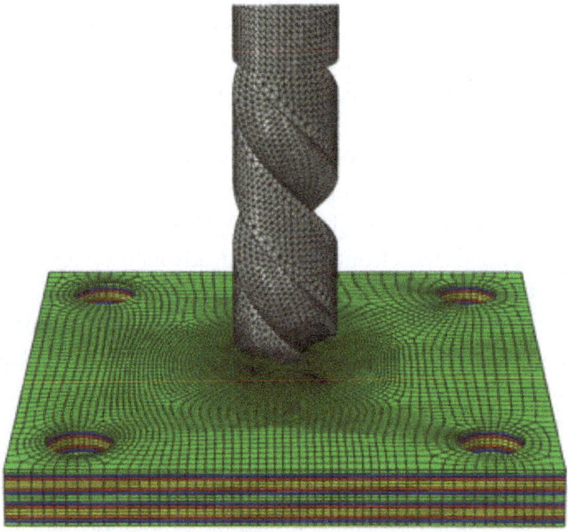

Fig. 15 Three dimensional finite element model of drilling of CFRP. Reprinted, with kind permission, from Isbilir and Ghassemieh (2012)

up to delamination initiation. The growth of the delamination was modelled with the mixed-mode energy criterion. They also adapted 3-D complex twist and step drill geometries. The models showed promising results for the optimisation of process parameters and tool geometry (see Fig. 16). Since the models used a basic Coulomb friction model without the heat effect, the authors addressed more detailed friction models and thermomechanical analysis for the discrepancies.

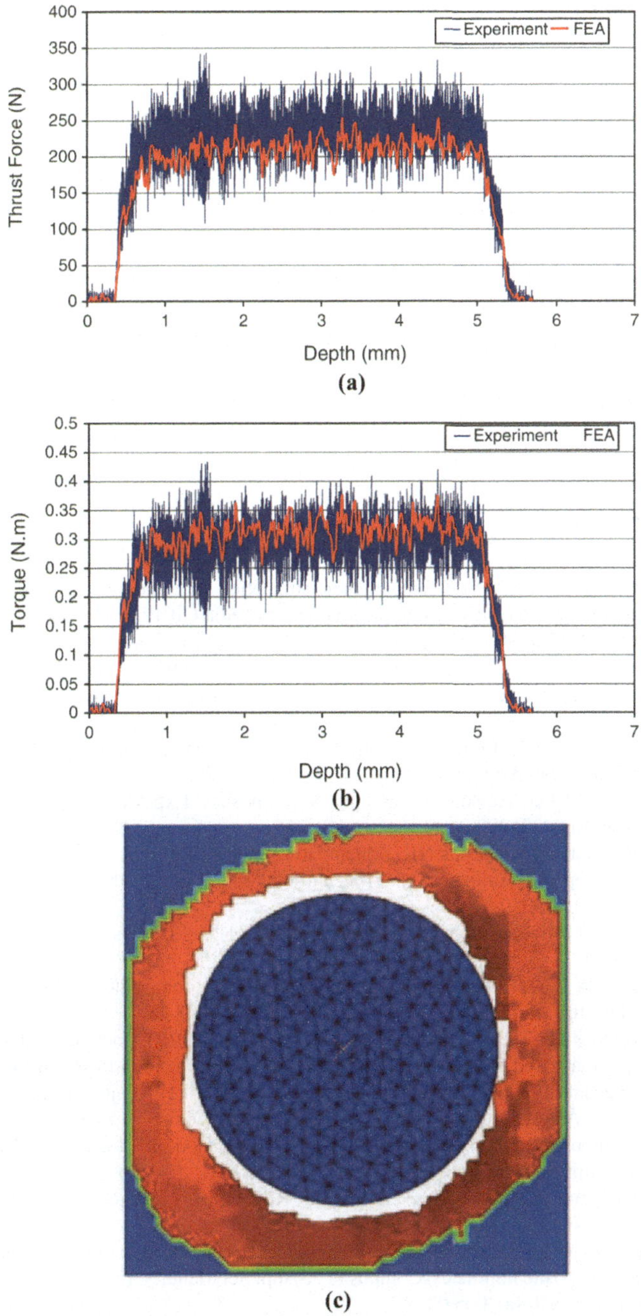

Fig. 16 Results of drilling of CFRP using FEM: **a** thrust force, **b** torque, **c** entrance delamination. Reprinted, with kind permission, from Isbilir and Ghassemieh (2014)

5 Conclusion

This chapter provides a brief overview on the analytical and numerical modeling approaches in machining processes of fiber reinforced composites based on the literature study. Constitutive law, damage mechanisms and modeling of damage for fiber reinforced laminated composites are explained comprehensively. Damage initiation criteria are reported for comparison reason.

It has been observed that majority of the studies are macroscale models. More refined models are required for better understanding of cutting and damage mechanisms in the fiber reinforced polymer composites and to predict machining outputs accurately. With the advances in computer technology and numerical methods, models will be developed towards nanoscale direction whereas computational costs will likely be less.

References

Abena A, Soo SL, Essa K (2015) A finite element simulation for orthogonal cutting of UD-CFRP incorporating a novel fibre-matrix interface model. Procedia CIRP 31:539–544. https://doi.org/10.1016/j.procir.2015.04.091

Abena A, Soo SL, Essa K (2017) Modelling the orthogonal cutting of UD-CFRP composites: development of a novel cohesive zone model. Compos Struct 168:65–83. https://doi.org/10.1016/j.compstruct.2017.02.030

Arola D, Ramulu M (1997) Orthogonal cutting of fiber-reinforced composites: a finite element analysis. Int J Mech Sci 39(5):597–613

Azzi VD, Tsai SW (1965) Anisotropic strength of composites. Exp Mech 5(9):283–288. https://doi.org/10.1007/BF02326292

Bhatnagar N, Ramakrishnan N, Naik NK, Komanduri R (1995) on the machining of fiber-reinforced plastic (FRP) composite laminates. Int J Mach Tools Manuf 35(5):701–716. https://doi.org/10.1016/0890-6955(95)93039-9

Budiansky B (1983) Micromechanics. Comput Struct 16(1):3–12. https://doi.org/10.1016/0045-7949(83)90141-4

Budiansky B, Fleck NA (1993) Compressive failure of fibre composites. J Mech Phys Solids 41(1):183–211. https://doi.org/10.1016/0022-5096(93)90068-Q

Calzada KA, Kapoor SG, DeVor RE, Samuel J, Srivastava AK (2012) Modeling and interpretation of fiber orientation-based failure mechanisms in machining of carbon fiber-reinforced polymer composites. J Manuf Process 14(2):141–149. https://doi.org/10.1016/j.jmapro.2011.09.005

Choi NS, Kinloch AJ, Williams JG (1999) Delamination fracture of multidirectional carbon-fiber/epoxy composites under mode I, mode II and mixed-mode I/II loading. J Compos Mater 33(1):73–100. https://doi.org/10.1177/002199839903300105

Davila CG, Camanho PP (2003) Failure criteria for FRP laminates in plane stress. NASA Langley Research Center; Hampton, VA, United States

Durão LMP, De Moura MFSF, Marques AT (2006) Numerical simulation of the drilling process on carbon/epoxy composite laminates. Compos Pt A-Appl Sci Manuf 37(9):1325–1333. https://doi.org/10.1016/j.compositesa.2005.08.013

Durão LMP, De Moura MFSF, Marques AT (2008) Numerical prediction of delamination onset in carbon/epoxy composites drilling. Eng Fract Mech 75(9):2767–2778

Everstine GC, Rogers TG (1971) A theory of machining of fiber-reinforced materials. J Compos Mater 5(1):94–106. https://doi.org/10.1177/002199837100500109

Guo DM, Wen Q, Gao H, Bao YJ (2011) Prediction of the cutting forces generated in the drilling of carbon-fibre-reinforced plastic composites using a twist drill. Proc Inst Mech Eng J Eng Manuf 226(1):28–42. https://doi.org/10.1177/0954405411419128
Hashin Z (1980) Failure criteria for unidirectional fiber composites. J Appl Mech 47(2):329–334. https://doi.org/10.1115/1.3153664
Hashin Z, Rotem A (1973) A fatigue failure criterion for fiber reinforced materials. J Compos Mater 7(4):448–464. https://doi.org/10.1177/002199837300700404
Ho-Cheng H, Dharan CKH (1990) Delamination during drilling in composite laminates. J Eng Indus 112(3):236–239. https://doi.org/10.1115/1.2899580
Hocheng H, Tsao CC (2003) Comprehensive analysis of delamination in drilling of composite materials with various drill bits. J Mater Process Technol 140(1–3):335–339
Hocheng H, Tsao CC (2005) The path towards delamination-free drilling of composite materials. J Mater Process Technol 167(2–3):251–264
Hocheng H, Tsao CC (2006) Effects of special drill bits on drilling-induced delamination of composite materials. Int J Mach Tools Manuf 46(12):1403–1416. https://doi.org/10.1016/j.ijmachtools.2005.10.004
Hoffman O (1967) The brittle strength of orthotropic materials. J Compos Mater 1(2):200–206. https://doi.org/10.1177/002199836700100210
Isbilir O, Ghassemieh E (2012) Finite element analysis of drilling of carbon fibre reinforced composites. Appl Compos Mater 19(3–4):637–656. https://doi.org/10.1007/s10443-011-9224-9
Isbilir O, Ghassemieh E (2013) Numerical investigation of the effects of drill geometry on drilling induced delamination of carbon fiber reinforced composites. Compos Struct 105:126–133
Isbilir O, Ghassemieh E (2014) Three-dimensional numerical modelling of drilling of carbon fiber-reinforced plastic composites. J Compos Mater 48(10):1209–1219. https://doi.org/10.1177/0021998313484947
Jain S, Yang DCH (1993) Effects of feedrate and chisel edge on delamination in composites drilling. J Eng Indus 115(4):398–405. https://doi.org/10.1115/1.2901782
Jain S, Yang DCH (1994) Delamination-free drilling of composite laminates. J Eng Indus 116(4):475–481. https://doi.org/10.1115/1.2902131
Khashaba UA (2004) Delamination in drilling GFR-thermoset composites. Compos Struct 63(3):313–327. https://doi.org/10.1016/S0263-8223(03)00180-6
Merchant ME (1945a) Mechanics of the metal cutting process. i. orthogonal cutting and a type 2 chip. J Appl Phys 16(5):267–275. https://doi.org/10.1063/1.1707586
Merchant ME (1945b) Mechanics of the metal cutting process. II. Plasticity conditions in orthogonal cutting. J Appl Phys 16(6):318–324. https://doi.org/10.1063/1.1707596
Nairn JA (2000) 2.12—matrix microcracking in composites. In: Kelly A, Zweben C (eds) Comprehensive composite materials. Pergamon, Oxford, pp 403–432. https://doi.org/10.1016/B0-08-042993-9/00069-3
Nayak D, Bhatnagar N, Mahajan P (2005) Machining studies of UD-FRP composites part 2: finite element analysis. Mach Sci Technol 9(4):503–528. https://doi.org/10.1080/10910340500398183
Niu K, Talreja R (2000) Modeling of compressive failure in fiber reinforced composites. Int J Solids Struct 37(17):2405–2428. https://doi.org/10.1016/S0020-7683(99)00010-4
Pinho ST, Davila CG, Camanho PP, Iannucci L, Robinson P (2005) Failure models and criteria for FRP under in-plane or three-dimensional stress states including shear non-linearity. NASA Langley Research Center, Hampton, VA, United States
Pinho ST, Darvizeh R, Robinson P, Schuecker C, Camanho PP (2012) Material and structural response of polymer-matrix fibre-reinforced composites. J Compos Mater 46(19–20):2313–2341. https://doi.org/10.1177/0021998312454478
Pipkin AC, Rogers TG (1971) Plane deformations of incompressible fiber-reinforced materials. J Appl Mech 38(3):634–640. https://doi.org/10.1115/1.3408866
Piquet R, Lachaud F, Ferret B, Swider P (2000) Étude analytique et expérimentale du perçage de plaques minces en carbone/époxy. Mécanique & Industries 1(1):105–111

Puck A, Schürmann H (1998) Failure analysis of FRP Laminates By Means Of Physically Based Phenomenological Models. This article represents the authors' contributions to a world-wide exercise to confirm the state-of-the-art for predicting failure in composites, organised by Hinton and Soden. Compos Sci Technol 58(7):1045–1067. https://doi.org/10.1016/S0266-3538(96)00140-6

Pwu HY, Hocheng H (1998) Chip formation model of cutting fiber-reinforced plastics perpendicular to fiber axis. J Manuf Sci Eng 120(1):192–196. https://doi.org/10.1115/1.2830100

Rahmé P, Landon Y, Lachaud F, Piquet R, Lagarrigue P (2011) Analytical models of composite material drilling. Int J Adv Manuf Technol 52(5):609–617. https://doi.org/10.1007/s00170-010-2773-5

Rao GVG, Mahajan P, Bhatnagar N (2007b) Micro-mechanical modeling of machining of FRP composites–Cutting force analysis. Compos Sci Technol 67(3–4):579–593

Rao GVG, Mahajan P, Bhatnagar N (2007a) Machining of UD-GFRP composites chip formation mechanism. Compos Sci Technol 67(11–12):2271–2281

Rentsch R, Pecat O, Brinksmeier E (2011) Macro and micro process modeling of the cutting of carbon fiber reinforced plastics using FEM. Procedia Eng 10:1823–1828

Sahraie Jahromi A, Bahr B (2010) An analytical method for predicting cutting forces in orthogonal machining of unidirectional composites. Compos Sci Technol 70(16):2290–2297. https://doi.org/10.1016/j.compscitech.2010.09.005

Takeyama H, Iijima N (1988) Machinability of glassfiber reinforced plastics and application of ultrasonic machining. CIRP Ann 37(1):93–96. https://doi.org/10.1016/S0007-8506(07)61593-5

Tsai SW, Wu EM (1971) A general theory of strength for anisotropic materials. J Compos Mater 5(1):58–80. https://doi.org/10.1177/002199837100500106

Tsao CC (2012) Effect of induced bending moment (IBM) on critical thrust force for delamination in step drilling of composites. Int J Mach Tools Manuf 59:1–5

Tsao CC (2006) The effect of pilot hole on delamination when core drill drilling composite materials. Int J Mach Tools Manuf 46(12):1653–1661. https://doi.org/10.1016/j.ijmachtools.2005.08.015

Tsao CC (2007a) Effect of deviation on delamination by saw drill. Int J Mach Tools Manuf 47(7):1132–1138. https://doi.org/10.1016/j.ijmachtools.2006.09.016

Tsao CC (2007b) Effect of pilot hole on thrust force by saw drill. Int J Mach Tools Manuf 47(14):2172–2176. https://doi.org/10.1016/j.ijmachtools.2007.05.008

Tsao CC, Hocheng H (2005a) Effect of eccentricity of twist drill and candle stick drill on delamination in drilling composite materials. Int J Mach Tools Manuf 45(2):125–130

Tsao CC, Hocheng H (2008) Effects of peripheral drilling moment on delamination using special drill bits. J Mater Process Technol 201(1):471–476. https://doi.org/10.1016/j.jmatprotec.2007.11.225

Tsao CC, Hocheng H (2005b) Effects of exit back-up on delamination in drilling composite materials using a saw drill and a core drill. Int J Mach Tools Manuf 45(11):1261–1270

Tsao CC, Hocheng H, Chen YC (2012) Delamination reduction in drilling composite materials by active backup force. CIRP Ann 61(1):91–94. https://doi.org/10.1016/j.cirp.2012.03.036

Zhang LC, Zhang HJ, Wang XM (2001) A force prediction model for cutting unidirectional fibre-reinforced plastics. Mach Sci Technol 5(3):293–305. https://doi.org/10.1081/MST-100108616

Zitoune R, Collombet F (2007) Numerical prediction of the thrust force responsible of delamination during the drilling of the long-fibre composite structures. Compos A Appl Sci Manuf 38(3):858–866. https://doi.org/10.1016/j.compositesa.2006.07.009

Zitoune R, Collombet F, Lachaud F, Piquet R, Pasquet P (2005) Experiment–calculation comparison of the cutting conditions representative of the long fiber composite drilling phase. Compos Sci Technol 65(3):455–466. https://doi.org/10.1016/j.compscitech.2004.09.028

Milling/Trimming of Carbon Fiber Reinforced Polymers (CFRP): Recent Advances in Tool Geometrical Design

S. A. Sundi, R. Izamshah, M. S. Kasim, M. F. Jaafar, and M. H. Hassan

Abstract Nowadays, Fiber Reinforced Polymers (FRP) materials namely carbon fiber-reinforced polymers (CFRPs) have been given such great attention by industries around the world as the end products material especially the aerospace industry due to the superior properties of the material themselves such as high-strength to the weight-ratio, high modulus-to-density ratio and high corrosion resistance. This chapter initiated to briefly describes the overview of one of the paramount processes in machining composite materials known as milling or trimming operation. In general, this chapter covers the basics of the CFRP material, fiber architectures, cutting tool materials, common types/geometries of cutting tool for milling/trimming CFRP material, a brief on the cutting tool manufacturing processes and wrapped-up with an exclusive recent advances reviews specifically emphasize on the impact of the cutting tool geometry in milling/trimming CFRP material. Ultimately, it is hoped that the contents of the chapter might be such a beneficial reference either to the academia or industrial practitioners who are engaged with composite materials in gaining initial ideas/overview related to the cutting tools and its geometries during milling/trimming of CFRP material.

Keywords Milling/trimming · CFRP material · Cutting tool geometry

S. A. Sundi (✉)
Faculty of Mechanical & Manufacturing Engineering Technology, Universiti Teknikal Malaysia Melaka (UTeM), Hang Tuah Jaya, 76100 Durian Tunggal, Melaka, Malaysia
e-mail: syahrul.azwan@utem.edu.my

S. A. Sundi · R. Izamshah · M. S. Kasim · M. F. Jaafar
Advanced Manufacturing Center (AMC), Universiti Teknikal Malaysia Melaka (UTeM), Hang Tuah Jaya, 76100 Durian Tunggal, Melaka, Malaysia

M. H. Hassan
School of Mechanical Engineering, Universiti Sains Malaysia, 14300 Nibong Tebal, Pulau Pinang, Malaysia

M. T. Hameed Sultan et al. (eds.), *Machining and Machinability of Fiber Reinforced Polymer Composites*, Composites Science and Technology,
https://doi.org/10.1007/978-981-33-4153-1_4

1 Introduction

As of today, Fiber Reinforced Polymers (FRP) composites are getting more popular to be used as components or end products for numbers of industries out there. One of the major impacted industries was aerospace industry. For instance, The *Boeing 787 Dreamliner* which was first flying in December 2009 exhibited gradual increases up to 50% in usage of composite material as the main structural components (Shennan 2014). The main reasons why the aerospace industry is slowly shifting from a metallic material to utilizing more composite materials as the main structural components in the aircraft's design are due to their high-strength to the weight-ratio, high modulus, high specific strength and high corrosion resistance which a few times stronger and better than metals.

Glass Fiber Reinforced Polymers (GFRP) and Carbon Fiber Reinforced Polymers (CFRP) are the two major FRPs used in aero-structural components. This chapter will only discuss milling/trimming related to the CFRP material as the main focus material.

Generally, it could be said that machining composite material is far different from machining homogeneous material such as metallic material, despite the fact that theoretically and technologically are the same used in metal machining. Although composite components are normally manufactured or fabricated to the near-net final shape, some machining processes are hard to be avoided. Milling/trimming and drilling/countersinking are among the compulsory machining processes which are usually applied as the post/finishing processes to achieve the desired dimensioning accuracy. Those processes are vital in ensuring the quality of the panels' final trimmed edges and holes for joining such as riveting are met according to the certain stated specification.

To deal with the uniqueness of the composite properties such as heterogeneous, anisotropic and interaction with the cutting tool during machining is such a hard situation to be understood. Machining might cause defect or flaw to the machined composite parts or also known as surface damages namely delamination, matrix cracking, matrix smearing, un-cut fiber, fiber pulled-out, and burned matrices. In addition, the abrasive nature of the fibers as well as the requirement needed to neatly shear or cut them off, brought the challenge of composite machining a step higher than metallic machining. Hence, to ensure the most optimum quality of the machined part, the best tool materials and tool geometrical design must be chosen accordingly (Sheikh-Ahmad et al. 2012; Mazumdar 2002; Rana and Fangueiro 2016). Therefore, there are various materials for cutting tools and geometrical design offered by tools manufacturers out there. The main reason for this option in tooling is the uniqueness of the composite properties which usually derived by various matrices, reinforcement materials, forms, and methods applied to prepare certain composites (López de Lacalle et al. 2009; López de Lacalle and Lamikiz 2010).

This chapter outlines a brief synopsis on the milling or trimming process of CFRP material. In addition, this chapter also covers the overall overviews of the cutting tool types and geometries utilized to perform the mentioned machining process. In the final section of the chapter, recent advances in milling or trimming CFRP materials mainly related to the impact of tool geometry towards machining performances are successfully critically reviewed and elaborated.

2 Overviews on CFRP

2.1 Types of Carbon Fiber

There are several ways in preparing carbon fibers from the polymeric precursor materials namely cellulose, pitch, polyvinylchloride and polyacrylonitrile (PAN). Among all, PAN is the most popular method used by most of the related industries today. PAN based carbon fibers have higher strength and modulus properties as compared to the pitch-based precursor carbon fibers (Morgan 2005). Figure 1 illustrates the normal processes in making polyacrylonitrile (PAN) based carbon fiber (Rana and Fangueiro 2016).

Carbon fibers are currently used primarily for aerospace industries due to their exceptionally high modulus-to-density ratio. Besides, the strength-to-density ratio is also very high. Other advantages of carbon fibers are their high electrical conductivity, high thermal conductivity, and low coefficient of thermal expansion. However, their high electrical conductivity possesses a problem in processing plants. Carbon dust or fine particles produced during the processing phase from carbon fibers may

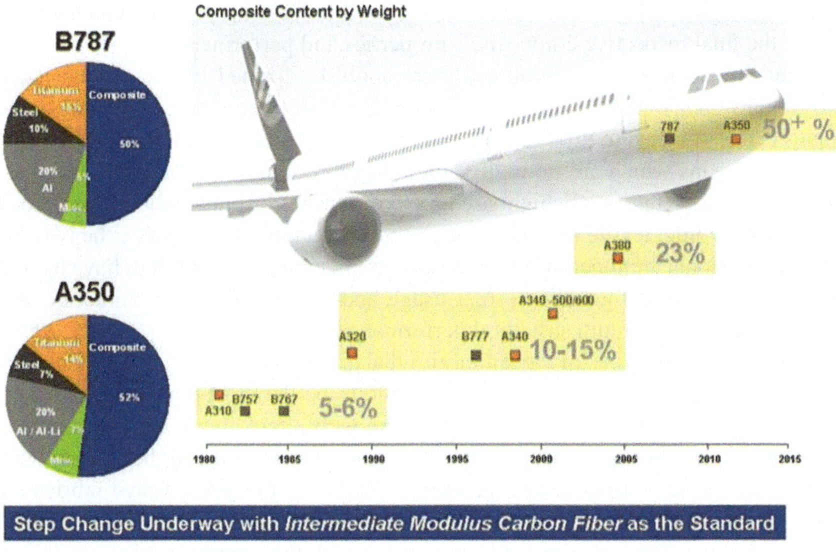

Fig. 1 Increased growth and content by weight of composite materials in aircrafts over the years (Shennan 2014)

short-circuit nearby electric motors and other electrical machines. Therefore, carbon-proofed/protected equipment is necessary to ensure these issues are resolved (Mallick 2018). Figure 2 illustrates the Processes of manufacturing carbon fibers from various precursors.

There are a few categories of CFRPs which are usually classified based on their modulus and tensile strength as listed below:

- Ultra-high modulus (UHM) = tensile modulus higher than 500 GPa
- High modulus (HM) fiber = modulus higher than 300 GPa and strength to modulus ratio <0.01
- Intermediate modulus (IM) = modulus up to 300 GPa and strength to modulus ratio >0.01
- Low-modulus Carbon fibers = modulus as low as 100 GPa and low strength (Fig. 2).

2.2 *Fiber Forms/Architecture*

The definition of fiber architecture is usually referred to as the arrangement of fibers in a composite. This arrangement shall determine the properties of the composite as a whole as well as the appropriate processes involved. Fiber orientation/stacking direction, fiber sequencing, fiber continuity, fiber crimping, and fiber interlocking are among the fiber architecture characteristics which usually affected the overall material properties of each composite. On the other hand, the void content, fiber wetting, fiber distribution, dry area and others are determined by the flow of the matrix through the fiber architecture during the processing period, which finally affects the final respective composite's properties and performances.

Fiber architecture for continuous fibers could be formed by one-dimensional, two-dimensional, or three-dimensional. A dedicated technique called pre-pregs techniques is usually used to produce one-dimensional architecture. Pre-pregs are fiber-reinforced resins which are normally cured under certain controlled heat and pressure to form an exceptionally high-strength to weight-ratio components (Corporation 2013). Meanwhile, textile manufacturing processes are used to produce the two and three-dimensional architectures. Each respective fiber architecture shall have its own characteristics. Therefore, the correct usage and design of fiber architecture will result in the most optimum structural performance of the composite (Mallick 2008).

Figure 3 exhibits types of woven fabrics that are normally used in manufacturing composite materials. The plain weave is found to be the most popular weave used to manufacture composite materials which the weft alternatingly crosses over and under the warp. This indirectly makes this type of weave as the highest crimp with the tightest fabric and poorest drape-ability. However, the plain weave fabrics are also the most resistant to the in-plane shear movement. On the other hand, the weft yarn arrangement for satin weaves is crossed over or skips a number of warps before it crosses under a single warp again which finally results in minimizing the crimp strengths and increasing flexibility and drape-ability of the fabrics. Twill weaves

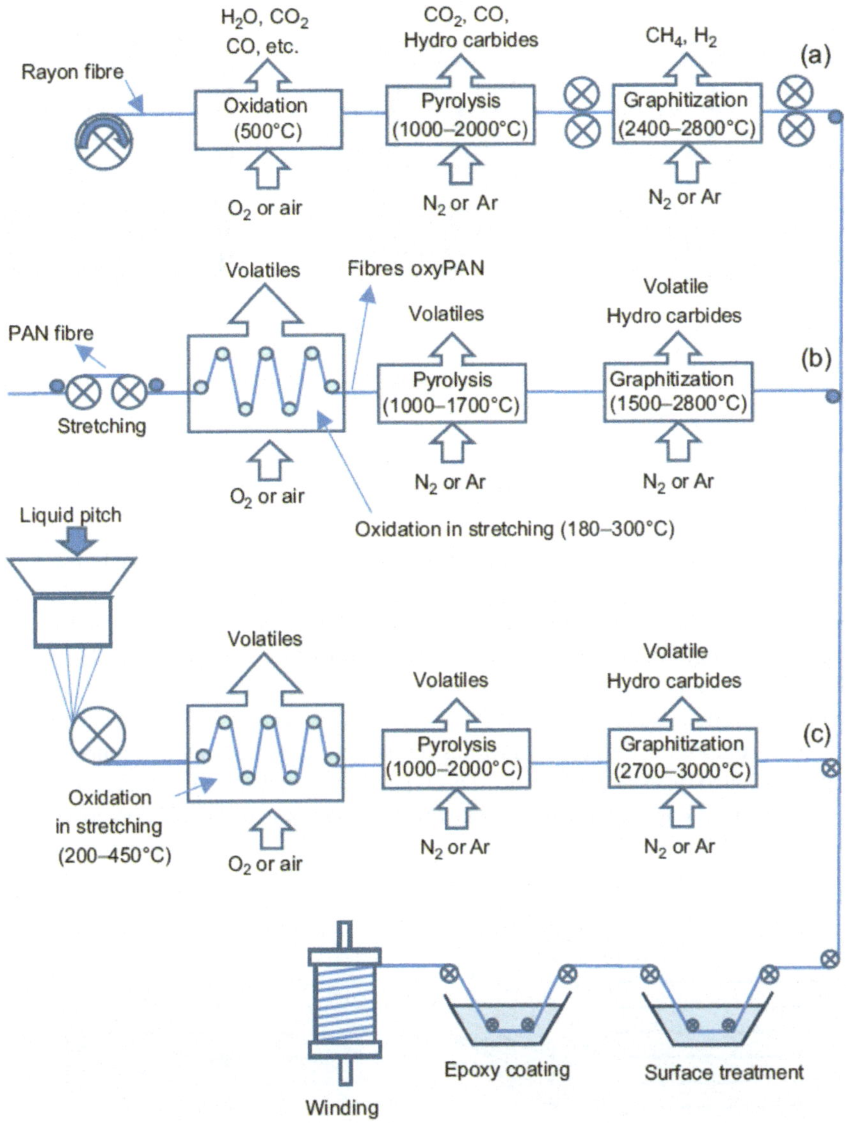

Fig. 2 Processes of manufacturing carbon fibers from various precursors; **a** rayon; **b** polyacrylonitrile (PAN); **c** pitch for oil (Rana and Fangueiro 2016)

are made in standard four-, five-, or eight-harness forms. The weave style can be varied according to the crimp and drape-ability. Low crimp gives better mechanical performance due to the straighter fibers carry greater loads. Meanwhile, the drapeable fabric is easier to lay up over complex form (Corporation 2013; Sheikh-Ahmad 2009).

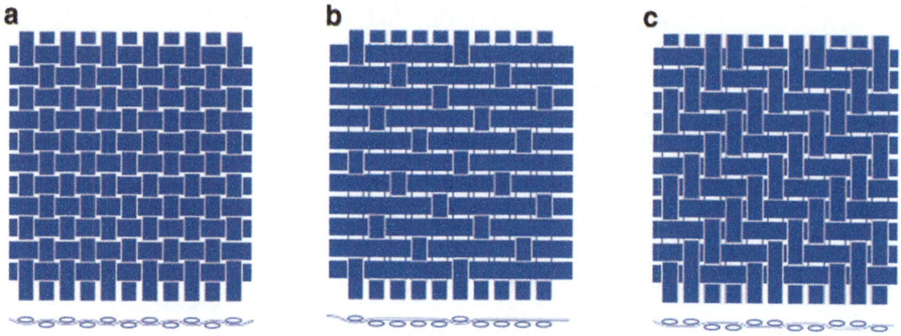

Fig. 3 Types of woven fabrics **a** plain weave (low drape-ability/high crimp); **b** satin weave (good drape-ability/low crimp); and **c** twill weave (average drape-ability/average crimp) (Corporation 2013)

2.3 *Laminated/Stacked Composite*

Laminated/stacked composite is the most common composite materials used by various industrial applications. Layer by layer of fibrous material is stacked or laminated together according to a certain direction based on a determined specification. A single direction of this type of composite is also known as Unidirectional (UD) composite. Meanwhile, various angles of stacking directions formed a composite called Multi-directional composite or also known as Cross-plied quasi-isotropic composite. Figure 4 illustrates the structure of laminated composites for both mentioned types of composites. This type of composite usually characterized by high in-plane strength and stiffness. However, the laminated/stacked composites are

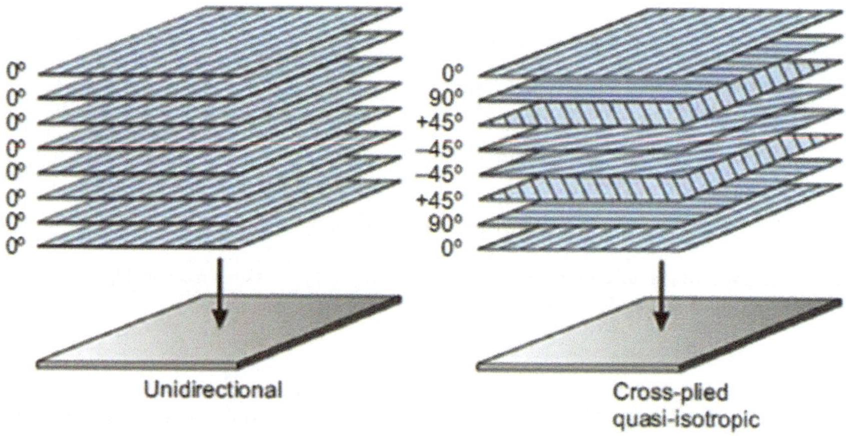

Fig. 4 Types of laminated/stacked composite; Uni-directional (UD)-Left; Multi-directional-Right (Rana and Fangueiro 2016)

considered quite poor strength via the-thickness direction which usually determined by flexural test (Rana and Fangueiro 2016).

2.4 *Sandwich Composite*

Sandwich composite classified to be a higher class of structural composites. The main advantages of this type of composite as compared to the one discussed previously (laminated/stacked composite) are better in a lightweight, higher stiffness and more durable. A sandwich composite comprises two outer sheets or plies that are separated by a thicker central structure called a core. Both parts are then adhesively bonded together to form a complete sandwich panel (Fig. 5). The outer sheets are usually made of a relatively stiffer and stronger material, such as fiber-reinforced plastics, aluminum alloys or other super alloying materials or plywood. Meanwhile, the core material is normally chosen from lightweight and low modulus of elasticity materials which typically laid under three categories namely rigid polymeric foams (phenolics, epoxy, polyurethanes), wood (balsa wood), and honeycombs. Structurally, the core materials are meant to provide continuous support for the outer sheets. Moreover, they must have sufficient shear strength to withstand transverse shear stresses and higher shear stiffness to resist the buckling effect of the overall composite. Honeycomb structure core is well-known to be used in an aerospace application. The unique design and structure of thin foils that are formed into interlocking hexagonal cells, with axes oriented perpendicular to the face planes, meet the higher strength and stiffness required by the respected application. In general, the strength and stiffness of honeycomb structures are measured based on size and the wall thickness of the honeycomb cells, as well as the material from which the honeycomb is made (William et al. 2010).

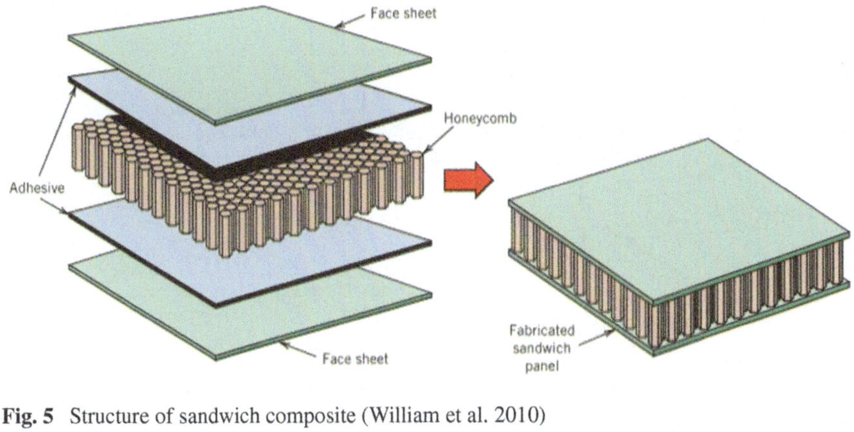

Fig. 5 Structure of sandwich composite (William et al. 2010)

3 Milling/Trimming of CFRP Material

Milling operation is defined as a removing material from a work piece by applying a rotating cutting tool which may consist of one or more active cutting edge. There are two (2) most common types of milling operation in machining FRPs material known as peripheral milling or also called as side trimming as well as an end milling process. Peripheral/side milling refers to the cutting condition which utilizes the body or diameter of the cutting tool. Other terms used by researchers and industries around the world which referring to the similar operation is edge/slot trimming/routing operation. Meanwhile, the end milling operation always refers to a machining process using the face or end section of the cutting tool (Sheikh-Ahmad 2009). Figure 6 distinguishes both mentioned milling operations.

3.1 *Cutting Tool Material*

A huge range of materials available today for the use of cutting tools to cater to various machining operations. These materials are usually classified according to the three (3) main characteristics namely the hardness, toughness and overall strength. Hardness and toughness are the two (2) vital characteristics to be evaluated in selecting the appropriate tool materials. Theoretically, the cutting tool's material should be having

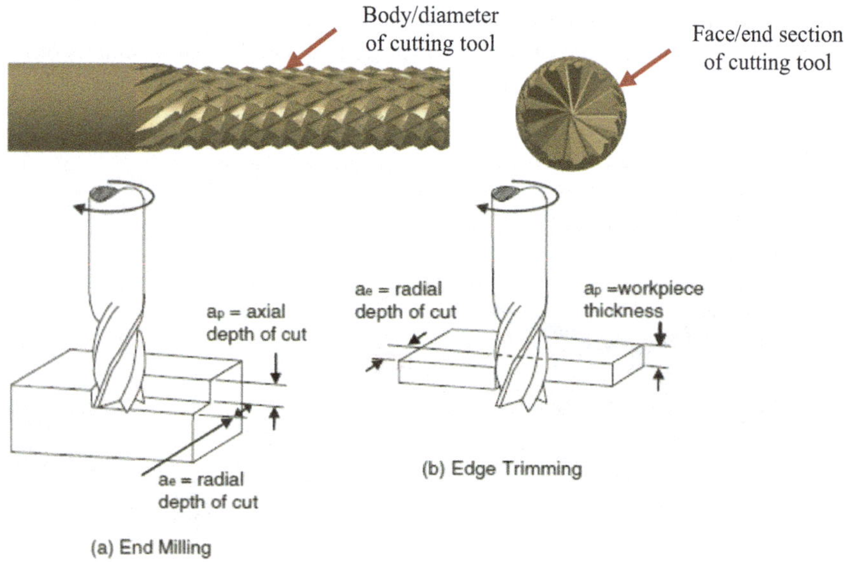

Fig. 6 Comparison between peripheral/edge and end milling (Sheikh-Ahmad 2009)

greater hardness properties than the part to be machined to allow the cutting or shearing off phenomenon occur efficiently.

Hardness refers to an ability of a material to localized plastic deformation caused by either abrasion or mechanical indentation. Meanwhile, the ability of a material to absorb energy before fracture or failure is known as toughness. The higher the toughness level of each material, the better it could resist the chipping and fracturing effect due to the vibration, chattering, runouts and other imperfections during machining. Both mentioned properties are changing in opposite directions for any specific tool materials. A recent trend in research and development (R&D) of tool materials is emphasizing on how to increase the toughness property of a material while ensuring the hardness property is kept constant.

The major properties of the most common cutting tool materials are shown in Fig. 7. The hardest tooling material obviously exhibited by polycrystalline diamond (PCD), however possesses the least toughness property as its sharp deformation occurs at around 600 °C. Contrary, high speed steel (HSS) material presented the best in toughness, but deforms around 700 °C as compared to the other types of tools material (Astakhov and Davim 2008).

3.1.1 Polycrystalline Diamond (PCD)

Polycrystalline diamond (PCD) is formed by a process of compacting high temperature and pressure of crystal diamonds with cobalt (Co) which react as the metallic binder to enhance the overall bonding strength due to a random orientation of the diamond crystals. Therefore, the main advantage of PCD compared to the single crystal diamond chain is its uniform mechanical properties which improves the hardness as well as the toughness of the material (Sheikh-Ahmad 2009). To manufacture PCD cutting tool, a mixing process of graphite and catalyst (typically nickel, Ni) which usually carried out under pressure of approximately at 7000 MP and temperature at 1800 °C to form a layer of diamond crystals before it being placed on a carbide substrate. In this process, the cobalt element from the tungsten substrates reacts as the binder to the diamond crystals which provides the required toughness to the polycrystalline diamond. Typical PCD tool materials could sustain from an abrasion resistance up to 500 times higher than the tungsten carbide material and has also higher thermal conductivity. To prolonged the tool life and increased productivity by adopting PCD tools in the real manufacturing situation, always offset the higher initial cost by lowering the unit cost of parts produced (Astakhov and Davim 2008).

3.1.2 High Speed Steel (HSS)

High speed steel (HSS) is one of the popular materials chosen to be the tool material due to its high toughness as compared to rest of the tool materials and an acceptable strength as well as moderate hardness (up to 68 HRC).

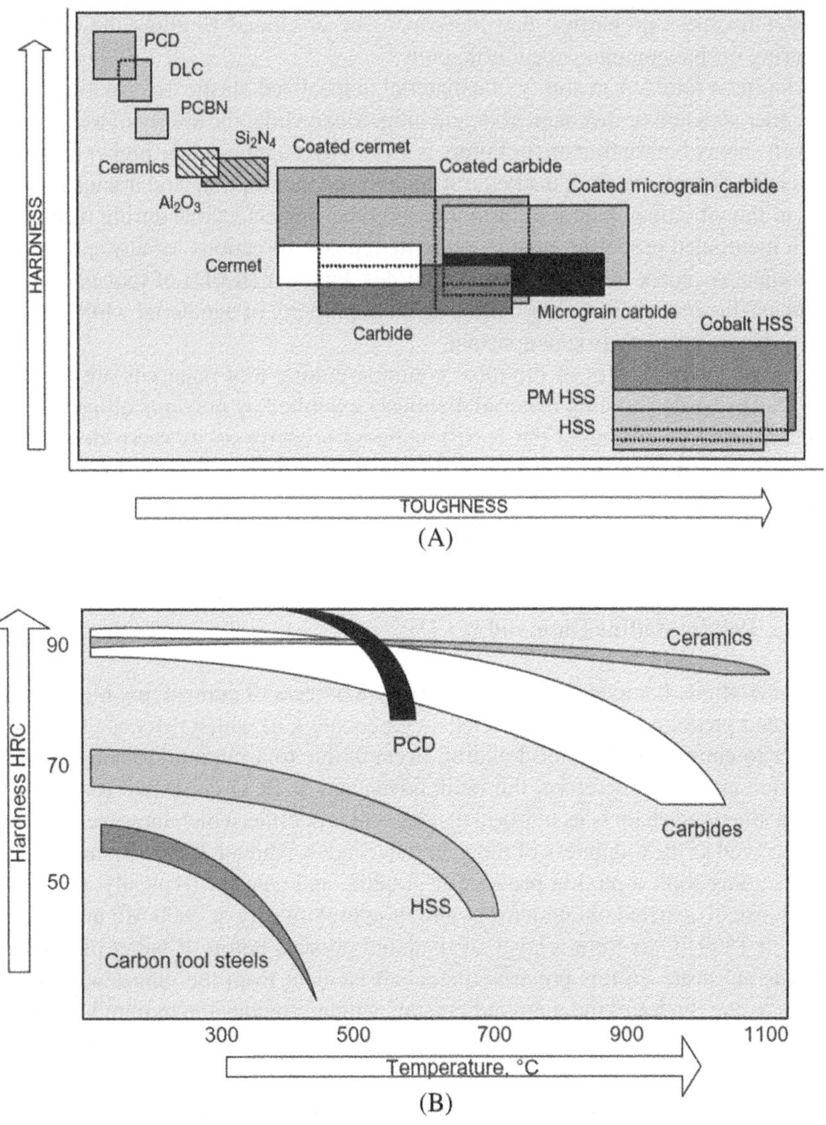

Fig. 7 **a** Overall hardness and toughness level for well-known cutting tool materials; **b** hardness of major tool materials in HRC versus temperature (Astakhov and Davim 2008)

Due to this characteristic, HSS easier to be ground into any desired cutting tool geometrical designs and sharp edges such as the most common geometry; helical helix end mills. This group of material could possibly be reground, heat-treated, and reused after being worn out. This is part of the cost-saving practices done by the existing industries out there. In terms of cost wise, HSS has been considered as low or more economical compared to the other tool materials due to the processes involved to manufacture this type of material which has been around for decades. However, the biggest drawback or weakness of HSS material is its inability to retain hardness at high temperatures (austenitic transformation temperatures), and thus not highly recommended for high-speed machining. Besides, due to their low hardness and moderate strength, this group of material is not suitable to be used to machine high abrasiveness materials such as FRPs and particulate-reinforced polymers especially in milling/trimming operation (Sheikh-Ahmad 2009). Nevertheless, HSS material is quite famous for drill types cutting tool for drilling of FRPs materials.

3.1.3 Cemented Carbide

Cemented carbides are well-known having such a superior wear resistance and high hardness. Tungsten carbides (WC) is found to be the main particle in the cemented carbide group. Besides, titanium (Ti), tantalum (Ta) and others combination of different particles are used to form cemented carbides family. The range of the particles' size used to fabricate cutting inserts are usually less than 0.8 μm for micro-grains, fine grains; 0.8–1.0 μm, medium grains; 1–4 μm, and more than 4 μm for coarse-grain. Cobalt content as the binder in each mixture of cemented carbide plays a significantly affect to the properties of carbide inserts. Normally, the amount of the cobalt used is between 3–20%, depending upon the required toughness and hardness as well as the application of the final inserts (Astakhov and Davim 2008).

Cold pressing process of carbide powders and mentioned metal binder (cobalt, Co) is utilized to produce carbide tool blanks which are usually manufactured to the near-net-shape. Sintering process is then taking place to sinter the tool blanks at temperatures in the range of approximately at 1,350–1,650 °C. The temperature applied is to melt the binder and indirectly creates bonding reaction between the metal binder and the carbide particles. The hardness of cemented carbides is basically determined by the hard carbide phase, while the metal binder provides the necessary toughness (Sheikh-Ahmad 2009).

As mentioned earlier, cobalt content determines the final toughness of a cutting tool. The more amount of cobalt particles added into the mixture increases the toughness of the cutting tool proportionally. However, the hardness and the overall strength of the cutting tool might be decreased. Having said that, the right combination of carbide and cobalt composition (grade), type of exterior coating layer and the coating technology applied could improve optimum productivity in metal cutting industry without the need to sacrifice the wear resistance.

3.2 *Types of Cutting Tool for Milling/Trimming CFRP Material*

Tool geometrical feature plays such a very important role in machining because it directly affects the final productivity of machining, tool life and surface quality/integrity of machined parts (Astakhov and Davim 2008). In general, common types of cutting tools used for milling/trimming of CFRP materials could be divided into two (2) namely polycrystalline diamond (PCD) inserts type which usually consists of solid carbide body with sintered PCD inserts. Another geometry of cutting tool is called solid carbide end mill which can be found as it is (uncoated) or with specific coating condition (coated). The coating layer usually added on the exterior surface of the cutting tool is meant to enhance the wear resistance or prolong the tool life of the cutting tool which indirectly impacted the yearly yields or productivity of the respective industry. The solid carbide end mill type is then divided into another two (2) other specific geometries known as helical spiral helix and router/burr/interlocking geometry. A family of solid (uncoated/coated) carbide end mill is illustrated in Fig. 8.

3.2.1 Polycrystalline Diamond (PCD) Tool

PCD tool usually consists of two main parts namely the PCD blanks/insert and the body/holder which usually made of carbide material. Non-traditional machining operations namely laser beam machining (LBM) or electrical discharge machining (EDM) are utilized to cut the sintered PCD blanks to the desired shape before they being brazed to a tungsten carbide body/holder. PCD tools (Fig. 9) are typically very expansive as compared to normal solid carbide end mills due to the technology and processes required to manufacture the tool such as extreme temperature and pressure needed for the sintering process. Besides, to perform the cutting and grinding process of the cutting edge is not an easy job to be done. Moreover, high technology and precision equipment are required to produce such high-quality tools. Nevertheless,

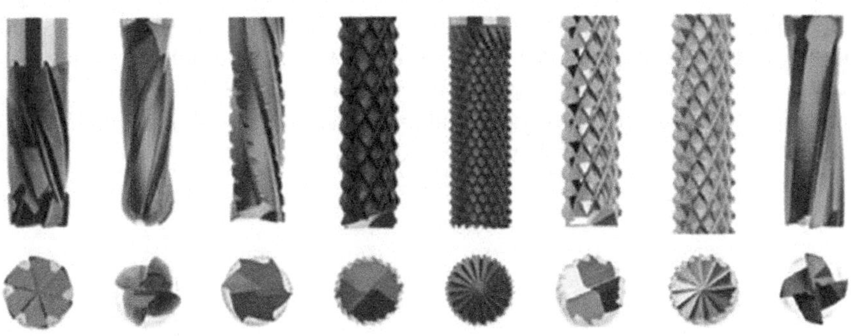

Fig. 8 Advanced tool design for composites machining (courtesy of Seco Tool)

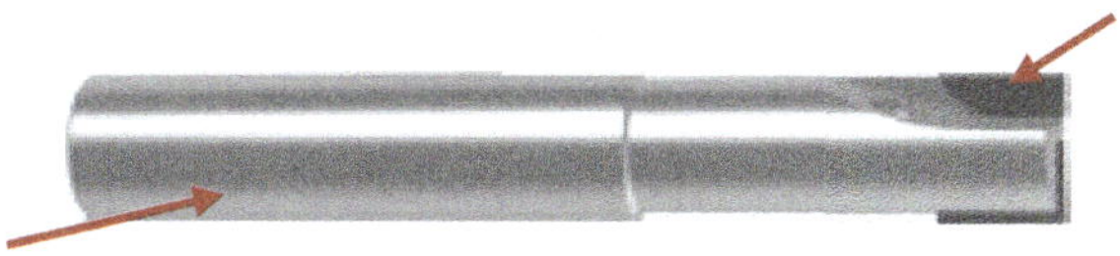

Fig. 9 PCD Tool (courtesy of Guhring KG)

Fig. 10 Helical helix end mill geometry (courtesy of Sandvik Coromant)

PCD tools could be more worth to invest compared to the solid carbide tools under certain optimum machining conditions due to its superior resistance to the wear rate.

3.2.2 Solid Carbide End Mill

The first type of solid or (uncoated/coated) carbide end mill geometry is known as Helical Helix tool (Fig. 10). This type of tool is very famous in cutting metallic materials. It has either right hand or left hand helical helix shape depending on the application. In metallic machining, helical helix tool geometry is very good to be used almost in every stage of machining phases; roughing, semi-finishing or finishing. As far as composite machining is concerned, researchers have been utilizing this type of tool geometry for years to search the best cutting conditions, methods, techniques and others approaches in machining composite materials.

The second type of solid or (uncoated/coated) carbide end mill geometry is called as burr or router tool. Other terminology being used by researchers around the world which referring to the similar tool geometry in milling/trimming composite materials are segmented helix, knurled, diamond interlocking and multi-tooth. Usually, this type of tool geometry has both right and left helix shape with certain angles. Number of flute/teeth for any designated tool is determined by the number of both helix (refer Fig. 11). This type of tool has been widely used by industries due to its economic cost as compared to the PCD tool and very efficient in neatly shearing/trimming FRPs materials. In addition, burr or router geometry has been also given a special attention by researchers around the world in searching the best method and approach to obtain the most optimum quality of the machined FRPs parts.

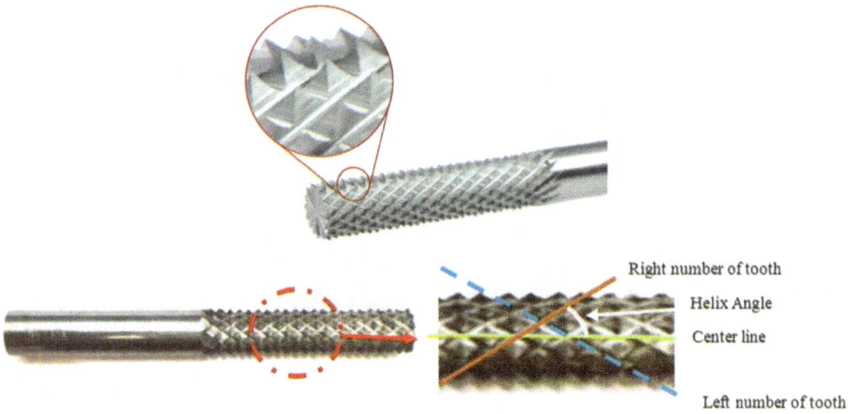

Fig. 11 Burr/router tool (courtesy of OSG) and details of geometrical features

3.3 Manufacturing/Grinding Processes of Solid Rod Cutting Tools

In cutting tool fabrication especially for solid rod type cutting tools, the CNC tool and cutter grinder are utilized to perform the grinding process by using various shapes and grinding wheel materials. The grinding process is part of the material removal processes by using an abrasive product (normally cylindrical wheel type) as the main cutting element or acts as the cutting tool. A rotating wheel is then controlled by certain cutting conditions namely the cutting speed (rotation per minute-rpm) and the feeding rate into contact with the respective part surface to be ground. Abrasive grains which are formed and bonded together by applying a specific binder begin to perform the machining or grinding process. The quality of the finished surface and the final dimensional accuracy of the ground part are partly determined by the type of abrasive wheel used besides the applied cutting conditions as stated earlier (El-hofy 2014). For solid carbide material, the only type of wheel which can be used to perform the grinding process is the diamond wheel due to the properties of diamond material which has higher hardness and strength as compared to tungsten carbide material. Meanwhile, Cubic Boron Nitride (CBN) grinding wheel's material is applied to perform the grinding process for a solid HSS rod.

A bare solid carbide rod is cut into the required overall length before the fabrication taking place. The desired tool's geometries are programmed conversationally via the software that came with the machine. Programmers may visualize a completed program and the overall grinding processes prior to the final execution. This will helps to prevent any unwanted incidents such as collisions or over-ground results. Furthermore, this process also determining the capability of the machine in producing any customization of the tool's geometries. The tool grinding process shall begin once the cycle start button is pressed and the cycle time of each tool produced will be depending on the complexity of the geometry as well as the cutting feed during

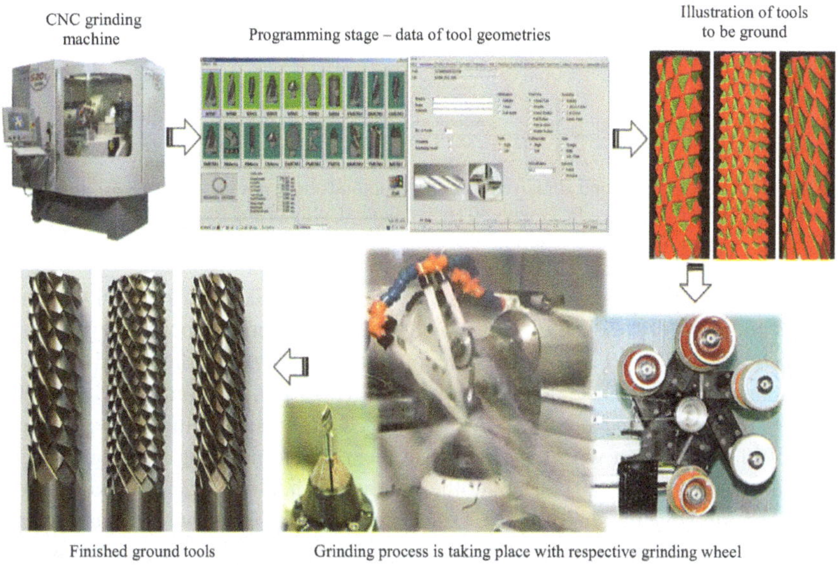

Fig. 12 Overall processes in cutting tool fabrication for solid rod type

grinding. Figure 12 summarizes the overall processes involve in fabricating a solid rod cutting tool utilizing the tool and cutter CNC grinding machine.

3.4 Recent Advances on the Effect of Tool Geometrical Design in Milling/Trimming of CFRP Materials

To date, researchers as well as industries around the world have primarily concentrated on researching and investigating the most optimum tool geometry in machining CFRP materials especially on milling/trimming operation. Koplev et al. and König et al. (1985) were among the earliest researchers who performed experimental studies in composite machining. They have recommended that routers with the "diamond cut"/burr tool geometry for glass and carbon fibers and opposed helical design for aramid fibers give the best results (König et al. 1985). Colligan and Ramulu (1992) initiated a research to study ply delamination effect during edge trimming graphite/epoxy laminates using two different tool geometry known as PCD and helical helix carbide end mill tool with some variation on the helix angle of both mentioned tools (Fig. 13). Authors revealed that the surface ply delamination appears to occur in three distinctly different types according to the fiber orientations (Colligan and Ramulu 1992). A few years later, Wang et al. (1995) experimented a study on the effect of resultant force with various PCD inserts tool geometries namely the rake angles; 0°, 5°, 10° and clearance angles; 7°, 17° (Fig. 13) (Wang et al. 1995). El-Hofy

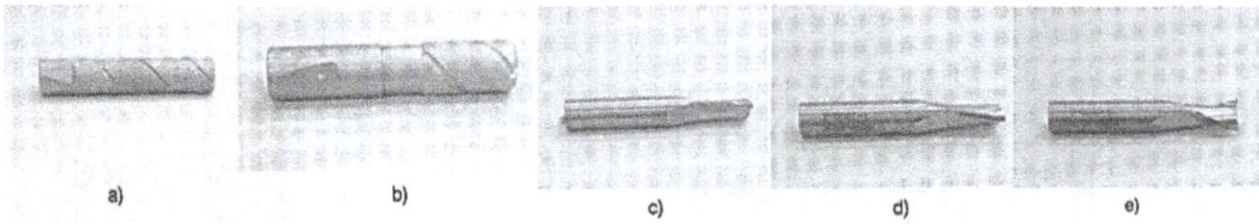

Fig. 13 **a** Coarse grit diamond abrasive **b** fine grit diamond abrasive **c** carbide with 10° helix angle **d** PCD with 10° helix angle **e** PCD with 30° helix angle (Colligan and Ramulu 1992)

et al. (2011) conducted a comparison study on Diamond-Liked-Carbon (DLC) coated carbide end mills and PCD tool. There were three (3) different types PCD grains size ranged from 1.26 to 30 μm which categorized into coarse, medium and fine were benchmarked against DLC coated carbide tool (Fig. 15). Authors found that PCD tools resulted certain acceptable surface quality without sacrificing the wear rate of the tool in comparison with the DLC coated carbide tool during slotting of CFRP material (El-Hofy et al. 2011). In another study by Chen et al. (2017) and Liu et al. (2017) reported that the staggered PCD cutter geometry (Fig. 14) which was designed by them indicated significant resistant on the wear rate and better burr suppression as compared to the normal 0° helix angle of PCD tool. The cutting edge of the staggered PCD cutter had an inclination angle of 5°, while the inclination angle of its adjacent cutting edge was in the opposite direction, and staggered along the circumferential direction of the tool (Liu et al. 2017; Chen et al. 2017). Earlier this year, Nguyen-Dinh et al. (2020) have proposed a newly designed of PCD tool known as the four serrated straight flutes tool (4SSF) (Fig. 16) which has been specially fabricated join-ventured with ASAHI Company. The main aim of the newly designed tool was to assist in reducing the impact on the harmful particles produced whilst milling/trimming of CFRP materials which could cause respiratory severe decease such as pulmonary alveoli. It was observed that the newly designed tool (4SSF), managed to reduce the dispersion of the particles in the air which indirectly resulted great reduction of the harmful particles in the air.

Davim and Reis (2005) performed a comparative study of two different helical helix end mills to investigate the effect of surface damages during milling of CFRP composite material. The first tool was a two-flute end mill with a helix angle of 30°,

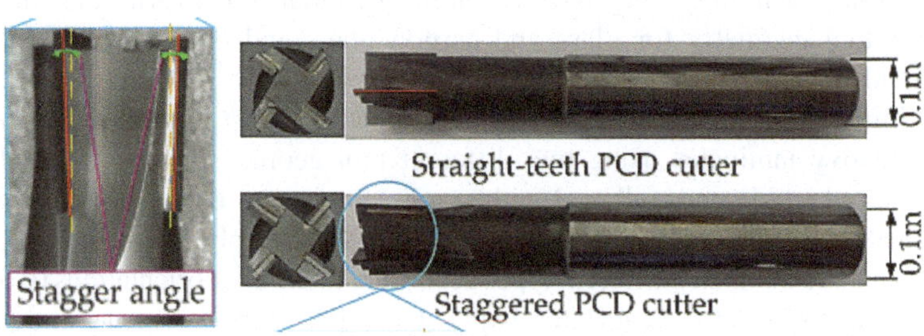

Fig. 14 Staggered PCD cutter details geometry (Chen et al. 2017)

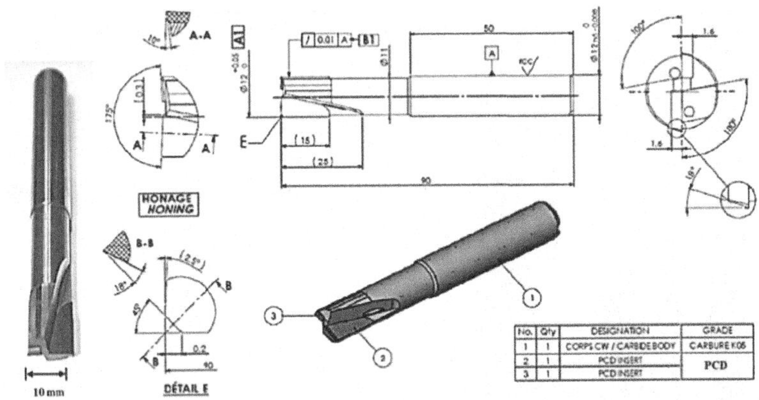

Element PCD grade	Reference number	Description
Coarse (CTM-302)	Seco 28108-928	Grain sizes varying from 2-30 μm. High abrasion resistance but relatively low chipping resistance
Medium (CTB-010)	Seco Reaming 28156-928*	Average grain size of ~ 6.7 μm. Good balance of abrasion and chipping resistance
Fine (CMX-850)	Seco 28155-928***	Grain size of 1.26 μm, which enables fabrication of cutters with finer cutting edges and high resistance to milling forces
WPC-102	Seco 0269269	Developed for woodworking applications with multi-layered PCD structure consisting of functionally graded PCD with diamond to prevent chipping

Fig. 15 Geometry and characteristics of PCD cutters (El-Hofy et al. 2011)

a rake angle of 10°/30°, a clearance angle of 9°. Meanwhile, the second tool was a six-flute helical helix end mill with a straight or 0° helix angle (Fig. 17). Authors discovered that the two-flute end mills with the specific geometries as mentioned earlier generated better surface quality with fewer damages than the six-flute helical helix end mill (Davim and Reis 2005). Uhlmann et al. (2016) studied on cutting strategies and high-speed cutting (HSC) for trimming CFRP material by using eight-flute end mills tungsten carbide material (Fig. 18). They revealed that the macro and micro specific design, as well as the tight tolerances on the geometrical features of the tool, were the ultimate key factors in ensuring the success of HSC milling (Uhlmann et al. 2016). Earlier before, Hosokawa et al. (2014) performed an investigation on the effect of Diamond-Like-Coating (DLC) and variation on helix angles tool which summarized that DLC coating did not really significant than tool geometry of the helical helix end mill. Authors have considered a high-helix angle of

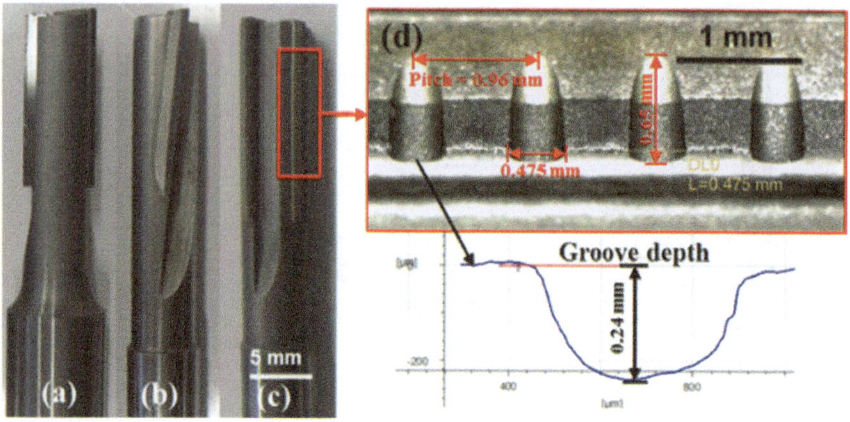

Fig. 16 **a** Two (2) straight-flutes (2SF), **b** two (2) helix-flutes (2HF), (c) four (4) serrated straight-flutes (4SSF), and **d** details of the grooves design od 4SSF tool (Nguyen-dinh et al. 2020)

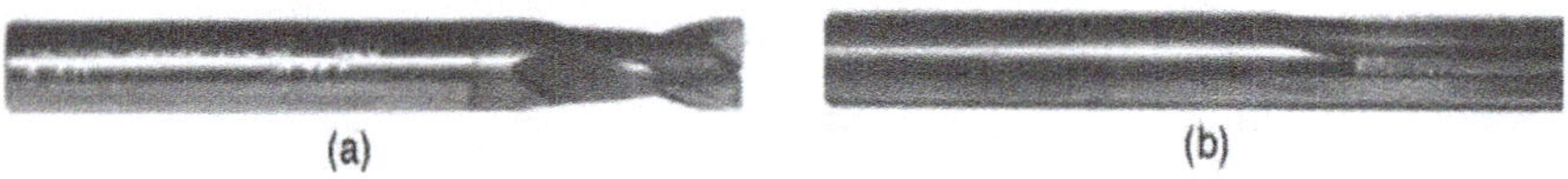

Fig. 17 **a** Two-flute helical helix end mill (helix angle; 30°, rake angle; 10°/30°, clearance angle; 9°), **b** Six-flute straight /0° helix angle end mill (Davim and Reis 2005)

Fig. 18 Details of cutting tools' geometry and the analyzed milling paths (Uhlmann et al. 2016)

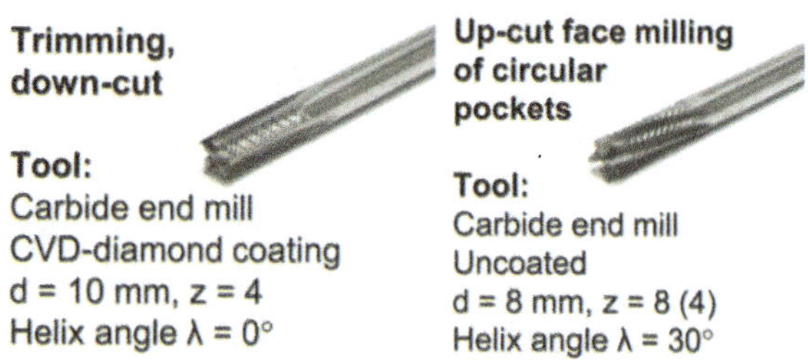

helical helix (60°) end mill as compare to the standard 30° end mill (Fig. 19) in their work (Hosokawa et al. 2014). In a different study, Can (2017) presented the surface roughness analysis of transverse and longitudinal directions by adopting arithmetic average (Ra) for vertical and incline milling position using three (3) different helical helix end mill geometries which were varied by the number of teeth/flutes (2, 4 and 6) (Fig. 20) for CFRP composite material. Inclined milling position was observed to be better than the vertical position in generating good surface roughness results at all cutting conditions. The authors reported that an increase in the number of teeth or flute decreases the deformation on the trimmed edges (Can 2017). In recent year, Wang et al. (2020) proposed a newly designed of tool geometry known as left-right end mill tool (Fig. 21) which proven in minimizing the trimmed surface quality

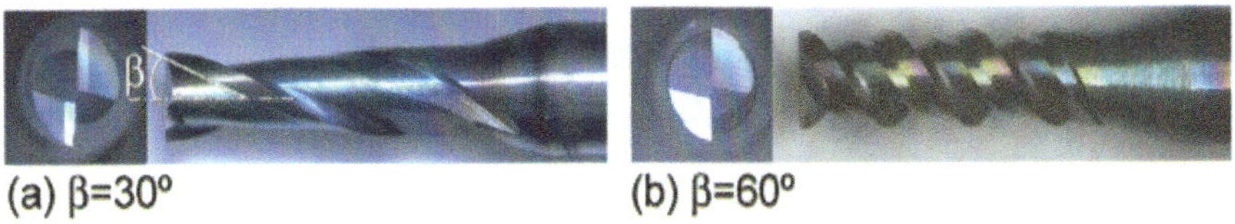

Fig. 19 DLC coated of normal and high-helix end mills with the specific helix angles (β) (Hosokawa et al. 2014)

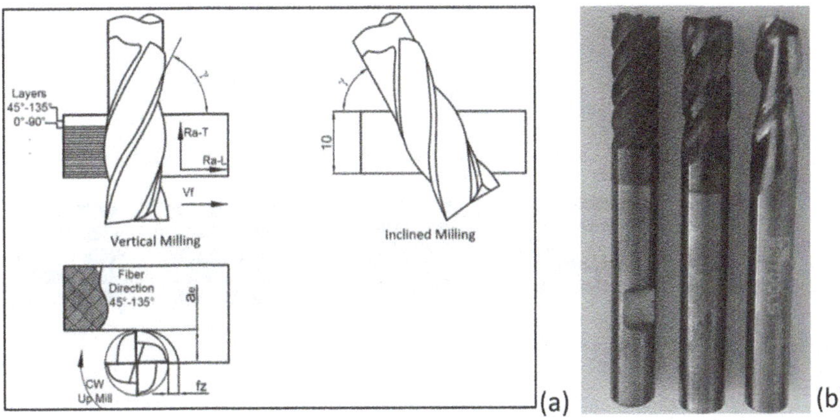

Fig. 20 **a** Details of vertical and incline milling positions, **b** 2-4-6 flutes helical helix end mill tools (Can 2017)

especially related to delamination as compare to standard helical helix tool with 30° degree helix angle at a small cutting depth of UD CFRP (Wang et al. 2020).

Janardhan et al. (2006) have performed an experimental investigation on the differentiation between up and down milling of CFRP material using a diamond interlocking/burr tool. The conventional or up-milling direction was identified to be better in terms of surface damages such as delamination and surface quality than climb or down-milling direction for CFRP material (Janardhan et al. 2006). An investigation work carried out by Duboust et al. reported that multiple teeth/burr geometry (diamond coated) tool (Fig. 22) produced an acceptable surface quality in comparison with polycrystalline diamond (PCD) tool although tested at high feed rate condition (Duboust et al. 2017). On the other hand, López de Lacalle and Lamikiz (2010) have performed a comprehensive study on the effect of various types of coating, the micro-grain of carbide and Cobalt content as well as various geometries of multi-tooth or burr tool (Fig. 23). Besides, they also performed a comparison study of machining performance between the burr tool and the PCD tool. PCD tools reported being not visible enough to be economically feasible concerning to their high price. Therefore, the multi-tooth milling tool or burr tool was concluded as the most recommendable for the trimming of several types of composite materials (López de Lacalle et al.

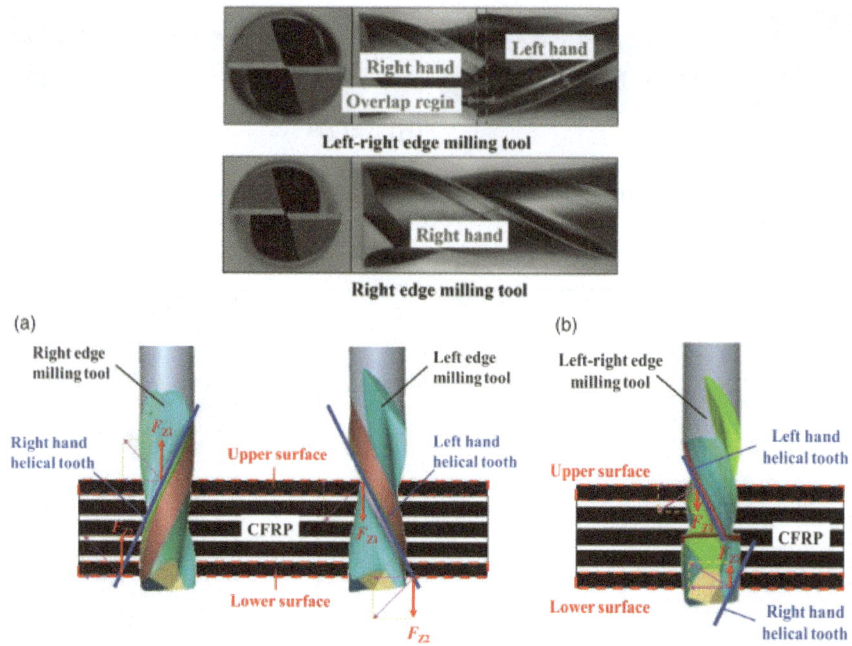

Fig. 21 **a** Right hand edge and left hand edge helical helix tool, **b** newly proposed tool; left-right edge helical helix tool (Wang et al. 2020)

Fig. 22 **a** Three (3)-flutes PCD tool **b** 12-flutes multiple teeth/burr (CVD coated) tool **c** three (3)-flutes PCD tool (Duboust et al. 2017)

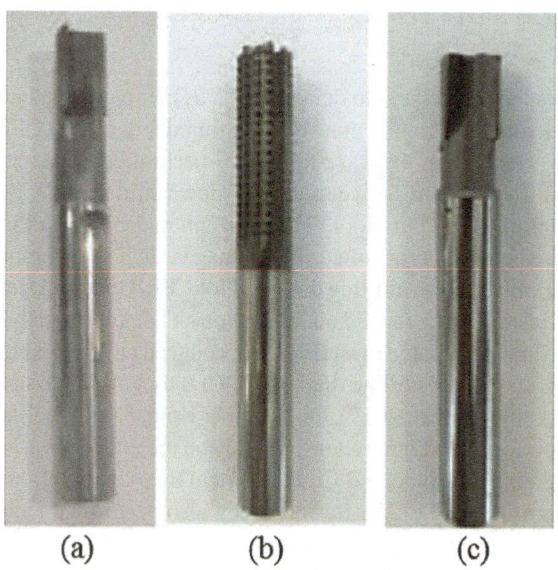

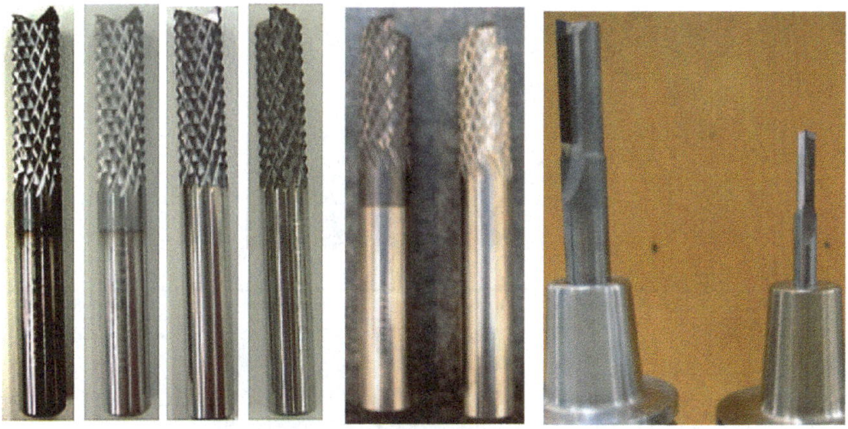

Fig. 23 Comparative study on the effect of coatings, micro grain, cobalt content and multi-tooth versus PCD tool (López de Lacalle et al. 2009)

2009; López de Lacalle and Lamikiz 2010). Bilek et al. initiated a research work focused on various types of tool geometries mainly router/burr and PCD type to investigate the effect of surface roughness, cutting forces and dimensional precision during the trimming of FRP materials. The authors have observed that due to certain specific characteristics, PCD (coated) tools indicated lower cutting efficiency compared to router/burr type tool. Therefore, an agreement achieved by the authors that the router/burr type tool is more recommendable for edge/side trimming of CFRP materials (Ondrej Bílek and Rusnáková 2016). Prakash et al. investigated the impact of the trimmed surface quality in high-speed milling of CFRP composites by varying three (3) type of tool geometries which categorized into two different detailed of burr type tool (T1 and T2) and helical helix geometry (T3) (refer Fig. 24). They reported that the first type of burr tool geometry; trapezoidal has produced an acceptable averaged surface roughness values (Ra) and lower cutting force without any delamination observed. While another type of burr tool known as pyramidal resulted in higher cutting force and averaged surface roughness value (Ra) than the first one. The final geometry; helical helix end mill type exhibited the highest averaged surface roughness value (Ra) as well as the cutting force. The delamination also identified to be the worst in comparison with another two types of router/burr geometry. It was presumed that the continuous flutes with a higher helical angle were the main reasons causing the pulling action of the extreme top and bottom plies of the laminate which results in delamination (Prakash et al. 2016).

An investigation by Haddad et al. (2013) considered two types of different geometries namely tungsten carbide uncoated and diamond coated tool of burr tool as well as four-flute diamond coated end mill tool (Fig. 25). The authors discovered that the cutting forces are proportionally influenced by the increase in the feed rate or cutting distance regardless of any type of tool geometry is used and at any cutting conditions. However, the higher the cutting speed decreases the cutting forces (Haddad

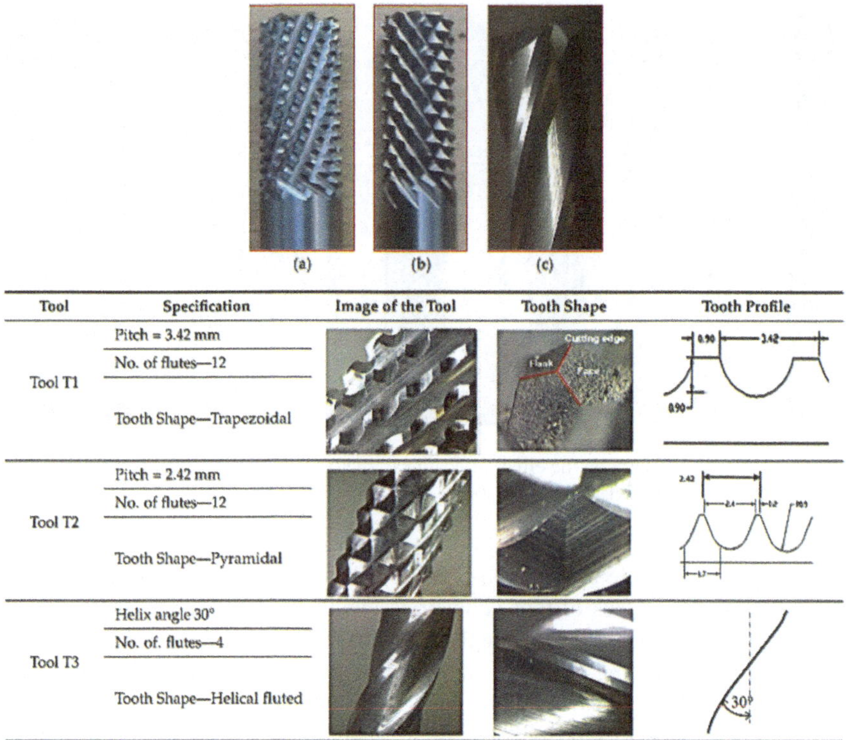

Fig. 24 Details of the investigated cutting tools (T1, T2 & T3) (Prakash et al. 2016)

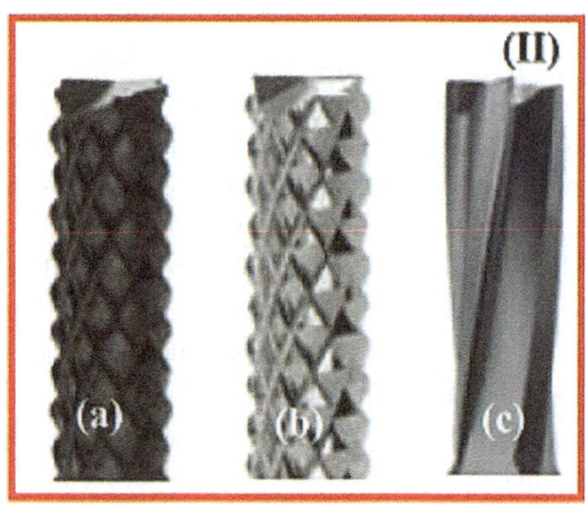

Fig. 25 **a** Burr tool (tungsten carbide) **b** Burr tool (diamond coated tungsten carbide) **c** Four-flute helical helix end mill (diamond coated) (Haddad et al. 2013)

et al. 2013). Meanwhile, Gara and Tsoumarev performed a series of a comparative study on a few different micro-grain of burr tools which categorized into fine, medium and coarse types (Fig. 26) in slotting CFRP material. They discovered that

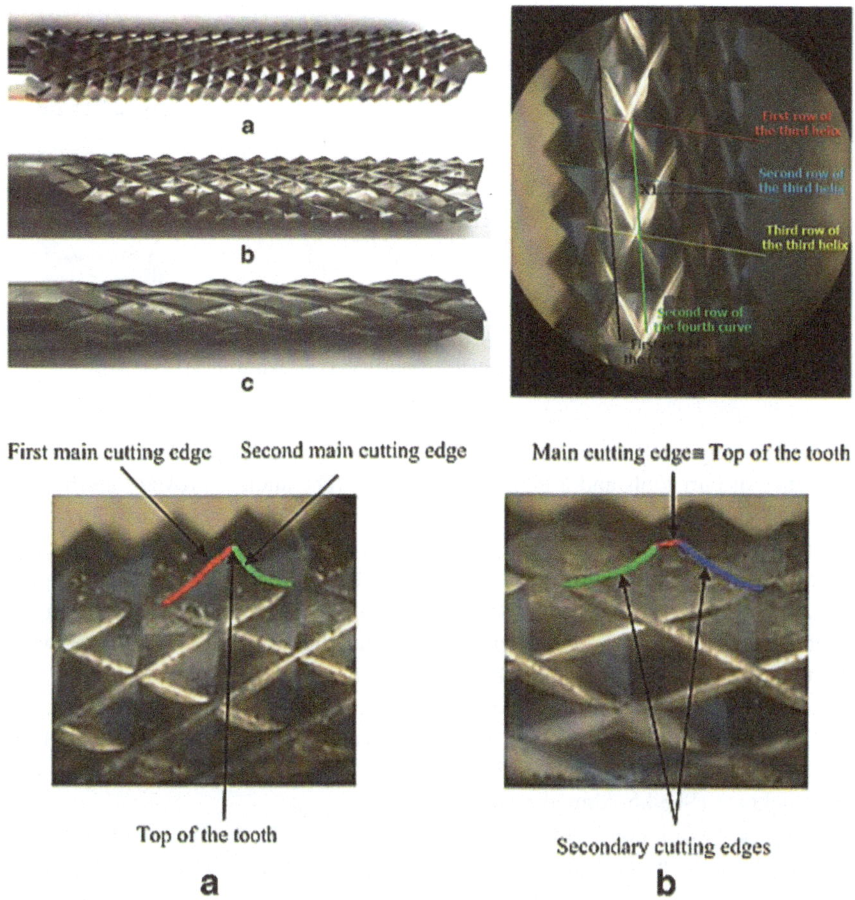

Fig. 26 Details burr tool geometries: **a** fine grain, **b** medium grain, **c** coarse grain (Gara and Tsoumarev 2016)

the cutting parameters had an insignificant influence on the transverse surface roughness. However, the tool geometry found to be the main role in determining the slotted surface quality in the transverse direction. The fine grain geometry of the burr tool was reported to be the most recommendable type of geometry to be used for the slotting of CFRP material as compared to the other two types of geometries (smooth and coarse) (Gara and Tsoumarev 2016, 2017). Authors have extended their work with the same three micro-grain carbide burr tools to observe the effect on the cutting and chip temperatures. The results revealed that the cutting tool geometry as well as the machining parameters had a great impact on the heat generation during slotting CFRP material. The most predominant factor that influenced the cutting temperature was the cutting speed followed by the feed per tooth (Haddad et al. 2013). Similarly, a research work initiated by Sheikh-Ahmad et al. (2018) also emphasized the

Cutter	Material	Cutting Flute	Geometry	
Burr Type I	CVD diamond coated carbide	Segmented flute	12 cutting flutes ($\phi_c = 15°$) 10 intersecting flutes ($\phi_i = 45°$)	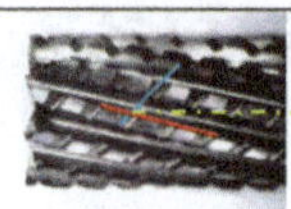
Burr Type II	TiAlN coated carbide	Burr	13 cutting flutes ($\phi_c = 30°$) 15 intersecting flutes ($\phi_i = 35°$)	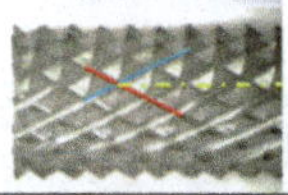
End Mill	PCD	Straight flute	2 flutes ($\phi_c = 0°$) Rake angle = 0° Clearance angle = 15°	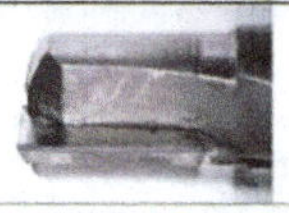

Fig. 27 Types of cutting tools and their details specification (Sheikh-Ahmad et al. 2018)

effect of cutting temperature generated by various tool geometrical designs namely two types of burr tools and a PCD tool (Fig. 27). The authors proved that the heat generated during the cutting/shearing process was mainly evacuated by the chips produces. Therefore, the size of the chips produced or also known as chip thickness determined the percentage of heat energy to be carried away during machining. In the case of a two-flute PCD tool, the thickness of the chips produced was the largest among all and the chips were then easily channeled out via the spacious gap between the two inserts as compared to the burr tools where the gap one flute to another was quite narrow. An increase in the feed rate may cause the increase of the chip thickness which indirectly increases the heat dissipated through the workpiece and the cutting tool to decrease and the portion going to the chips to increase due to the increase in chip thickness (Sheikh-Ahmad et al. 2018).

A research work conducted by Ashworth et al. (2019) to observe the effect of trimming process by two different machines tools namely robotic arm/articulated robotic system and 5-Axis milling CNC on the surface defects with two different tool geometries; burr tool type and herringbone intersection or usually known as compression router (Fig. 28). Authors discovered that tool geometry plays an important role in generating different surface matrices as well as the total power of the machining process but it does not influence the flexural strength directly (Ashworth et al. 2019). In another study by Sundi revealed that the tool geometrical features namely the number of flute/teeth of router/burr tool geometry may significantly influence the trimmed surface finish in edge trimming of CFRP composite material (Sundi et al. 2019a, 2020a). In addition, authors also have proven the effect of machining parameters or cutting conditions namely cutting speed, V_c and feed rate, V_f on the surface quality in edge trimming of a specific CFRP material in their separate research works (Sundi et al. 2019b, 2020b, c). A comparison study of multi-tooth (MT) or burr tool and up-down or compression router (UD) (Fig. 29) on the cryogenic condition was successfully examined by Cunningham et al. (2018). Authors discovered that cryogenic machining improves the average surface roughness as well as delamination length for both mentioned tool geometries (Cunningham et al. 2018). In more recent research carried out by Geier and Pereszlai (2020), reported that cutting

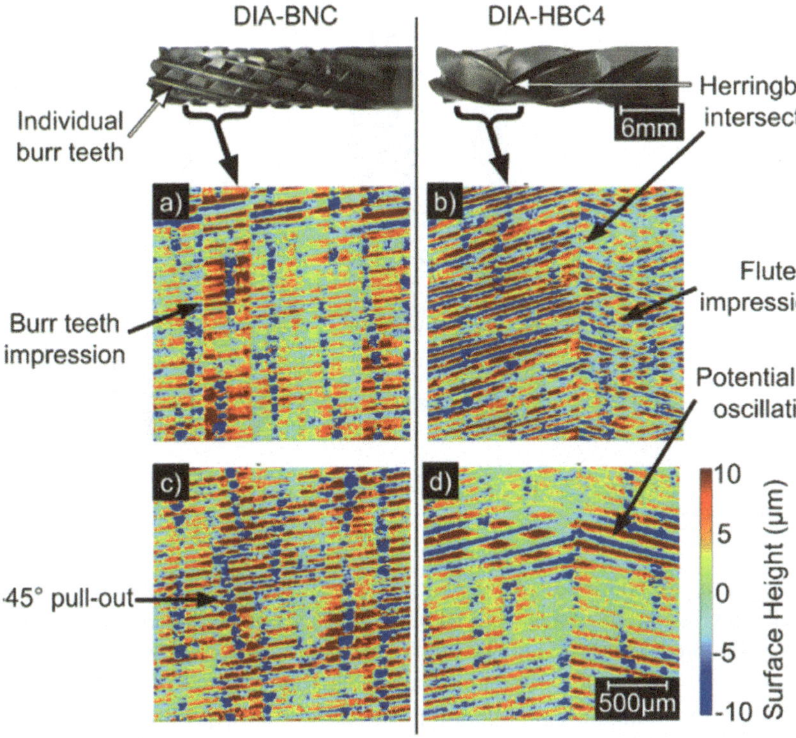

Fig. 28 Details of the tool geometries and the trimmed surfaces topographical results (Ashworth et al. 2019)

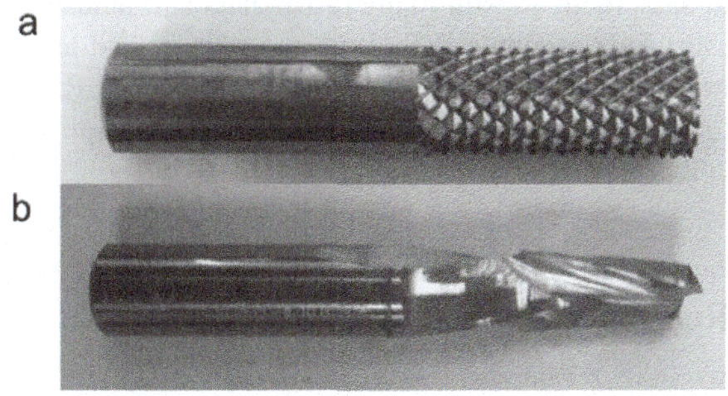

Fig. 29 **a** Multi-tooth/burr tool geometry; **b** up-down/compression router tool (Cunningham et al. 2018)

tool geometry and type of machining has a significant influence on the characteristics of surface roughness. Authors have analyzed two measurement methods namely contact profilometer and confocal microscope utilizing various tool geometries which include standard compression/burr end mill, coarse tooth compression/burr end mill, medium tooth compression end mill and helical helix end mill (Fig. 30) (Geier and Pereszlai 2020).

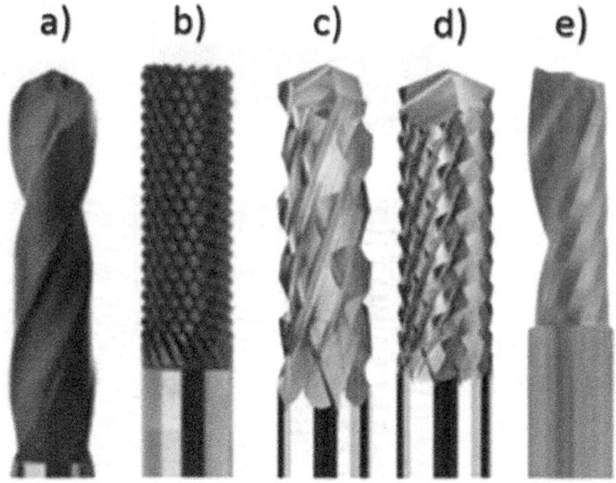

Fig. 30 **a** Ø11.138 mm double point angle twist drill (by Seco) **b** Ø10 (mm) burr tool (by Seco); **c** Ø10 mm burr tool (coarse tooth by Fraisa) and **d** Ø10 mm burr tool (medium tooth by Fraisa); **e** Ø10 mm helical helix end mill (by Tivoly) (Geier and Pereszlai 2020)

4 Conclusion

The overall chapter addresses a brief synopsis on the milling or trimming process of CFRP material. An overview of CFRP material is described in the earlier section comprises the processes involved in manufacturing carbon fiber as well as the fiber architectures. Moreover, this chapter also covers the fundamentals in a milling operation, cutting tool materials, cutting tool geometries in milling/trimming of CFRP material and a short elaboration on the cutting tool manufacturing processes. Deeper attention has been given to the sections related to the cutting tools. A comprehensive recent advances review, mainly emphasized on the impact of the tool geometrical design in milling/trimming CFRP material towards machinability has proven that to date, there are still many researchers and industries around the world striving in every single minute to investigate, evaluate and analyze the best approach/method to machine one of the most unique materials; CFRP besides determining the perfect tool geometrical design to be utilized in milling/trimming CFRP material without compromising the machining quality.

References

Ashworth S et al (2019) Effects of machine stiff ness and cutting tool design on the surface quality and flexural strength of edge trimmed carbon fibre reinforced polymers. Compos Part A 119(May 2018):88–100

Astakhov VP, Davim JP (2008) Tools (geometry and material) and tool wear. In: Machining. Springer, London

Can A (2017) Effect of edge trimming parameters on surface quality of carbon fiber reinforced polymer composites. J Sci Eng 17:302–311

Chen T, Wang D, Gao F, Liu X (2017) Experimental study on milling CFRP with staggered PCD cutter. Appl Sci 7(9):934

Colligan K, Ramulu M (1992) The effect of edge trimming on composite surface plies. Manuf Rev 5(4):274–283

Corporation H (2013) HexPly prepreg technology

Cunningham CR, Shokrani A, Dhokia V (2018) Edge trimming of carbon fibre reinforced plastic. Procedia CIRP (Hpc):6–10

Davim JP, Reis P (2005) Damage and dimensional precision on milling carbon fiber-reinforced plastics using design experiments. J Mater Process Technol 160(2):160–167

Duboust N et al (2017) An optical method for measuring surface roughness of machined carbon fibre-reinforced plastic composites. J Compos Mater 51(3):289–302

El-hofy HA (2014) Fundamentals of machining processes; conventional and nonconventional processes, 2nd ed. CRC Press, Taylor & Francis Group

El-Hofy MH, Soo SL, Aspinwall DK, Sim WM, Pearson D, Harden P (2011) Factors affecting workpiece surface integrity in slotting of CFRP. Procedia Eng 19:94–99

Gara S, Tsoumarev O (2016) Effect of tool geometry on surface roughness in slotting of CFRP. Int J Adv Manuf Technol 86(1–4):451–461

Gara S, Tsoumarev O (2017) Optimization of cutting conditions in slotting of multidirectional CFRP laminate. Int J Adv Manuf Technol 95:3227–3242

Geier N, Pereszlai C (2020) Analysis of characteristics of surface roughness of machined CFRP composites. Period Polytech Mech Eng 64(1):67–80

Haddad M, Zitoune R, Eyma F, Castanie B (2013) Influence of tool geometry and machining parameters on the surface quality and the effect of surface quality on compressive strength of carbon fibre reinforced plastic. Mater Sci Forum 763:107–125

Hosokawa A, Hirose N, Ueda T, Furumoto T (2014) High-quality machining of CFRP with high helix end mill. CIRP Ann Manuf Technol 63(1):89–92

Janardhan P, Sheikh-ahmad J, Cheraghi H (2006) Edge trimming of CFRP with diamond interlocking tools edge trimming of CFRP with diamond interlocking tools. SAE Int 2006-01-31(Sept)

König W, Wulf C, Graß P, Willerscheid H (1985) Machining of fibre reinforced plastics. CIRP Ann Manuf Technol 34(2):537–548

Liu G, Qian X, Chen H, Gao F, Chen T (2017) Development of a staggered PCD end mill for carbon fiber reinforced plastic. Appl Sci 7(3):245

López de Lacalle A, Lamikiz LN (2010) Milling of carbon fiber reinforced plastics. Adv Compos Mater 83–86:49–55

López de Lacalle N, Lamikiz A, Campa FJ, Valdivielso AF, Etxeberria I (2009) Design and test of a multitooth tool for CFRP milling. J Compos Mater 43(26):3275–3290

Mallick PK (2008) Fiber-reinforced composites: materials, manufacturing and design. CRC Press, Taylor & Francis Group

Mallick PK (2018) Processing of polymer matrix composites. CRC Press, Taylor & Francis Group

Mazumdar SK (2002) Composites manufacturing. In: Materials, product and process engineering, vol 32, no 1. CRC Press

Morgan P (2005) Carbon fibers and their composites

Nguyen-dinh N, Hejjaji A, Zitoune R, Bouvet C, Salem M (2020) New tool for reduction of harmful particulate dispersion and to improve machining quality when trimming carbon/epoxy composites. Compos Part A 131(Feb):105806

Ondrej Bílek MZ, Rusnáková S (2016) Cutting-tool performance in the end milling of carbon-fiber-reinforced plastics. Mater Tehnol 50(5):819–822

Prakash R, Krishnaraj V, Zitoune R, Sheikh-Ahmad J (2016) High-speed edge trimming of CFRP and online monitoring of performance of router tools using acoustic emission. Materials (Basel) 9(10):798

Rana S, Fangueiro R (2016) Advanced composite materials for aerospace engineering: processing, properties and applications. Woodhead Publishing WP

Sheikh-Ahmad JY (2009) Machining of polymer composites. Springer

Sheikh-Ahmad J, Urban N, Cheraghi H (2012) Machining damage in edge trimming of CFRP. Mater Manuf Process 27(7):802–808

Sheikh-Ahmad J, Almaskari F, Hafeez F (2018) Thermal aspects in machining CFRPs: effect of cutter type and cutting parameters. Int J Adv Manuf Technol 100(9–12):2569–2582

Shennan C (2014) Carbon fiber composites in wind energy: challenges and solutions. Hexcel Corporation

Sundi S, Izamshah R, Kasim M, Mohd Amin A, Kumaran T (2019a) Influence of router tool geometry on surface finish in edge trimming of multi-directional CFRP material. IOP Conf Ser Mater Sci Eng 469:012026

Sundi S, Izamshah R, Kasim M, Abdullah MK (2019b) Effect of machining parameters on surface quality during edge trimming of multi-directional CFRP material: Taguchi method. IOP Conf Ser Mater Sci Eng 469:012095

Sundi SA, Izamshah R, Kasim MS, Jaafar MF, Hassan MH (2020a) Surface roughness and tool wear in edge trimming of carbon fiber reinforced polymer (CFRP): variation in tool geometrical design. Lect Note Mech Eng 2:402–408

Sundi SA, Izamshah R, Kasim MS (2020b) Effect of machining parameters on surface roughness in edge trimming of carbon fiber reinforced plastics (CFRP). Tribol Online 15(2):53–59

Sundi SA, Izamshah R, Kasim MS, Jaafar MF, Hassan MH (2020c) Surface roughness and cutting forces during edge trimming of multi-directional carbon fiber reinforced polymer (CFRP). Lect Note Mech Eng 2:409–415

Uhlmann E, Richarz S, Sammler F, Hufschmied R (2016) High speed cutting of carbon fibre reinforced plastics. Procedia Manuf. 6:113–123

Wang DH, Ramulu M, Arola D (1995) Orthogonal cutting mechanisms of graphite/epoxy composite. Part I: Unidirectional laminate. Int J Mach Tools Manuf 35(12):1623–1638

Wang F, Yan J, Zhao M, Wang D, Wang X, Hao J (2020) Surface damage reduction of dry milling carbon fiber reinforced plastic/polymer using left–right edge milling tool, no 2

William J, Callister D, Rethwisch DG (2010) Material science and engineering, 8th ed. Wiley

Comprehensive Study on Tool Wear During Machining of Fiber-Reinforced Polymeric Composites

Sikiru Oluwarotimi Ismail, Shoaib Sarfraz, Misbah Niamat, Mozammel Mia, Munish Kumar Gupta, Danil Yu Pimenov, and Essam Shehab

Abstract The use of fiber reinforced polymeric (FRP) composites has increased rapidly, especially in many manufacturing (aerospace, automobile and construction) industries. The machining of composite materials is an important manufacturing process. It has attracted several studies over the last decades. Tool wear is a key factor that contributes to the cost of the machining process annually. It occurs due to sudden geometrical damage, frictional force and temperature rise at the tool-work interaction region. Moreover, tool wear is an inevitable, gradual and complex phenomenon. It often causes machined-induced damage on the workpiece/FRP composite materials. Considering the geometry of drill, tool wear may occur at the flank face, rake face and/or cutting edge. There are several factors affecting the tool wear. These include,

S. O. Ismail (✉)
Department of Engineering, School of Physics, Engineering and Computer Science, Centre for Engineering Research, University of Hertfordshire, Hatfield AL10 9AB, UK
e-mail: s.ismail3@herts.ac.uk

S. Sarfraz
Manufacturing Department, School of Aerospace, Transport and Manufacturing, Cranfield University, Cranfield, Bedfordshire MK43 0AL, UK

M. Niamat
Mechanical Engineering Department, Muhammad Nawaz Sharif University of Engineering and Technology, Multan 66000, Pakistan

M. Mia
Department of Mechanical Engineering, Imperial College London, South Kensington, London, UK

M. K. Gupta
Key Laboratory of High Efficiency and Clean Mechanical Manufacture, School of Mechanical Engineering, Shandong University, Jinan, People's Republic of China

D. Y. Pimenov
Department of Automated Mechanical Engineering, South Ural State University, Lenin Prosp. 76, Chelyabinsk 454080, Russia

E. Shehab
Mechanical and Aerospace Department, School of Engineering and Digital Sciences, Nazarbayev University, Nur-Sultan, Kazakhstan

M. T. Hameed Sultan et al. (eds.), *Machining and Machinability of Fiber Reinforced Polymer Composites*, Composites Science and Technology,
https://doi.org/10.1007/978-981-33-4153-1_5

but are not limited to, drilling parameters and environments/conditions, drill/tool materials and geometries, FRP composite compositions and machining techniques. Hence this chapter focuses on drilling parameters, tool materials and geometries, drilling environments, types of tool wear, mechanisms of tool wear and methods of measurement of wear, effects of wear on machining of composite materials and preventive measures against rapid drill wear. Conclusively, some future perspectives or outlooks concerning the use of drill tools and their associated wears are elucidated, especially with the advancement in science and technology.

Keywords FRP composites · Machining/drilling · Tool/drill wear · Measurement · Mechanism

1 Introduction

The significance of drilling process has been increased widely both technologically and commercially in recent years (Groover 2007). Tool wear is a main factor contributing to the cost of a machining process. Drill wears are responsible for the roundness of holes, burr formation, surface roughness; hence, they directly affect the quality of machined hole (Abu-Mahfouz 2005).

Wear has been commonly defined as the amount of matter lost by the drill tool (Ertunc et al. 2001; Iliescu et al. 2010). Tool wear occurs in diverse and varied ways being a complex occurrence (Dimla et al. 1997) and gradual process (Astakhov and Davim 2008). There are a number of factors dependently contribute to the wear of cutting tool: cutting speed, depth of cut and feed rate (cutting parameters), machine-tool characteristics, workpiece materials and cutting fluids (Astakhov and Davim 2008). Carbon fibers are highly non-homogeneous, discontinuous, brittle, abrasive, and anisotropic in nature and have limited plastic deformation, as such an excessive abrasive wear is located along the cutting edges of twist drills, especially on the uncoated carbide types (Faraz et al. 2009). Consequently, as it prolongs, roundness and bluntness of the drill occur, as depicted in Fig. 1. When the drill has not been used, the maximum wear, VB_{max} is zero (null). Immediately after 560 holes have been drilled, the VB_{max} increased to 235 μm, later rendered the drill ineffective after drilling 610 holes.

Moreover, the rise in cutting speed causes an increase in both thrust force and torque (cutting forces). They which adversely increase the tool wear, especially the flank wear during high speed drilling of carbon fiber reinforced polymer (CFRP), irrespective of the tool coatings (Murphy et al. 2002). In addition, chip formation involves a brittle fracture, which is mainly based on the alignment of the composite fibers, with powdery and dust-like chips rampant while stock on the drill causes wear. The thermal conductivity of drill materials strongly determines their wear rate, and it has been reported that wear decreases with higher thermal conductivity of the drill materials (Sakuma et al. 1985).

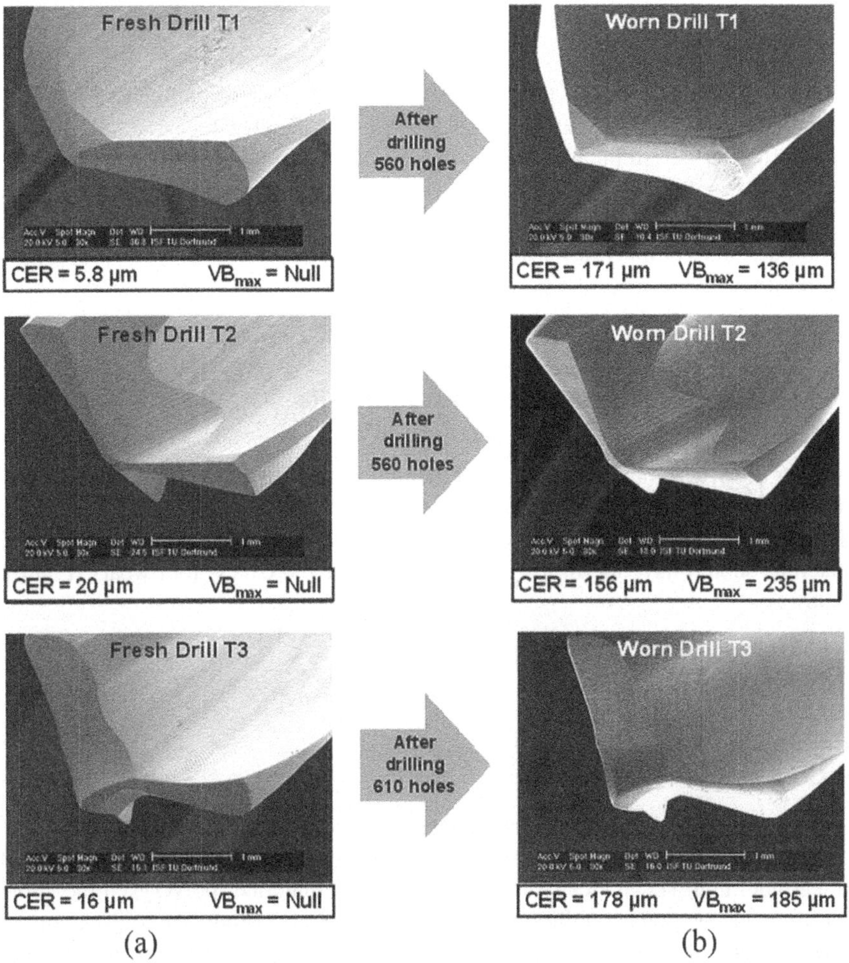

Fig. 1 Scanning electron microscope (SEM) images of **a** sharp cutting edges and **b** smoothly rounded/worn heads (Faraz et al. 2009)

The resultant effects of tool wear include an increase in drilling forces, degraded surface finish, increase in temperature, drilled hole inaccuracy and tool breakage, poor hole roundness, burr formation and centering issue (Abu-Mahfouz 2005). When drill wear reaches an unacceptable level under certain drilling conditions, it causes production loss, later it damages the work and machine tool, if not properly controlled.

2 Drilling Parameters

Selection of drilling parameters and proper setting of these parameters are very important, because input factors, such as drill bit diameter, feed rate and cutting speed affect or determine tool wears. Jayabal and Natarajan (2010) established an optimal scheme of drilling parameters, which offered minimum values of tool wear for drilling process of composites. Genetic algorithm and Nelder-Mead optimisation techniques were employed for analysis and empirical modelling of the parameters and their levels.

3 Drilling Tool Materials and Geometries

Drilling tools with better physical and thermal properties ensure better interaction of tool with workpiece/fiber reinforced polymeric (FRP) composites with minimal chances of rapid tool wear. For drilling of various FRP composite, selected tool materials should have the following properties to avoid excess tool wears (Lantrip 2008):

- Good toughness.
- Fair hardness.
- Good wear resistance.
- Chemical non-reactive.
- Good thermal conductivity.

A range of materials are available and they have been commonly used for manufacturing of various drill bits for composite drilling. These include, but are not limited to, carbide, high speed steel, super hard material (diamond, polycrystalline cubic boron nitride or polycrystalline diamond), cobalt and coating materials: titanium nitride (TiN), black oxide and titanium carbonitride cobalt (TiCN), among other substances. Different studies have compared these tool materials for minimum tool wear, as subsequently discussed.

Aluminum titanium nitride (AlTiN) coated drill enhances the hardness of the drills and provides solid lubrication to curb bluntness of the drill bit. Moreover, it can be used at higher temperatures, as AlTiN is a high temperature alloy, as reported by Sriraman et al. (2015). In another study, Park et al. (2012) examined drilling of CFRP and titanium stacks with uncoated tungsten carbide (WC) drill and boron aluminum magnesium (BAM) coated WC drill. During this process, tool surface was observed by X-ray energy dispersive spectroscopy (XRD), SEM and confocal laser scanning microscope (CLSM). A lower flank wear was observed for drilling with coated drill, when compared with an uncoated drill. Moreover, after drilling 20 holes, the BAM coat was damaged and tool wear increased after drilling 40 holes. Hence, the coated tools exhibited a longer tool life. Edge roundness is a dominant wear during drilling of CFRP materials. CFRP composites are brittle in nature, stagnation region confronting cutting edge propagates dullness at the cutting edge, as investigated by Wang et al. (2013).

In addition to coated drills, drill bits with special geometrical design are gaining attention. Drill geometry is mainly based on the following different characteristics (Xu et al. 2019):

- Angles (helix angle and point angle).
- Edge preparation (chamfer or round).
- Drill shape (Twist or helical).

These geometrical variables are closely related globally for outstanding performance of special drills, considering better quality of machined surface and minimum tool wear. Senthil Kumar et al. (2013) studied tool wear by machining 100 holes separately with two drill point angles of 118° and 130°. Drill with point angle of 130° showed lower tool wear. According to Garrick (2007), the diamond coated drill bits with special K-land design resulted in a longer tool life than other comparative conventional drills.

4 Drilling Environments

Application of right types of lubricant or cooling media helps to minimise tool wear during drilling process of composite materials. In drilling, the cutting fluids can reduce friction at tool-chip interface point, minimise heat generation at drilling point and facilitate the removal of chips produced during the process. Moreover, the cutting fluid/coolant mainly performs following functions when drilling composite materials (Xu et al. 2019):

- Elimination of chips sticking on drill bit flute.
- Reduction of force generation on tool-chip interaction point.
- Reduction in heat generation that enhances tool wear.

Many experimental studies have examined and concluded that tool life can be significantly enhanced by using cooling media during drilling of composite materials. For instance, for drilling CFRP composite, Xia et al. (2016) experimentally investigated that cryogenic cooling has a significant effect on reducing the outer corner wear, roundness in the cutting edge of cutting tool and surface quality of machined hole. Uncoated tools were used for the experiments. Moreover, comparison of both drilling environments (cryogenic cooling and dry condition) showed that a lower drill wear due to thermal damage is observed in cryogenic cooling.

The successful application of cutting fluids largely depends on the method of application of the cutting fluids during drilling process. Three commonly known fluid supply methods include flood cooling, wet drilling and mist spraying (Sen et al. 2019).

Moving forward, Tashiro et al. (2011) compared two drilling environments, dry process drilling and water mist cooling. Although, water mist cooling process exhibited a lower thrust force, however, chips produced were harder. Moreover, comparison based upon the number of holes drilled in both environments showed that a longer tool life is possible in dry process when compared with the mist cooling process. A sequence of experiments was conducted for the evaluation of three application methods of cutting fluids: flooding, misting and spraying. Drilling with spray mist method offered better results of tool wear, when compared with other two methods of cutting fluid application (Shyha et al. 2011).

In addition, the use of excessive cutting fluids has damaging effects on environment and health of the people in the working area. To overcome these issues, near dry machining appeared as an emerging technique. Minimum quantity lubrication (MQL), also called near dry machining, has gained attention in composite materials machining. MQL is a sustainable process that considers environmental and economic aspects (Xu et al. 2019). MQL is the application of minimal amount of biodegradable oil droplets along with compressed air at too-chip interaction point to minimise the heat produced (temperature rise) during machining, which consequently prolongs the life span of tools (Rahim and Sasahara 2011). For examples, Krolczyk et al. (2019) and Sen et al. (2019) recently and extensively reviewed the cooling lubrication techniques applied in machining. MQL was suggested as such a method of cooling-lubrication that has a great potential to replace conventional cooling techniques, because MQL did not only reduce the tool wear and surface roughness, but also it is less dangerous to environment and human health. MQL has attract wide industrial applications and become a famous technique worldwide, because of less consumption of lubricant and its better results than those obtained by traditional flood cooling technique in terms of tool/drill wear and quality of surface characteristic of machined part (Brinksmeier et al. 1999). In another study, Brinksmeier and Janssen (2002) performed a number of experiments and reported that tool wear can be minimised using internal supply of MQL during drilling. Similarly, minimum drill wear has been observed when drilling of CFRP/Ti6Al4V stacks under MQL cooling condition (Xu et al. 2019).

Also, experimental investigation into drilling of composite-titanium compound in a MQL environment has been performed. The results indicate lower drill wear and improved surface quality with a MQL drilling. However, when Xu et al. (2019) compared MQL and dry drilling, dry drilling produced lower thrust force and hole cylindricity. Wang et al. (2018) experimentally observed that the application of coolant mist through the outlet located at the lower most position of secondary cutting-edge resulted in great reduction of drill wears. In a comparative experimental study, drilling in MQL produced decrease of 22% (lower) flank wear when compared with dry and flood coolant and decrease of 30%/lower flank wear than in compressed air environment. Besides, MQL in combination with lower oil flow and higher air flow rates offered a maximum tool life (Iskandar et al. 2014). Bhattacharyya and Horrigan (1998) employed a liquid nitrogen as cooling media when drilling Kevlar composite samples. Under the cryogenic cooling, a lower tool wear was observed. Additionally, protruded, fuzzy and uncut fiber defects were completely removed with the application of cryogenic coolant. This facilitated the entry and exit of drill and consequently, reduced the chances of tool wear and hence tool life was increased (Xia 2014).

Kannan et al. (2018) investigated into drilling of CFRP under different environments (conventional flood drilling, dry and compressed air and MQL). Experimental results revealed that the dry cooling offered lower thrust force when compared with other cooling environments, while flood lubrication showed highest thrust force. MQL drilling presented a better machining performance, longer tool life and superior surface quality than dry drilling. MQL and dry drillings offered good 110 and 80 holes with coated WC drill, respectively.

5 Types of Drill Wear

There are many types of drill wear based on parts of the drill where such wear occurs, as shown in Figs. 2 and 3. Drill wears and their causative mechanisms are subsequently elucidated.

5.1 *Crater Wear*

Crater wear arises from erosion of the drill just under the cutting edge, occurring mostly on the tool rake face through diffusion mechanism. Wearing action is accelerated by higher temperatures and stresses induced by tool-chip interaction. It has a parabolic relationship with feed rate and distance along the drill lips and linear relationship with cutting speed (Choudhury and Raju 2000). The wear is usually measured by the depth or area coved by the crater (Groover 2007). Crater wears are

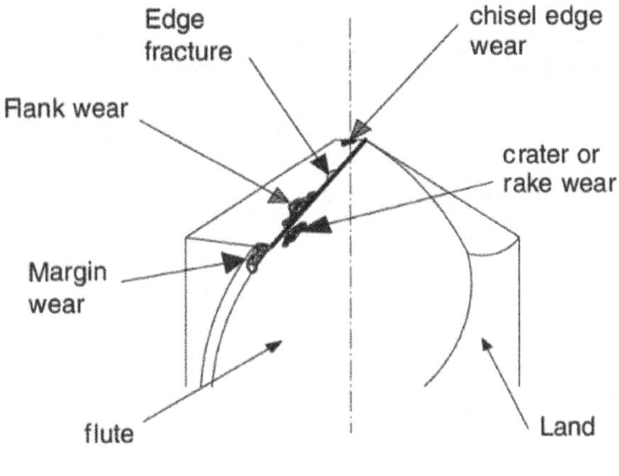

Fig. 2 A schematic representation of various types of wear on twist drill (Abu-Mahfouz 2005)

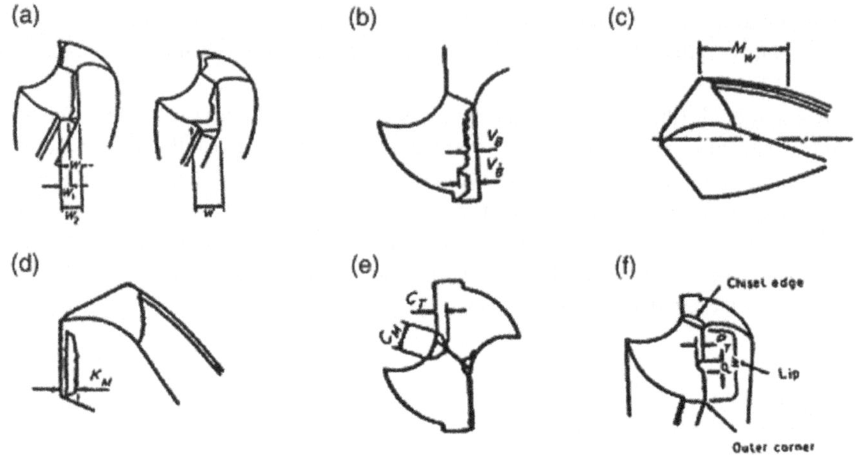

Fig. 3 Various types of wear: **a** outer corners wear; **b** flank wear; **c** margin wear; **d** crater wear; **e** chisel edge wear and **f** chipping at lip (Ertunc et al. 2001)

attributed to the diffusion process. Application of protective coating on tool surface is an effective way of slowing the diffusion process, hence, minimising crater wear.

5.2 Flank Wear

Flank wear arises on the flank (relief) face of the tool. This wear is a progressive process and a commonly used indication for detecting the degree of condition of drill wear or as basis for tool wear (Abu-Mahfouz 2005). Both elevated temperature and intimate friction contact at drill tool-work interface cause flank wear. It increases with increasing delamination factor (Khashaba et al. 2010) and is predominant at a low speed. The summarised basic causes, mechanisms, types, consequences of flank type of wear are later shown in Fig. 5. Basically, tool flank and rake faces are the two important zones where tool wear occurs in a drill (Liu et al. 2000).

5.3 Fracture or Cutting-Edge Chipping

An increase in wear leads to an increase in torque and thrust force. Failure to endure the rise in progressive wear causes drill edge chipping or fracture/breakage. Irrespective of the types of drill wear, they are caused by composite-drill high interface temperature and friction, fiber orientations, anisotropic and abrasive nature of CFRP composite materials, drilling variables/parameters, especially high cutting speed. Mostly, friction and temperature wears are considered as abrupt wears and these two

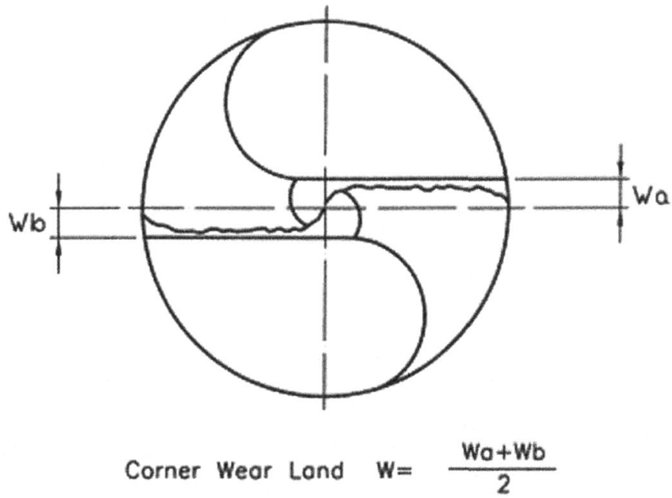

Fig. 4 Description of a drill corner wear land, depicting its simple analysis and features (Liu et al. 2000)

wears are undesirable tool failure modes rather than gradual wear which results in a longer tool life.

5.4 Other Types of Wear

- *Margin or Corner Wear*: It occurs due to the misalignment of the drill and heat caused by both insufficient coolant and high spindle speed.
- *Chisel Wear*: It is very rampant at chisel point of the cutting tool (drill).
- *Outer Corner Wear*: It occurs on the outer corner of the drill tool lips (Ertunc et al. 2001). The drill corner wear land can be simply analysed, as indicated in Fig. 4. Where *W*, *Wa* and *Wb* represent the drill corner wear land, right and left cutting edges, respectively.

6 Mechanisms and Determination of Twist Drill Wear During Drilling of Composite Materials

6.1 Mechanisms of Wear

Tool wear during conventional composite machining is an unavoidable phenomenon, since production of chips through plastic deformation is required before occurrence of cutting (drilling) operation, under cutting forces (thrust and torque). This occurs after interaction between cutting tool (drill) and workpiece (composite material), leading to a high friction and interface temperature. These two factors caused the atoms of the cutting tool to gain great kinetic energy, resulting to tool wear as the particles of the cutting tool tends to move away.

In other words, wear occurs due to removal of some parts of material from the surface of cutting tools which is due to cutting tool-workpiece physical (mechanical) and chemical interactions (Ertunc et al. 2001). Classification of wear mechanism according to Astakhov and Davim (2008) are shown in Fig. 5 and subsequently explained.

- *Abrasion:* This is a thermo-dynamic wear mechanism, commonly called mechanical wear. Usually, hard abrasive particle present in workpiece/FRP composite causes scratching and eliminating small portions from the tool. Flank wear majorly occurs by the abrasion wear.
- *Adhesion*: It happens due to friction, pressure and high temperature. It occurs at tool rake face and chip interface point. As newly formed chip flows along the tool, it removes a portion of tool surface with it and causes erosion on the surface.

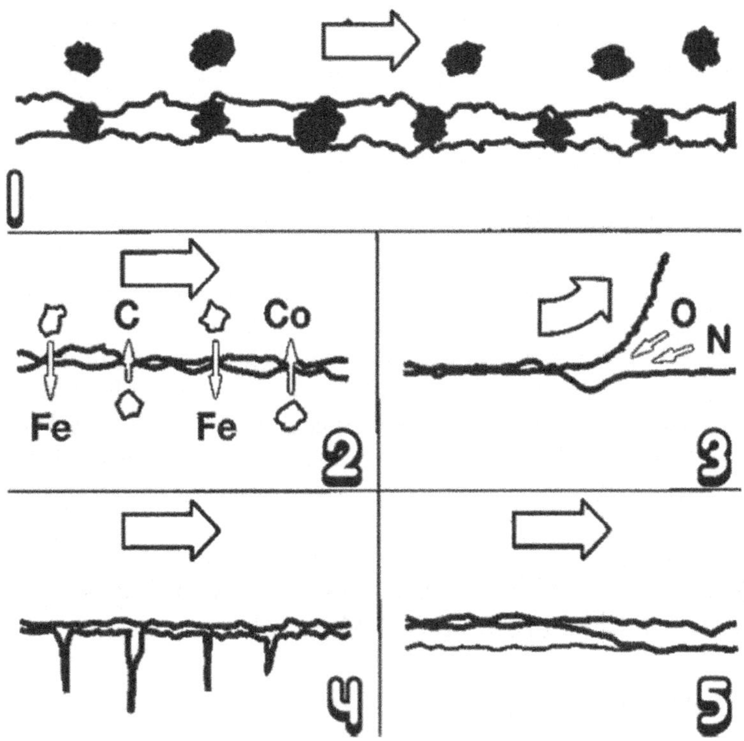

Fig. 5 Wear mechanisms, showing (1) abrasion, (2) diffusion, (3) oxidation, (4) fatigue and (5) Adhesion (Astakhov and Davim 2008)

- *Diffusion*: This is a thermo-chemical wear process, commonly called chemical wear, whereby atoms of hard material (drill tool) diffuse into the soft material matrix (workpiece) at high temperature. It is strongly dependent on temperature. Diffusion is a significant cause of crater wear.
- *Oxidation:* It occurs due to the chemical reaction between the drill tool face and oxygen. It mainly happens on the rake side of the tool. Oxidation layer being weaker layer easily scratches off, revealing new surface for further reactions.
- *Fatigue:* Fatigue occurs at high pressure when two surfaces slide on each other due to friction, tensile and compressive forces causing surface crack. Fatigue mainly results in flank wear.

Furthermore, the most commonly encountered mechanisms are abrasion and adhesive wears, but abrasion and adhesion mechanisms of wear cannot completely describe wear in cutting tools (Ertunc et al. 2001). Tool wear being a convoluted phenomenon depends on many parameters, such as nature of cutting tool and/or workpiece, cutting conditions, contact stress, vibration in machine tool and cutting edge temperature. In addition, the tribological study reveals that wear rate depends on temperature, friction, pressure, speed, nature of workpiece, tool geometry and drill materials. Increase in friction causes an increase in temperature, due to high kinetic energy gained by the drill atoms, resultantly, the wear increases. Excessive wear results into the failure of a drill.

In addition, with an increase in the cutting forces, the flank wear, VB increases, as depicted in Fig. 6. VB affects the surface roughness. Drill wear passes through several processes before ends in a plastic deformation which causes failure or rupture.

Besides, it is necessary to compare wear rate during machining of synthetic and sustainable natural or plant FRP composite samples. It is evident that the drill wear rate during machine of CFRP composite materials is higher when compared with that of hemp FRP counterpart (Ismail et al. 2016). A null or negligible wear was observed on the high speed steel twist drill diameters of 5 and 10 mm after drilling 32 and 64 holes of hemp FRP composite samples, respectively, compared with a minimum (greater) drill wear that was observed with CFRP counterparts, under same drilling parameters, conditions and environment.

More also, tool wear phenomenon is usually a slower and progressive during drilling operation, but catastrophic and abrupt in case of tool failure as well as cutting edge breakage. Considering Fig. 7, wear of drill bit starts immediately it is used for operation, known as initial wear, W_o before stable wear region, but with faster rate once it becomes dull. It finally reaches an accelerated failure stage, where it is inefficient in cutting (Ertunc et al. 2001). In stable region, the slope of the wear curve is controlled by the cutting conditions and properties of work materials. With increasing hardness of work materials, drill wear also increases accordingly. Similarly, an increase in cutting speed causes an increase in drill wear ratio.

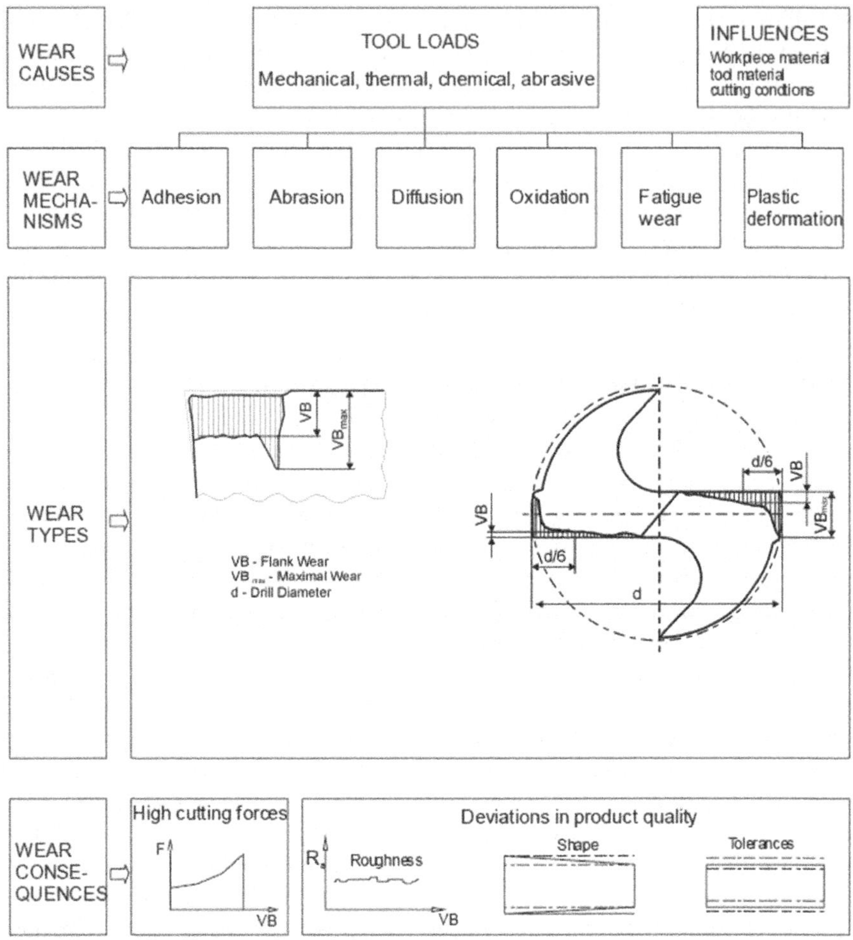

Fig. 6 An outline of the causes, mechanisms, types and consequences of tool wear (Dolinšek and Kopač 2006)

6.2 Drill Tool Life

Tool life is expressed as a time span for which a drill can be used for machining until its catastrophic wear occurs. In production, using tool till or to a stage of catastrophic failure is usually avoided, as it demands re-sharpening the tool and can impair the quality of the machined surfaces (Groover 2007). Tool life expectancy can also be predicted empirically, using Taylor's equation, as expressed in Eq. (1).

$$V_c T^n = C \tag{1}$$

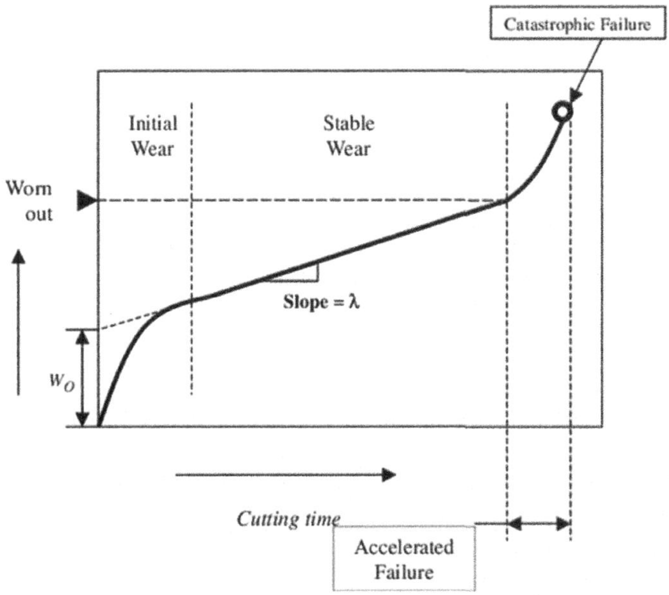

Fig. 7 An illustration of drill flank wear evolution and its basic features (Abu-Mahfouz 2005)

where V_c represents the cutting speed (m/min), T denotes the tool life (min) and n stands for exponent, as shown in Table 1 for different cutting tool materials. The exponent depends on drilling parameters, while C designates a constant. Some of these terms can be obtained from past published works or experimental results.

Equation (1) can be represented graphically, as shown in Fig. 8.

Table 1 The values of exponent for various cutting tool materials (Astakhov and Davim 2008)

Tool material	High speed steel	Cemented carbides	Ceramics
n	0.1–0.2	0.2–0.5	0.5–0.7

Fig. 8 Graph of log V against log T

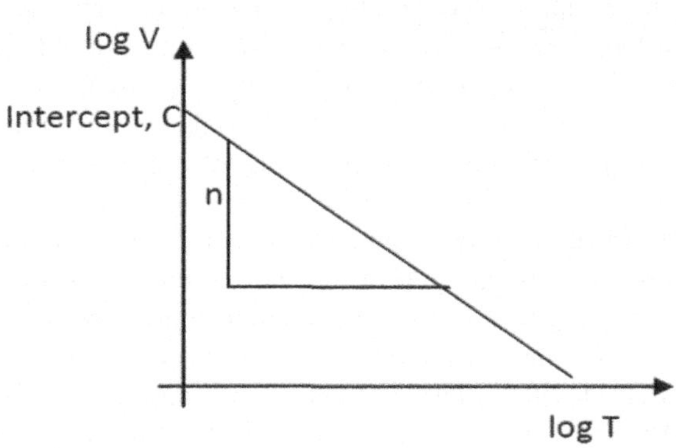

For the estimation of the feed rate and depth of cut, Taylor's equation can be generally expressed as Eq. (2).

$$V_c T^n D^x S^y = C \quad (2)$$

where D and S stand for the depth of cut and feed rate, respectively. The exponents, x and y, are obtained experimentally.

Although, tool life prediction can be made based on flank wear, using Taylor's equation. However, in industry, it is difficult and time consuming to estimate the flank wear (Groover 2007). The following are some important measures to predict a particular tool life:

- Visual inspection of tool failure.
- Failure of cutting edge due to fracture failure, temperature failure or gradual failure.
- Scratch test by the operator to check the abnormalities.
- Changes in the typical sound of the operation.
- Stringy chips formation.
- Rougher machined surface.
- Significant increase in power consumption during operation.
- Total drilling time that the drill has been used for drilling.
- The number of holes made by the drill or number of parts machined by the tool.

6.3 Wear Determination or Measurement Techniques

During elastic deformation of a material, energy is required in the form of strain energy to control the atomic bonds. The plastic deformation rate is fast for a machining process, such as drilling, at a high strain rate. Almost all the energy used is converted into heat, which causes a rise in temperature, leading to earlier wears and eventually tool failure. The temperature increases with speed when machining materials, such as ferrous and high strength materials include CFRP composites (Choudhury and Raju 2000). Szwajka and Trzepieciński (2016) determined the effect of varying cutting speed on the life span of tool. The results showed that the useful life of tool decreased with an increasing cutting speed, and the wear was more dominant on tool flank. In another study, Lin and Chen (1996) evaluated the effect of cutting speed on drilling of CFRP composite material. It was observed that the tool wear occurred more rapidly at higher speeds rather than at lower speeds. Drilling of CFRP-Ti stack composite material showed that Ti and CFRP cutting promoted WC flank wear and depreciated the drill cutting edge, respectively, as confirmed or observed through SEM and CLSM examinations (Beal et al. 2011).

Moving forward, the maximum flank wear, VB_{max} was carefully measured, using a modern instrument (Mitotoyo tool measuring microscope, type MF workstation) coupled with an integrated desktop computer and a built-in-camera installed with AnalySIS, a commercial digital image processing software (Faraz et al. 2009).

Furthermore, the accruing magnitude of cutting edge roundness (CER) was measured with an application of a commercial optical fringe microscope workstation, known as GFM MikroCAD system. The same technique was used by Shyha et al. (2010), but a toolmaker's microscope (WILD M3Z) with attached Nikon EOS 400D digital camera was used and managed by a digital imaging software called Omnimet 8.7.

In addition, failure in drill tool can be monitored by signal processing techniques. Non-linear relationships as well as recognition pattern from noisy complex data of drill wear have been carried out using a multivariate nonlinear analytical tool, called artificial neural network (ANN) algorithms; learning vector quantization (LVQ) and the fuzzy learning vector quantization (FLVQ). FLVQ was found more efficient than LVQ in assessing size of flank wear (Abu-Mahfouz 2005). Also, algorithm for synthesis of polynomial networks (ASPNs) was used by Liu et al (2000) for online prediction of corner wear in drill bits during drilling operations while application of intelligent tool condition monitoring (TCM) systems using ANN was reported by Dimla et al. (1997). Govekar and Grabec (1994) demonstrated a Kohonen type, a self-organising neural network (NN) technique to classify flank wear in drills subject to cutting forces. Tansel et al. (1993), utilised NN, type-ART2 and wavelet transformation techniques to examine the tool wear just at the beginning of failure of tool in micro-drilling operation. The results showed that wavelet monitoring technique was better, especially for automatic monitoring.

Moreover, different measurement techniques can be further applied to measure different types of tool wear. For instance, a universal tool room microscopy can be employed for the estimation of the width of flank wear and analysis of worn surfaces. An investigation through SEM was recommended by Sathish and Raj (2012). Rawat and Attia (2009) experimentally investigated tool wear mechanism and cutting forces affecting the part surface. Chipping and abrasion wears were the dominant mechanisms controlling the drill damage. Li and Tso (1999) estimated the state of drill wear, using values of motor currents for both spindle and feed. It was a new regression technological and fuzzy taxonomy method, involving numbers of cutting parameters: feed rate, diameter of drill and speed. Regression analysis models were established through experimental study to predict when a tool is to be replaced.

Furthermore, the profiles of worn cutting edges of the tool have been observed using the confocal microscope. The detailed wear patterns were studied by superior magnification pictures, which were taken with aid of a SEM (Wang et al. 2013). Periodic recordings of drill weight loss was used by Khashaba et al. (2010) to carry out machinability analysis, effect of drill wear when drilling woven GFR/epoxy composites. Kim et al. (2002) measured the power consumed by spindle motor to accurately estimate the drill wear. The estimated wear error was lower than 0.02 mm, proving the reliability of the real-time error since drill replacement of 0.18 mm is required in flank wear.

6.4 Preventive Measures Against Rapid Drill Wear

In a quest to have an accurate drilled hole for composite parts, the tool/drill should be in a good condition. Drill wear can be reduced by using optimal setting of drilling parameters, good tool materials and efficient coatings, lubrication or MQL technique as well as suitable coolants or cutting fluids. Coolants could be compressed air (dry type) and mist coolants (wet type). Their various practical applications have been earlier discussed compressively in Sect. 4. The Use of wet or liquid coolants are often discouraged during drilling of natural/plant (bast) FRP composites, due to the hydrophobic nature of plant (flax, sisal, palm tree, bamboo, hemp, jute, to mention but a few) fibers. Though, liquid coolants are very effective towards reduction of high drill-composite friction, interface temperature and consequently the tool wear during drilling operation, but the drawbacks of some wet/liquid coolant include high cost, poor application and most importantly an increased moisture absorption (swollen) by the plant FRP biocomposite materials. It further causes a poor/weak fiber-matrix interfacial adhesion, delamination and most importantly, decrease in mechanical properties of biocomposites. Therefore, an application of a compressed air to reduce high tool-workpiece interface temperature and drill wear is advisable when drilling natural/plant FRP biocomposites, as used by Wang et al. (2019). However, wet coolants have a good application during drilling of a selected few synthetic (carbon and glass) FRP composite materials.

7 Concluding Remarks and Future Perspectives

7.1 Conclusions

Tool wear during conventional composite machining is an inevitable phenomenon. The abrasive, heterogeneous, brittle and anisotropic properties of FRP composite materials/work piece support rapid tool wear rate. Tool wear is a complex occurrence, because its mechanisms involve several branches of science and engineering. Wear of tools differ from one FRP composite to another. Drill wears faster when drilling synthetic (carbon and glass) FRP composite materials than natural/plant FRP counterparts. In addition, tool wear depends on numerous factors, such as types of composites, drilling parameters, tooling materials, tool/drill geometries, to mention but a few. Different types of wear and their mechanisms have been extensively discussed. Both the region where wear occurs and method of occurrence determine the type of wear.

Also, several techniques of determining and analysing wear have been well elucidated within this chapter. Though, wear remains an unavoidable occurrence, but wear rate can be reduced and/or tool life can be maintained by using suitable coolants (either dry or wet/liquid type), coated tools, optimal process parameters, efficient drill design and geometry, among others. The effectiveness of coolant depends on types (compositions) of FRP composites and methods of application.

7.2 Future Outlooks

Cutting tools, such as drill bits are very important in manufacturing industries. This is because holes are indispensably required to assemble, join or couple composite components. The widest users of cutting tools during machining of several composites include, but are not limited to, transportation (aerospace, marine and automobile), telecommunication and construction industries. As these sectors grow and develop, the application of cutting tools increases to meet the ever-increasing and insatiable needs of human, using composite parts. Also, in an attempt to maximise productivity and profitability continuously, the pursuit of having an optimal drill with excellent tool life and reduced wear rate remains a continuous exercise.

The effects of tool wear include high power/energy consumption and capital involvement, increased machining-induced damage and high number of part rejects as well as accidents and deaths. These unwanted consequences of tool wear will be either drastically reduced or totally eradicated with advancement in technology/engineering and/or advent of several modern, sophisticated and state-of-the-art machine centres and software packages. These technologies include computer numerical control (CNC), non-conventional machining (abrasive/waterjet, laser, electric discharged, ultrasonically-assisted types, among others), additive manufacturing/3D printing and robot in manufacturing, to mention but a few.

References

Abu-Mahfouz I (2005) Drill flank wear estimation using supervised vector quantization neural networks. Neural Comput Appl 14(3):167–175

Astakhov VP, Davim JP (2008) Tools (geometry and material) and tool wear. In: Davim JP (ed) Machining, Chapter 2. Springer, London, pp 29–57

Beal A, Kim DDW, Park KH, Kwon P (2011) A comparative study of carbide tools in drilling of CFRP and CFRP-Ti stacks. In: ASME 2011 international manufacturing science and engineering conference, vol 1. ASMEDC, pp 145–152

Bhattacharyya D, Horrigan DPW (1998) A study of hole drilling in Kevlar composites. Compos Sci Technol 58:267–283

Brinksmeier E, Janssen R (2002) Drilling of multi-layer composite materials consisting of carbon fibre reinforced plastics (CFRP), titanium and aluminum alloys. CIRP Ann 51:87–90

Brinksmeier E, Walter A, Janssen R, Diersen P (1999) Aspects of cooling lubrication reduction in machining advanced materials. Proc Inst Mech Eng Part B: J Eng Manuf 213:769–778

Choudhury S, Raju G (2000) Investigation into crater wear in drilling. Int J Mach Tools Manuf 40:887–898

Dimla DE, Lister PM, Leighton NJ (1997) Neural network solutions to the tool condition monitoring problem in metal cutting—a critical review of methods. Int J Mach Tools Manuf 37:1219–1241

Dolinšek S, Kopač J (2006) Mechanism and types of tool wear; particularities in advanced cutting materials. J Achiev Mater Manuf Eng 19:11–18

Ertunc HM, Loparo KA, Ocak H (2001) Tool wear condition monitoring in drilling operations using hidden Markov models (HMMs). Int J Mach Tools Manuf 41:1363–1384

Faraz A, Biermann D, Weinert K (2009) Cutting edge rounding: An innovative tool wear criterion in drilling CFRP composite laminates. Int J Mach Tools Manuf 49:1185–1196

Garrick R (2007) Drilling advanced aircraft structures with PCD (poly-crystalline diamond) drills. In: Aerospace technology conference and exposition proceedings. SAE technical paper series. https://doi.org/10.4271/2007-01-3893

Govekar E, Grabec I (1994) Self-organizing neural network application to drill wear classification. J Eng Ind 116:233–238

Groover MP (2007) Fundamentals of modern manufacturing: materials processes, and systems. Wiley, New Jersey

Iliescu D, Gehin D, Gutierrez ME, Girot F (2010) Modeling and tool wear in drilling of CFRP. Int J Mach Tools Manuf 50:204–213

Iskandar Y, Tendolkar A, Attia MH, Hendrick P, Damir A, Diakodimitris C (2014) Flow visualization and characterization for optimized MQL machining of composites. CIRP Ann 63(1):77–80

Ismail SO, Dhakal HN, Popov I, Beaugrand J (2016) Comprehensive study on machinability of sustainable and conventional fibre reinforced polymer composites. Eng Sci Technol Int J 19:2043–2052

Jayabal S, Natarajan U (2010) Optimization of thrust force, torque, and tool wear in drilling of coir fibre-reinforced composites using Nelder-Mead and genetic algorithm methods. Int J Mach Tools Manuf 51:371–381

Kannan S, Pervaiz S, Ghoshal A (2018) An experimental investigation of inclined hole drilling for CFRP under various lubrication techniques. In: Proceedings of advanced manufacturing. American society of mechanical engineers international mechanical engineering congress and expedition, November 9–15, 2018, vol 2, pp 1–10, Pittsburgh, Pennsylvania, USA

Khashaba UA, El-Sonbaty IA, Selmy AI, Megahed AA (2010) Machinability analysis in drilling woven GFR/epoxy composites: Part II—Effect of drill wear. Compos A Appl Sci Manuf 41:1130–1137

Kim HY, Ahn JH, Kim SH, Takata S (2002) Real-time drill wear estimation based on spindle motor power. J Mater Process Technol 124:267–273

Krolczyk GM, Maruda RW, Krolczyk JB, Wojciechowski S, Mia M, Nieslony P, Budzik G (2019) Ecological trends in machining as a key factor in sustainable production—a review. J Clean Prod 218:601–615

Lantrip J (2008) New tools needed: new cutting tool materials are required for new generation airframes. Cut Tool Eng Mag 72–84

Li X, Tso SK (1999) Drill wear monitoring based on current signals. Wear 231:172–178

Lin SC, Chen IK (1996) Drilling carbon fibre-reinforced composite material at high speed. Wear 194:156–162

Liu H, Lee B, Tarng Y (2000) In-process prediction of corner wear in drilling operations. J Mater Process Technol 101(1–3):152–158

Murphy C, Byrne G, Gilchrist MD (2002) The performance of coated tungsten carbide drills when machining carbon fibre-reinforced epoxy composite materials. Proc Inst Mech Eng Part B: J Eng Manuf 216:143–152

Park KH, Kwon P, Kim D (2012) Wear characteristic on BAM coated carbide tool in drilling of composite/titanium stack. Int J Precis Eng Manuf 13(7):1073–1076

Rahim EA, Sasahara H (2011) Investigation of tool wear and surface integrity on MQL machining of Ti-6AL-V using biodegradable oil. Proc Inst Mech Eng Part B: J Eng Manuf 225(9):1505–1511

Rawat S, Attia H (2009) Wear mechanisms and tool life management of WC–Co drills during dry high speed drilling of woven carbon fibre composites. Wear 267:1022–1030

Sakuma K, Seto M, Taniguchi M, Yokoo Y (1985) Tool wear in cutting carbon-fibre-reinforced plastics: the effect of physical properties of tool materials. Bull Jpn Soc Mech Eng 28:2781–2788

Sathish RU, Raj RL (2012) Applying wear maps in the optimisation of machining parameters in drilling of polymer matrix composites—a review. Res J Recent Sci 1:75–82

Sen B, Mia M, Krolczyk GM, Mandel AK, Mondal SP (2019) Eco-friendly cutting fluids in minimum quantity lubrication assisted machining: a review on the perception of sustainable manufacturing. Int J Precis Eng Manuf-Green Technol 1–32

Senthil Kumar M, Prabukarthi A, Krishnaraj V (2013) Study on tool wear and chip formation during drilling carbon fibre reinforced polymer (CFRP)/titanium alloy (Ti6Al4V) stacks. Procedia Eng 64:582–592

Shyha I, Soo S, Aspinwall D, Bradley S (2010) Effect of laminate configuration and feed rate on cutting performance when drilling holes in carbon fibre reinforced plastic composites. J Mater Process Technol 210(8):1023–1034

Shyha IS, Soo SL, Aspinwall DK, Bradley S, Perry R, Harden P, Dawson S (2011) Hole quality assessment following drilling of metallic-composite stacks. Int J Mach Tools Manuf 51(7–8):569–578

Sriraman P, Prasanna G, Pradeep S, Praveen Raja D (2015) A review on experimental optimisation in drilling of cfrp composite. Int J Recent Trends Eng Technol 10:1–4

Szwajka K, Trzepieciński T (2016) Effect of tool material on tool wear and delamination during machining of particleboard. J Wood Sci 62(4):305–315

Tansel IN, Mekdeci C, Rodriguez O, Uragun B (1993) Monitoring drill conditions with wavelet based encoding and neural networks. J Mach Tools Manuf 33:559–575

Tashiro T, Fujiwara J, Inada K (2011) Drilling of CFRP/Ti-6Al-4V stacks. Adv Mater Res 325:369–374

Wang X, Kwon PY, Sturtevant C, Kim DDW, Lantrip J (2013) Tool wear of coated drills in drilling CFRP. J Manuf Process 15(1):127–135

Wang F, Qian B, Jia Z, Cheng D, Fu R (2018) Effects of cooling position on tool wear reduction of secondary cutting edge corner of one-shot drill bit in drilling CFRP. Int J Adv Manuf Technol 94(9):4277–4287

Wang D, Onawumi PY, Ismail SO, Dhakal HN, Popov I, Silberschmidt VV, Roy A (2019) Machinability of natural-fibre-reinforced polymer composites: conventional *vs* ultrasonically-assisted machining. Compos A Appl Sci Manuf 119:188–195

Xia T (2014) Investigation of drilling performance in cryogenic drilling on CFRP composite laminates. Master's dissertation, Department of Mechanical Engineering, University of Kentucky, USA

Xia T, Kaynak Y, Arvin C, Jawahir IS (2016) Cryogenic cooling-induced process performance and surface integrity in drilling CFRP composite material. Int J Adv Manuf Technol 82:605–616

Xu J, Ji M, Chen M, Ren F (2019) Investigation of minimum quantity lubrication effects in drilling CFRP/Ti6Al4V stacks. Mater Manuf Process 34(12):1401–1410

Milling Behavior of Injection Molded Short Fiber-Reinforced Green Composites

K. Debnath, M. Roy Choudhury, G. Surya Rao, and R. N. Mahapatra

Abstract Milling is a machining operation which is widely and frequently used to produce surface profile by removing excess material with the help of milling cutter. However, milling of fiber-reinforced green composites is quite challenging due to its anisotropic and heterogeneous characteristics. This chapter present and discuss the milling behavior of developed short fiber-reinforced green composites. The biodegradable polymer namely poly(lactic) acid has been reinforced with short natural fiber to fabricate the green composites using the direct injection molding technique. The effect of milling process parameters such as depth of cut, feed rate, and spindle on the performance parameters such as induced forces, temperature, surface roughness, and burr height have been studied by performing response surface methodology (RSM). The milling operations have been carried out using an end mill cutter of 6 mm in diameter. The relative significance of the process parameters has also been studied by performing analysis of variance (ANOVA). The three-dimensional surface plots and two-dimensional contour plots have been constructed to understand the mutual effect of process parameters on the responses. The optimization of process parameters has been also carried out for obtaining better milled surface in the developed composites.

Keywords Green composites · Milling · Forces · Temperature · Surface roughness · Burr formation · Response surface methodology

1 Introduction

Composites are heterogeneous material composed of two or more materials with different properties and having properties superior to the individual constituents. Composite materials have numerous applications in different sectors as these materials possess excellent properties such as high specific strength and stiffness, ease of

---

K. Debnath (✉) · M. R. Choudhury · G. S. Rao · R. N. Mahapatra
Department of Mechanical Engineering, National Institute of Technology Meghalaya, Shillong 793 003, India
e-mail: debnath.iitr@gmail.com

M. T. Hameed Sultan et al. (eds.), *Machining and Machinability of Fiber Reinforced Polymer Composites*, Composites Science and Technology,
https://doi.org/10.1007/978-981-33-4153-1_6

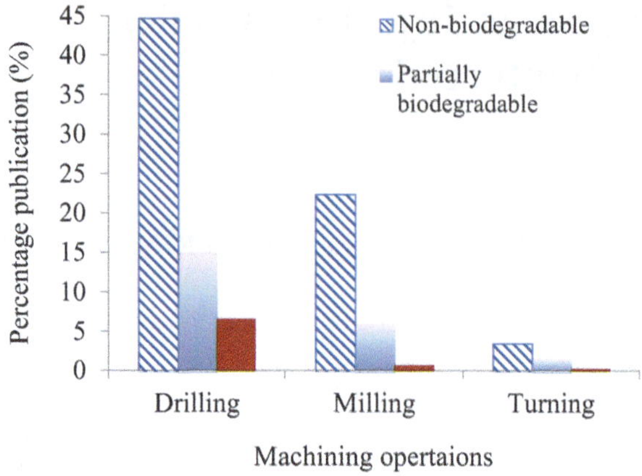

Fig. 1 Summary of research publications

design flexibility, and high chemical and corrosion-resistant to name a few. A new type of composites called green composites which are fully-biodegradable has been developed due to increase in the environmental concern. This type of composites consists of biodegradable polymer and natural fibers and can easily be fabricated to near net shape. The secondary manufacturing process (machining, joining, etc.) of these materials is an indispensable operation to produce an intricate shape product and to obtain the structural integrity. However, the machining operation is a challenging task due to the anisotropic and heterogeneous nature of this composite. Therefore, the study of the machinability of this composite is an area of paramount importance. Figure 1 presents a summary of works carried out in the past decade to investigate the different machining operations in fiber-reinforced polymer composites. The percentage contribution in terms of the total number of research publications is taken as reference data. Many works have been carried out in the past decade to study the machining behavior of nonbiodegradable synthetic fiber-reinforced composites. A few researches are available on partially biodegradable composites. The partially biodegradable composites consist of nonbiodegradable polymer and natural fiber. But a handful works available on the machining behavior of fully-bridgeable or green composites. The figure also shows that the milling and turning operations have gained less importance in the past decade as compared to drilling operation. Milling is an important and commonly used machining operation and thus the milling behavior of biodegradable green composites needs to be investigated to further expand the application spectrum of this composite.

2 Literature Review

Milling is a machining operation used for removing excess materials to obtain high-quality surface with close dimensional tolerance. However, a number of challenges are associated with the milling of fiber-reinforced composites. The common defects that occur during milling of fiber-reinforced composites are surface damage, formation of rough surfaces and burrs. These damages of the machined surface detrimentally affect the life of the product (Patel et al. 2018). Some investigations were carried out by the researchers to minimize the challenges associated with the milling of fiber-reinforced composites. Wang et al. (2016a) investigated the quality of surface produced during milling of carbon fiber-reinforced composites and correlated with induced temperature and forces generated during milling. Rajmohan et al. (2019) investigated the induced damage like fiber debonding, internal cracks, tool wear, and fiber pullout during milling of fiber-reinforced composites. Jia et al. (2018) experimentally studied the influence of induced temperature on cutting force and surface quality during milling of carbon fiber-reinforced composites. Davim et al. (2004) studied the effect of different resin materials on the induced forces and damages formed during milling of glass fiber-reinforced composites. The minimum damage was found to be associated with the milling of unsaturated polyester-based composites. Davim and Reis (2005) conducted milling of carbon fiber-reinforced composites with different milling cutters and concluded that six-flute end mill produces more damage as compared to two-flute end mill. Palanikumar et al. (2006), Puw and Hocheng (1993), Koplev et al. (1983), Islam et al. (2015), Hintze et al. (2015) Ghafarizadeh et al. (2016), and Wang et al. (2016b) studied the influence of fiber orientation angle on the surface quality during milling of synthetic fiber-reinforced composites. The researchers have concluded that good surface finish can be achieved at a fiber orientation angle of 0°. It was also stated that the surface roughness increases as the orientation angle of fiber increases. Sorrentino and Turchetta (2014) concluded that during milling of carbon fiber-reinforced composite a good quality surface can be achieved at a higher feed rate. On the contrary, it was reported that the damages associated with milling can be minimized by lowering the feed rate and increasing the spindle speed (Erkan et al. 2013; Jenarthanan et al. 2016; Raj et al. 2012; Jenarthanan and Jeyapaul 2013; Kiliçkap et al. 2015). Rusinek (2010) found a non-linear relationship of induced force with the feed rate. However, contributions of feed rate and depth of cut in altering induced force are 54% and 45%, respectively (Azmi et al. 2013). Sreenivasulu (2013) stated that surface roughness and delamination occur during milling of glass fiber-reinforced composites is mainly affected by the depth of cut and cutting speed, respectively. Razfar and Zadeh (2009) optimized the process parameters during milling of glass fiber-reinforced composites by performing artificial neural network (ANN) and genetic algorithm (GA). Some studies conducted to understand the milling characteristic of partially biodegradable composites. Babu and Kasu (2012) and Babu et al. (2013) studied the milling characteristics of different natural fiber-reinforced composites and found that composite reinforced with jute fiber offers more damage. The high-quality milled surface was produced in hemp fiber-reinforced composites. The study on milling characteristics of banana fiber-reinforced composites revealed that the delamination is minimum at lower feed and higher speed (2017). Balasubramanian et al. (2016) developed a fuzzy model in milling of natural fiber-reinforced composites to evaluate the thrust force and torque generated during machining. Sankar et al. (2015) and Çelik et al. (2019) stated that the thrust force induced during milling of jute fiber-reinforced composite was influenced by cutting speed and depth of cut. The feed rate and cutting speed were found to be the most significant factor during milling of kenaf fiber-reinforced composite (Harun et al. 2015). Chegdani et al. (2015) investigated the tribological behavior of different natural fiber composites while performing milling operation. It was found that bamboo fiber-reinforced composite produces a good surface as compared to other types of short fibers chosen for the purpose of investigation.

It can be observed from the above discussion that most of the researches is related to the milling of nonbiodegradable and partially-biodegradable composites. Negligible research has been carried out to study the milling characteristic of fully-biodegradable green composites. In this study, a green composite composed of biodegradable polymer (PLA) and bagasse fiber has been fabricated using an injection molding machine. The response surface methodology (RSM) has been applied to perform the experimental investigation. The milling parameters such as feed rate, depth of cut, and spindle speed have been optimized to obtain the minimum induced forces, temperature, surface roughness, and burr height. The relative significance of each parameter and their effects on the output responses have been obtained by performing analysis of variance (ANOVA).

3 Material and Methods

3.1 Fabrication of Composites

PLA in pellets form was supplied by Natur Tec India Pvt. Ltd, Chennai, India. The glass transition temperature, crystalline melting temperature, and density of PLA are 55–60 °C, 145–160 °C, and 1.24 g/cm^3. The short bagasse fiber has been used as a reinforcing material. The fabrication of composite specimen has been carried out using an injection molding machine. The heating temperature has been adjusted based on the melting temperature of PLA with the help of four heaters located at four positions of the process line in the injection molding machine. The required pressure of 65 psi has been applied by the inbuilt system at a speed of 50 mm/s. The cooling time of the specimen in the die has been set at 20 s. The length and width of the fabricated specimen are 125 mm and 12.7 mm, respectively. The injection molding machine and the developed green composites are shown in Fig. 2.

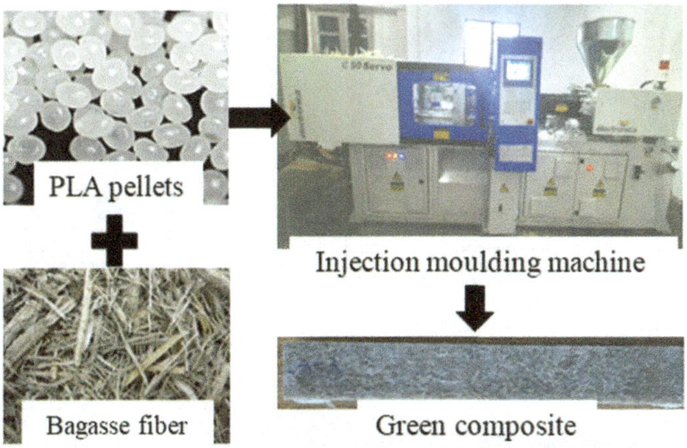

Fig. 2 Fabrication of green composites

3.2 Measurements

A knee type vertical milling machine (Baltiboi Ltd., BFV5) has been used to perform the milling operation on the developed composites. All the milling operations have been conducted using an HSS end mill cutter of 6 mm in diameter. The induced forces during milling have been recorded and analyzed using a dynamometer connected with a multi-channel charge amplifier (Kistler, 5070A) and data acquisition system (Kistler, 5697A1). The dynamometer has been fixed on the machine table above which the composite specimen has been fixed with the help of a delicate fixture. The sampling rate was set as high as 1000 Hz for detecting the minute variation in the force signals. A thermographic camera (Testo, 885-2 Set) has been used to indicate the induced temperature during the milling operation. The resolution of the infrared camera is 320 $\times$ 240 pixels, thermal sensitivity is <30 mK, and accuracy is ± 2 °C. The surface roughness of the milled specimen has been evaluated using a surface roughness tester (Taylor Hobson, Surtronic S-128) at a cut-off length of 0.25 mm. The returning speed and measuring speed of the surface roughness tester have been set at 1.5 and 1 mm/s, respectively. Each experiment was repeated three times and the average value was considered for the analysis purpose while measuring the roughness of the milled surface. The burr height has been measured using the tool maker's microscope (Mitutoyo, No. 152–390) with the least count of 0.005 mm and measuring range of 0–25 mm, respectively. The experimental setup and different equipment used in this study are shown in Fig. 3.

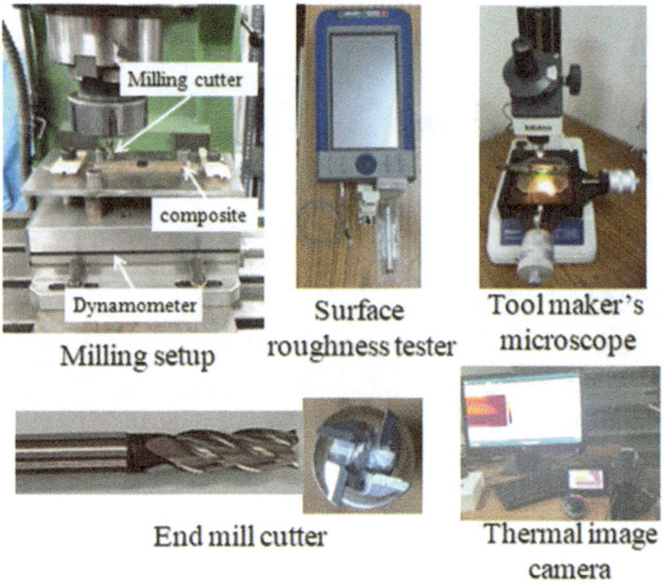

Fig. 3 Milling setup and other measuring equipment

3.3 *Design of Experiment*

In this study, the milling characteristics of the developed composites have been investigated by performing the experiments. The experiments were designed based on the response surface methodology. The Box-Behnken statistical screening design has been used to statistically optimize the process parameters and evaluate the main effects, two-way interaction effects, and square effects of the process parameters on the responses. A total of 15 experiments have been carried out to find the effect of different process parameters (depth of cut, feed, and spindle speed) on the output responses (forces, temperature, surface roughness, and burr height). A second-order polynomial model has been constructed by three factors having three corresponding levels. ANOVA has been conducted to provide the statistical basis of the experimental data. The three-dimensional surface plots and two-dimensional contour plots have been also developed to understand the mutual effect of process parameters. The process parameters have been optimized to obtain better milling performance by minimizing the output responses. The input process parameters and their corresponding levels are given in Table 1.

Table 1 Milling parameters and their corresponding levels

Factors	Levels		
	−1	0	1
A: Depth of cut (mm)	1	1.5	2
B: Feed (mm/min)	112	160	224
C: Spindle speed (RPM)	710	1400	2000

4 Results and Discussion

The experimental layout and results of output responses are presented in Table 2. The milled green composite specimens are shown in Fig. 4. The milling operation has been carried out for a distance of 3.5 mm from the edge of each specimen.

4.1 Effect of Input Parameters on Induced Forces

The forces induced during milling are important to analyze as higher force results in damage to the milled surface. The dynamometer used to sense the forces during milling provides all three orthogonal forces (F_x, F_y, and F_z). The maximum values of

Table 2 Experimental layout and results

Run no	A: Depth of cut (mm)	B: Feed rate (mm/min)	C: Spindle speed (RPM)	Induced forces (N)	Induced temperature (°C)	Burr height (mm)	Surface roughness (μm)
1	2.0	160	2000	66.150	113.1	6.85	1.134
2	1.5	160	1400	82.340	73.8	6.08	1.033
3	2.0	112	1400	79.450	66.8	6.00	1.000
4	1.0	160	710	87.440	49.7	5.00	0.866
5	1.5	112	710	82.620	49.3	4.86	0.733
6	1.5	224	2000	73.890	116.6	6.90	1.167
7	2.0	160	710	89.963	60.8	5.07	0.900
8	1.5	224	710	115.43	64.0	5.67	0.967
9	1.5	160	1400	82.340	73.8	6.08	1.033
10	1.5	160	1400	82.340	73.8	6.08	1.033
11	1.0	112	1400	66.720	60.9	5.06	0.966
12	1.5	112	2000	60.360	81.2	6.60	1.060
13	1.0	160	2000	63.060	95.0	6.70	1.133
14	2.0	224	1400	102.48	114.3	6.30	1.166
15	1.0	224	1400	87.070	75.9	6.25	1.066

Fig. 4 Milled composite specimens

these forces have been considered for each set of experiments. The resultant (F_w) of these three force components has been calculated using Eq. (1). This resultant force (F_w) is considered as the induced force during milling of green composites.

$$F_w = \sqrt{F_x^2 + F_y^2 + F_z^2} \quad (1)$$

where, F_x, F_y, and F_z are in Newton (N).

The force signals have been recorded at various combinations of feed rate, spindle speed, and depth of cut. F_x, F_y, and F_z have been recorded at a depth of cut of 1.5 mm, feed rate of 160 mm/min, and a spindle speed of 1400 RPM are shown in Fig. 5. The figure clearly shows that the signals of orthogonal force components are quite

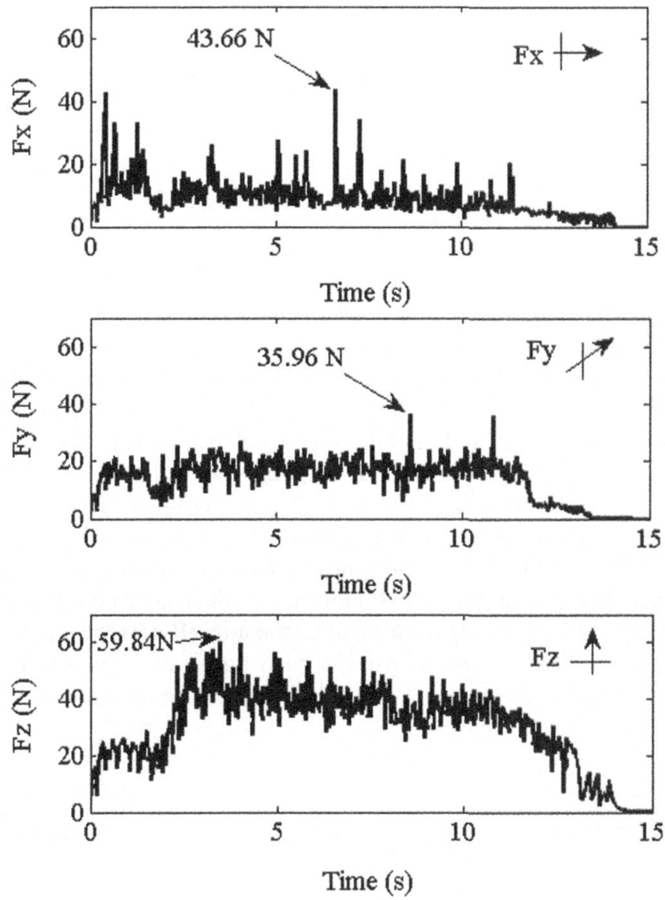

Fig. 5 Orthogonal components of induced forces during milling of green composites

Table 3 Analysis of variance for induced forces

Source	DF	Adj. SS	Adj. MS	F-Value	P-Value
Model	9	2941.31	326.81	17.20	**0.003**
Linear	3	2716.83	905.61	47.66	**0.000**
A	1	142.41	142.41	7.49	**0.041**
B	1	1006.39	1006.39	52.96	**0.001**
C	1	1568.03	1568.03	82.52	**0.000**
Square	3	129.60	43.20	2.27	0.198
A × A	1	21.57	21.57	1.13	0.335
B × B	1	59.28	59.28	3.12	0.138
C × C	1	39.48	39.48	2.08	0.209
2-way interaction	3	94.88	31.63	1.66	0.288
A × B	1	1.80	1.80	0.09	0.771
A × C	1	0.08	0.08	0.00	0.951
B × C	1	93.01	93.01	4.89	0.078
Error	5	95.01	19.00		
Lack of fit	3	95.01	31.67	*	*
Pure error	2	0.00	0.00		
Total	14	3036.32			$R^2 = 0.09687$

different from each other and the magnitude of the force in the z-direction is more as compared to the other two force components.

ANOVA has been performed to obtain the statistical significance of the experimental data as presented in Table 3. The main effects of the process parameters have been found to be significant as the p-values are less than 0.05 at a confidence interval of 95%. However, two-way interaction effects and square effects are not significant as p-values are greater than 0.05. But it can be observed that square and two-way interaction coefficients are important to make the model significant. From the main effects, it can be seen that F-value is higher for factor C (spindle speed). This indicates that spindle speed has the highest influence on the induced forces followed by feed and depth of cut. The coefficient of determination (R^2) has also been found to be quite high as 0.9687%. This indicates a good fitness of the experimental data. The quadratic regression equation generated using the experimental data is given by Eq. (2).

$$\begin{aligned}\text{Force} &= 41.9 + 32.8\text{A} - 0.084\text{B} + 0.0214\text{C} - 9.67\text{A}^2 + 0.001278\text{B}^2 \\ &\quad - 0.000008\text{C}^2 + 0.0239\text{A} \times \text{B} + 0.00044\text{A} \times \text{C} - 0.000133\text{B} \times \text{C}\end{aligned} \tag{2}$$

The three-dimensional surface plot and two-dimensional contour plots of the induced force are shown in Fig. 6. The figure gives a clear idea of the combined

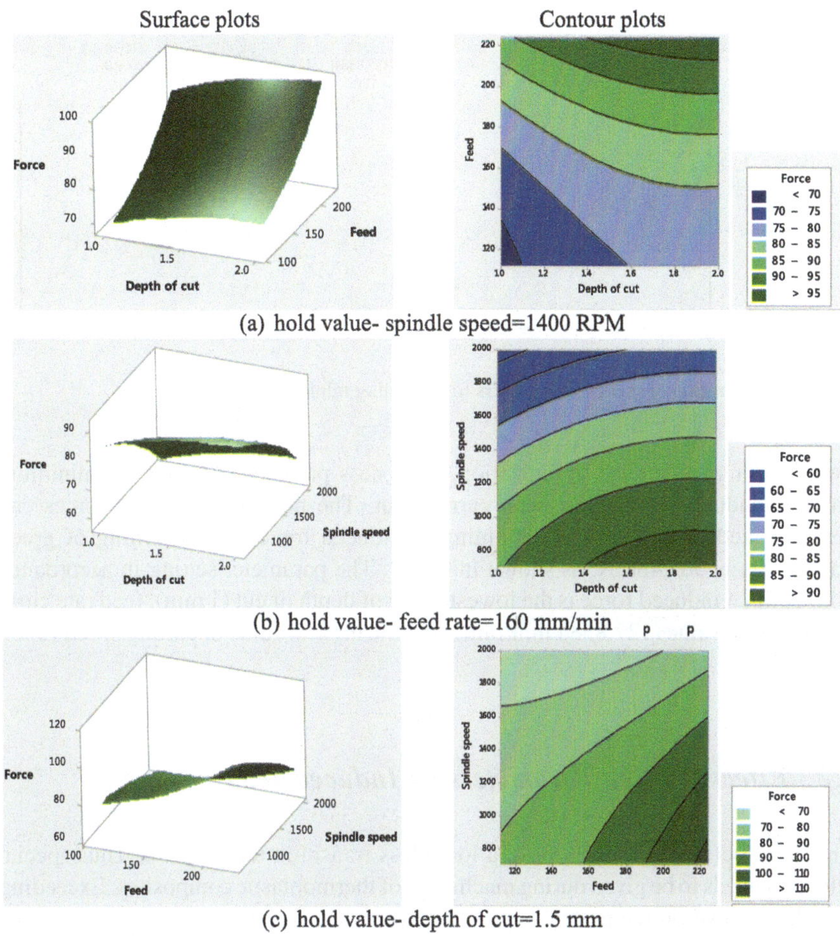

Fig. 6 Response surface plots of induced force with respect to mutual effects of process parameters **a** feed rate versus depth of cut, **b** spindle speed versus depth of cut, and **c** spindle speed versus feed rate

effect of any two process parameters on response variable. The response contour plot is the mapping of induced force coded with a different color. It can be observed that minimum induced force can be obtained at a lower level of feed rate and depth of cut (Fig. 6a). The milling cutter removes more amount of material at a higher feed rate which results in higher induced forces. Also, cutting force increases as the depth of cut increases. This is attributed to more mechanical contact between the milling cutter and the material at the higher depth of cut. The forces induced corresponding to the higher level of spindle speed has been found to be less irrespective of the corresponding level of the depth of cut and feed rate (Fig. 6b, c). Milling at higher spindle speeds results in more heat generation due to which softening of the polymer occurs. The removal of softened polymer became relatively easy which results in

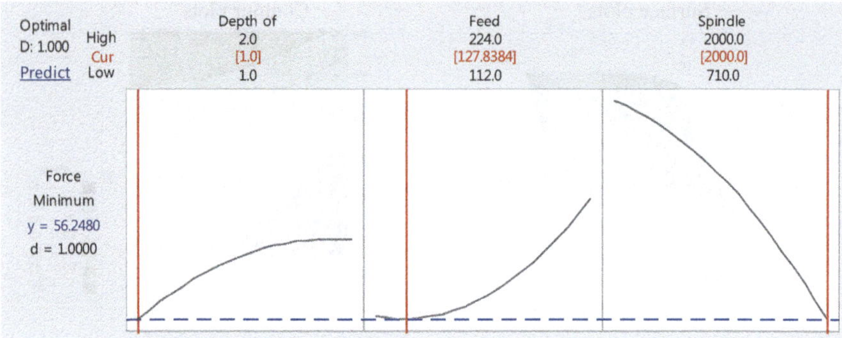

Fig. 7 Optimization of process parameters for obtaining minimum induced force

lower cutting force. The optimization of process parameters to get the minimum value of induced forces has been carried out. The minimum induced forces can be calculated using Eq. 2. The minimum induced force during milling of green composites is 56.2480 N, as shown in Fig. 7. The parameter setting in accordance with optimal induced force is the lowest value of depth of cut (1 mm), feed rate close to the lowest value (127.8384 mm/min), and the highest value of cutting speed (2000 RPM).

4.2 *Effect of Input Parameters on Induced Temperature*

The thermoplastic polymers have a low glass transition temperature. Thus special attention needs to be given during machining of thermoplastic composites. Exceeding the glass transition temperature of the polymer during machining may result in an overall decrease in the strength of the composites. The quality of the milled surface is also significantly affected by the temperature generated during machining. Thus it is pertinent to analyze the influence of milling parameters on the temperature induced during milling of green composites. The thermal imaging camera has been used to record the in-situ temperature that generates during the milling operation. The recorded thermographs have been further analyzed to calculate the temperature of the hotspot using the data capture program (TH71-717, 2.0C). The maximum temperature attained during each experiment has been considered for the study. The induced temperature during the milling operation is graphically presented in Fig. 8.

ANOVA has been performed to obtain the statistical significance of the experimental data (Table 4). The main effects of the process parameters and two-way interaction of A $\times$ B and B $\times$ C have been found to be significant as the p-values are less than 0.05 at a confidence interval of 95%. However, the square effects are not significant as p-values are greater than 0.05. But square and two-way interaction are important coefficients to make the model significant. From the main effects, it can be

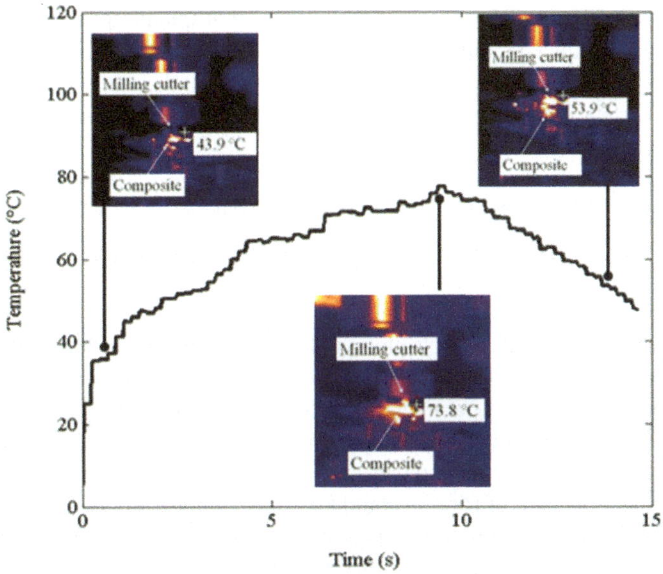

Fig. 8 Temperature versus time plot (depth of cut: 1.5 mm, feed rate: 160 mm/min, and spindle speed: 1400 RPM)

Table 4 Analysis of variance for induced temperature

Source	DF	Adj. SS	Adj. MS	F-Value	P-Value
Model	9	6861.26	762.36	55.11	**0.000**
Linear	3	6405.18	2135.06	154.33	**0.000**
A	1	675.28	675.28	48.81	**0.001**
B	1	1584.84	1584.84	114.56	**0.000**
C	1	4145.05	4145.05	299.62	**0.000**
Square	3	72.65	24.22	1.75	0.272
A × A	1	52.62	52.62	3.80	0.109
B × B	1	13.33	13.33	0.96	0.371
C × C	1	15.90	15.90	1.15	0.333
2-way interaction	3	383.43	127.81	9.24	**0.018**
A × B	1	264.06	264.06	19.09	**0.007**
A × C	1	12.25	12.25	0.89	0.390
B × C	1	107.12	107.12	7.74	**0.039**
Error	5	69.14	13.83		
Lack of fit	3	69.17	23.06	*	*
Pure error	2	0.00	0.00		
Total	14	6930.43			$R^2 = 0.99$

seen that F-value is higher for factor C. This means that spindle speed has the highest influence on the induced temperature followed by the feed rate and depth of cut. The R^2 has also been found to be quite high as 0.99 which indicates good fitness of the experimental data. The following quadratic regression equation has been generated using the experimental data:

$$\begin{aligned}\text{Temperature} &= 133.2 - 83.0A - 0.582B - 0.0104C + 15.10A^2 + 0.000606B^2 \\ &\quad + 0.000005C^2 + 0.2902A \times B + 0.00543A \times C + 0.000143B \times C \end{aligned} \tag{3}$$

The three-dimensional surface plots and two-dimensional contour plots of the induced temperature are shown in Fig. 9. From the response surface plots, it can be observed that all the process parameters viz. feed rate, spindle speed, and depth of cut affect the induced temperature. The repetitive interaction of the milling cutter with the composite at a higher level of the process parameters results in more friction. The heat generated as a result of friction could not transfer efficiently to the surroundings due to the low thermal conductivity of the composite constituents. The optimization of process parameters to get the minimum value of induced temperature has been carried out. The minimum temperature induced during milling of green composites is 46.0812 °C, as shown in Fig. 10. The control parameter setting in accordance with optimal induced temperature is the depth of cut close to the center point (1.5455 mm), lowest value of feed (112 mm/min), and lowest value of cutting speed (710 RPM).

4.3 *Effect of Input Parameters on Surface Roughness*

The evaluation of surface roughness of the milled surface is an important consideration as it significantly affects the overall quality of the final product. The composites may prone to several damages like fiber pull-out, delamination, and fiber breaking during milling. The roughness of the milled surface for each experimental trial has been measured three times and the average value has been considered for analysis purposes. Figure 11 represents the graph obtained by surface roughness tester showing the deviations from the mean line. The arithmetic deviation from the mean line is denoted by R_a (μm).

ANOVA has been performed to obtain the statistical significance of the experimental data, as given in Table 5. The main effects of the process parameters, the square effect of C^2, and the interaction effect of B × C have been found to be significant as the p-values are less than 0.05 at a confidence interval of 95%. The coefficients of square and two-way interaction are important to make the model significant. From the main effects, it can be seen that F-value is higher for factor C. This simply means spindle speed is the most significant factor followed by feed rate and depth of cut. The value of R^2 has also been found to be quite high (98.99%). This indicates a

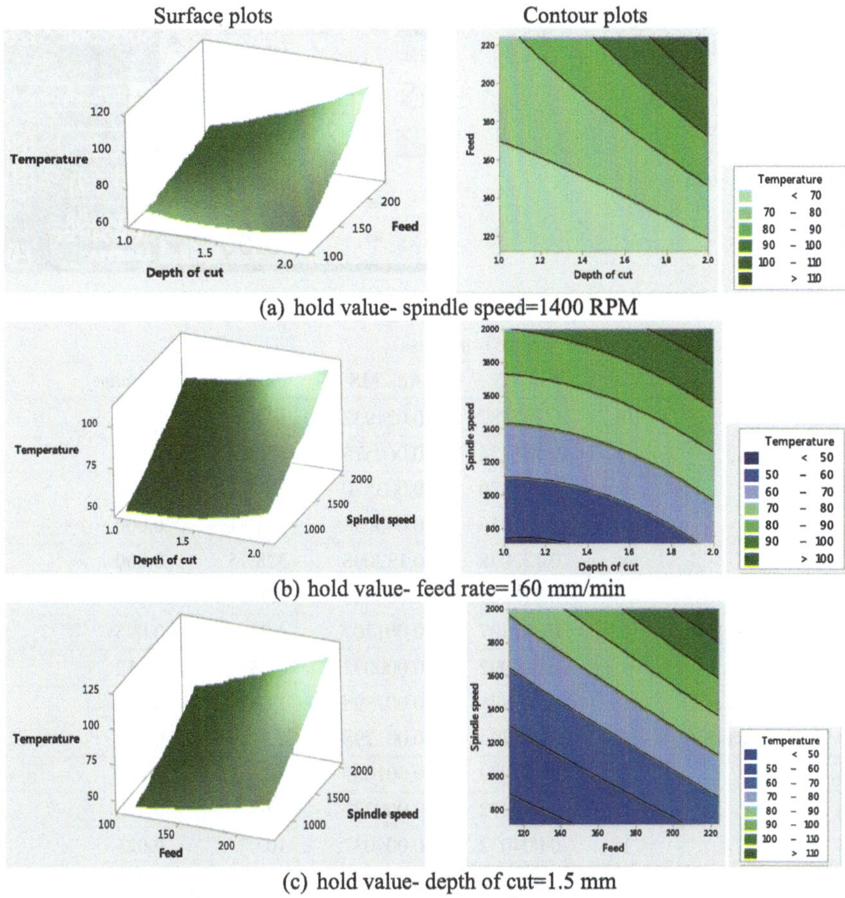

Fig. 9 Response surface plots of induced temperature with respect to mutual effects of process parameters **a** feed rate versus depth of cut, **b** spindle speed versus depth of cut, and **c** spindle speed versus feed rate

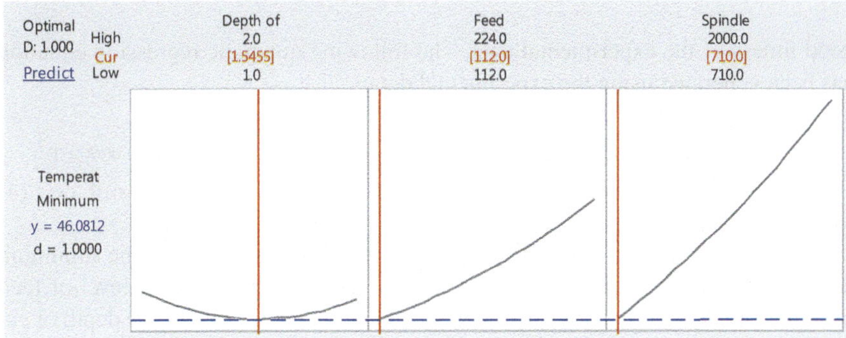

Fig. 10 Optimization of process parameter for obtaining minimum induced temperature

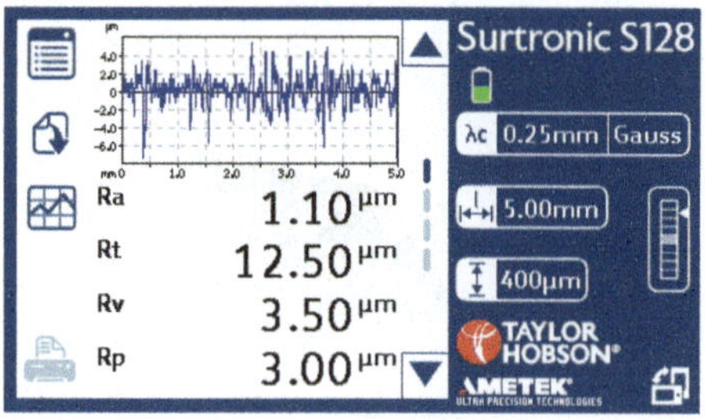

Fig. 11 Roughness of the milled surface (depth of cut: 1.5 mm, feed rate: 160 mm/min, and spindle speed: 1400 RPM)

Table 5 Analysis of variance for surface roughness

Source	DF	Adj. SS	Adj. MS	F-Value	P-Value
Model	9	0.197387	0.021932	54.48	**0.000**
Linear	3	0.181724	0.060575	150.48	**0.000**
A	1	0.003570	0.003570	8.87	**0.031**
B	1	0.046056	0.046056	114.41	**0.000**
C	1	0.132098	0.132098	328.15	**0.000**
Square	3	0.010269	0.003423	8.50	**0.021**
A × A	1	0.001707	0.001707	4.24	0.095
B × B	1	0.000092	0.000092	0.23	0.652
C × C	1	0.007898	0.007898	19.62	**0.007**
2-way interaction	3	0.005394	0.001798	4.47	0.070
A × B	1	0.001089	0.001089	2.71	0.161
A × C	1	0.000272	0.000272	0.68	0.448
B × C	1	0.004032	0.004032	10.02	**0.025**
Error	5	0.002013	0.000403		
Lack-of-fit	3	0.002013	0.000671	*	*
Pure error	2	0.000000	0.000000		
Total	14	0.199400			$R^2 = 98.99\%$

good fitness of the experimental data. The following quadratic regression equation has been generated using the experimental data:

$$\begin{aligned}\text{Surface roughness} &= 0.313 - 0.280\text{A} + 0.00220\text{B} + 0.000687\text{C} + 0.0860\text{A}^2 - 0.000002\text{B}^2 \\ &\quad - 0.000000\text{C}^2 + 0.000589\text{A} \times \text{B} - 0.000026\text{A} \times \text{C} - 0.000001\text{B} \times \text{C} \quad (4)\end{aligned}$$

The response plots of surface roughness are shown in Fig. 12. The minimum surface roughness of the milled surface can be obtained at the lower level of feed rate and spindle speed irrespective of the depth of cut (Fig. 12a–c). The depth of cut has less contribution to the roughness of the surface produced during milling. The temperature generated at the machining zone is quite high at higher feed rate and

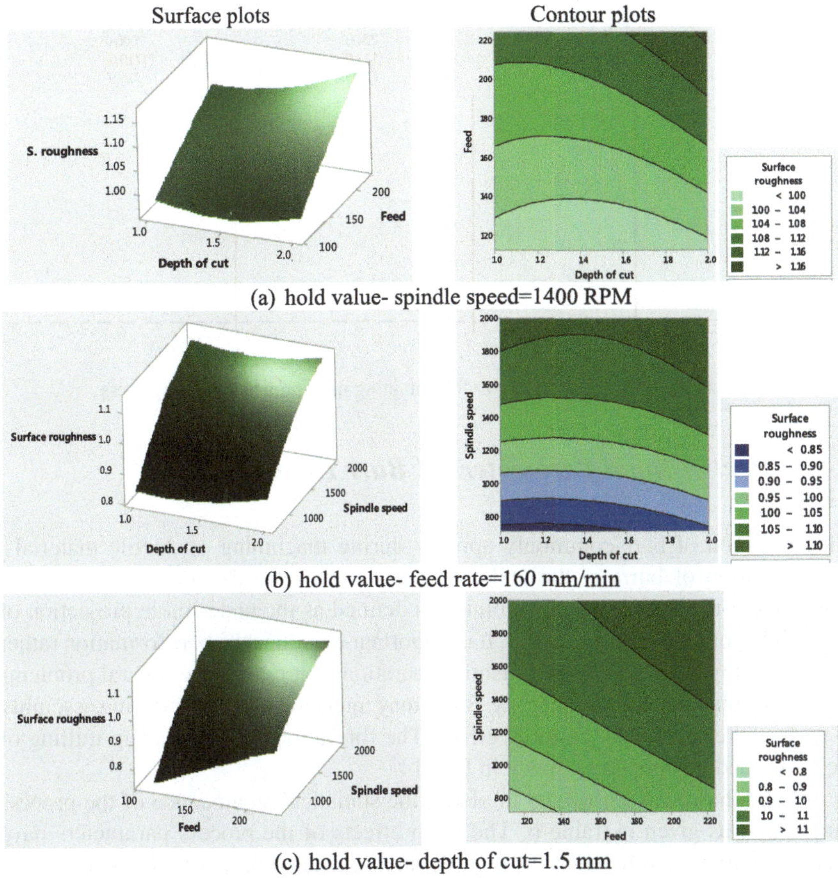

Fig. 12 Response surface plots of surface roughness with respect to the mutual effects of process parameters **a** feed rate versus depth of cut, **b** spindle speed versus depth of cut, and **c** spindle speed versus feed rate

spindle speed. The polymer becomes soft and results in uneven material removal during milling. Hence, a lower value of feed rate and spindle speed is preferred to obtain a smooth milled surface. The optimization of process parameters to get the minimum value of surface roughness has been carried out. The minimum surface roughness during milling of green composites is obtained as 0.7437 μm, as shown in Fig. 13. The control parameter setting in accordance with the optimal surface roughness is the depth of cut close to the center point (1.3535 mm), lowest value of feed rate (112 mm/min), and lowest value of cutting speed (710 RPM).

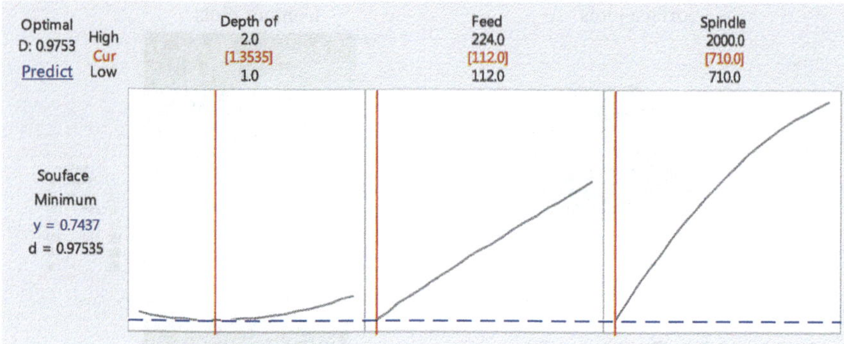

Fig. 13 Optimization of process parameter for obtaining minimum surface roughness

4.4 *Effect of Input Parameters of Burr Height*

The formation of burr commonly appears during machining of ductile materials. The formation of burrs at the end of a cut is similar to the formation of chips. The formation of burr during machining is defined as the undesirable projection of material beyond the specified limit. It is important to reduce the burr formation rather than deburring in a subsequent finishing operation as it may cause several problems for product quality and functionality. Burrs may interfere as a barrier during assembly of parts and can cause a jamming effect. The formation of burrs during milling of the developed composite is shown in Fig. 14.

ANOVA has been performed to obtain the statistical significance of the process parameters, as given in Table 6. The main effects of the process parameters have been found to be significant as the p-values are less than 0.05 at a confidence interval of 95%. However, two-way interaction effects and square effects have been found to be insignificant. But square and two-way interaction are important coefficients to make the model significant. From the main effect, it can be seen that F-value is higher for factor C which means spindle speed has the highest influence on the burr formation followed by feed rate and depth of cut. The higher value of R^2 indicates

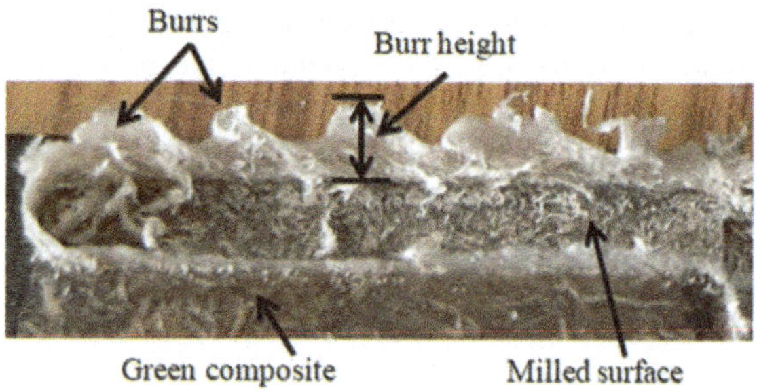

Fig. 14 Burr formation in milled surface (depth of cut: 1.5 mm, feed rate: 160 mm/min, and spindle speed: 1400 RPM)

Table 6 Analysis of variance for milling induced temperature

Source	DF	Adj. SS	Adj. MS	F-Value	P-Value
Model	9	6.56986	0.72998	29.28	**0.001**
Linear	3	6.22832	2.07611	83.26	**0.000**
A	1	0.18301	0.18301	7.34	**0.042**
B	1	0.84500	0.84500	33.89	**0.002**
C	1	5.20031	5.20031	208.55	**0.000**
Square	3	0.07688	0.02563	1.03	0.455
A × A	1	0.07237	0.07237	2.90	0.149
B × B	1	0.00519	0.00519	0.21	0.667
C × C	1	0.00452	0.00452	0.18	0.688
2-way interaction	3	0.26465	0.08822	3.54	0.104
A × B	1	0.19803	0.19803	7.94	0.037
A × C	1	0.00160	0.00160	0.06	0.810
B × C	1	0.06502	0.06502	2.61	0.167
Error	5	0.12468	0.02494		
Lack-of-fit	3	0.12468	0.04156	*	*
Pure error	2	0.00000	0.00000		
Total	14	6.69458			$R^2 = 0.9814$

a good fitness of the experimental data. The quadratic regression equation has been generated using the experimental data is given by Eq. (5).

$$\begin{aligned}\text{Burr height} = & -1.47 + 3.23\text{A} + 0.0265\text{B} + 0.001978\text{C} - 0.560\text{A}^2 - 0.000012\text{B}^2 \\ & - 0.000000\text{C}^2 - 0.00795\text{A} \times \text{B} + 0.000062\text{A} \times \text{C} - 0.000004\text{B} \times \text{C}\end{aligned} \tag{5}$$

The response plots of burr height are shown in Fig. 15. The minimum burr height corresponds to the lower level of depth of cut, feed rate, and spindle speed. The forces induced are less at lower feed rate and depth of cut. The partially deformed chips during milling bend in the direction of the cutting velocity and results in the formation of burr at a higher induced force. Also, some melted polymer slicked to the milling surface resulting in formation of burrs at higher spindle speed due to high heat generation. The optimization of process parameters to get the minimum burr height has been carried out. The optimized process parameters offer the minimum burr height assuming that the response variable is related to independent variables according to Eq. 5. The minimum burr height during milling of green composites is obtained as 4.2550 mm, as shown in Fig. 16. The control parameter setting in accordance with optimal burr height is the lowest value of depth of cut (1 mm), lowest value of feed rate (112 mm/min), and lowest value of cutting speed (710 RPM).

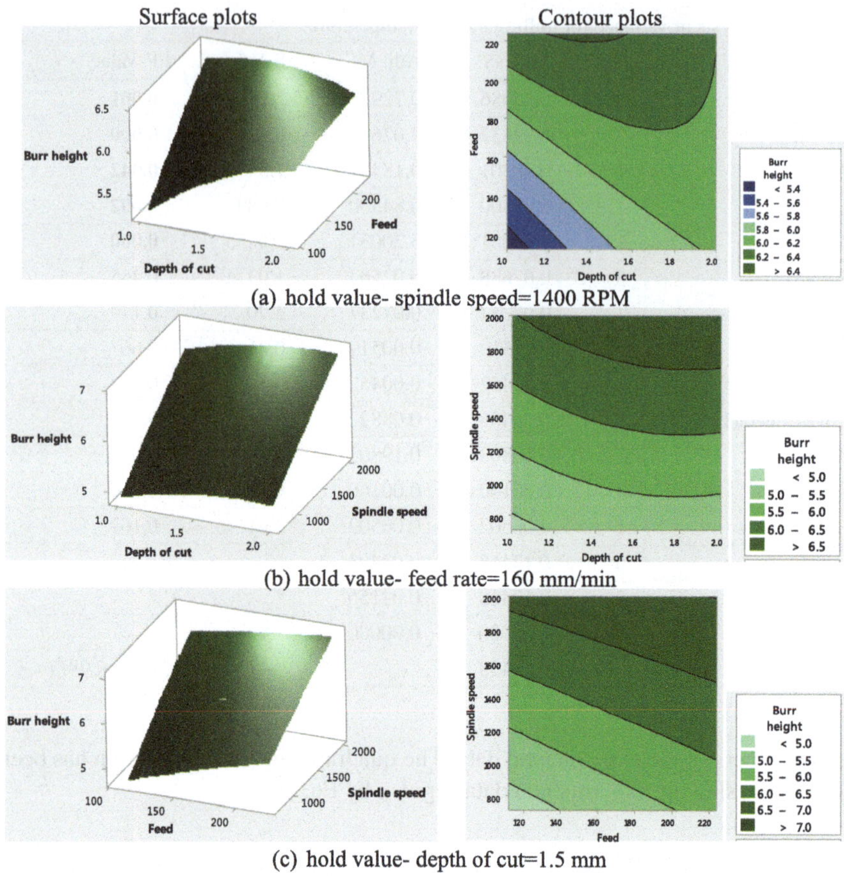

Fig. 15 Response surface plots of burr height with respect to the mutual effects of process parameters **a** feed rate versus depth of cut, **b** spindle speed versus depth of cut, and **c** spindle speed versus feed rate

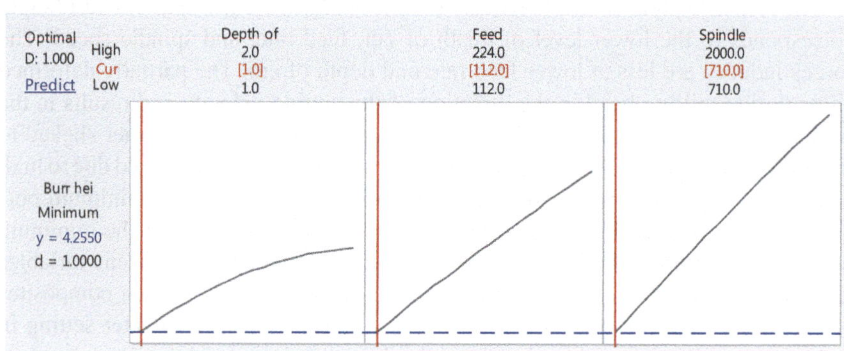

Fig. 16 Optimization of process parameter for obtaining of minimum burr height

5 Conclusions

In this study, green composites have been developed using an injection molding machine. The milling characteristic of the developed green composites has been evaluated experimentally using RSM. The process parameters have been optimized to obtain the lower value of induced forces, temperature, surface roughness, and burr height. The conclusions can be drawn from the experimental investigation are presented below:

1. ANOVA of the experiment data to obtain the induced force reveals that the main effects of the process parameters are significant. The interaction effects and square effects were found to be insignificant. The minimum value of induced force (56.2480 N) was found to be obtained at the lowest value of depth of cut (1 mm), feed rate close to the lowest value (127.8384 mm/min), and highest value of cutting speed (2000 RPM).
2. The main effects of the process parameters and two-way interaction of A × B and B × C were found significant from ANOVA of induced temperature. The minimum induced temperature obtained during milling of green composites was 46.0812 °C at the optimum setting i.e., depth of cut close to the center point (1.5455 mm), lowest value of feed (112 mm/min), and lowest value of cutting speed (710 RPM).
3. The main effect of process parameters, the square effect of C^2, and the interaction effect of B × C were found to be significant from the ANOVA of surface roughness. The minimum surface roughness obtained during milling of green composites was 0.7437 μm. The control parameter setting in accordance with the optimal surface roughness was depth of cut closed to the center point (1.3535 mm), lowest value of feed rate (112 mm/min), and lowest value of cutting speed (710 RPM).
4. The main effects of the process parameters were found to be significant for obtaining burr height. The two-way interaction effects and square effects were found to be insignificant. The minimum burr height obtained during milling of green composites was 4.2550 mm. The control parameter setting in accordance with optimal burr height is the lowest value of depth of cut (1 mm), lowest value of feed rate (112 mm/min), and lowest value of cutting speed (710 RPM).

References

Azmi AI, Lin RJT, Bhattacharyya D (2013) Machinability study of glass fibre-reinforced polymer composites during end milling. The Int J Adv Manuf Technol 64(1–4):247–261

Babu GD, Kasu Y (2012) Determination of delamination of milled natural fiber reinforced composites. Int J Eng Res Technol 1(8):1–6

Babu GD, Babu KS, Gowd BUM (2013) Effect of machining parameters on milled natural fiber-reinforced plastic composites. J Adv Mech Eng 1(1):1–12

Balasubramanian K, Sultan MTH, Cardona F, Rajeswari N (2016) Machining analysis of natural fibre reinforced composites using fuzzy logic. IOP Conf Ser Mater Sci Eng 152(1):012051

Çelik YH, Kilickap E, Kilickap Aİ (2019) An experimental study on milling of natural fiber (jute)-reinforced polymer composites. J Compos Mater 53(22):3127–3137
Chegdani F, Mezghani S, El Mansori M, Mkaddem A (2015) Fiber type effect on tribological behavior when cutting natural fiber reinforced plastics. Wear 332:772–779
Davim JP, Reis P (2005) Damage and dimensional precision on milling carbon fiber-reinforced plastics using design experiments. J Mater Process Technol 160(2):160–167
Davim JP, Reis P, Antonio CC (2004) A study on milling of glass fiber reinforced plastics manufactured by hand-lay up using statistical analysis (ANOVA). Compos Struct 64(3–4):493–500
Erkan Ö, Işık B, Çiçek A, Kara F (2013) Prediction of damage factor in end milling of glass fibre reinforced plastic composites using artificial neural network. Appl Compos Mater 20(4):517–536
Ghafarizadeh S, Lebrun G, Chatelain JF (2016) Experimental investigation of the cutting temperature and surface quality during milling of unidirectional carbon fiber reinforced plastic. J Compos Mater 50(8):1059–1071
Harun AB, Haron CHBC, Ghni JBA, Mokhtar SB, Ting ST (2015) Study the effect of milling parameters on surface roughness during milling kenaf fibre reinforced plastic. Adv Environ Biol 9(13):46–53
Hintze W, Cordes M, Koerkel G (2015) Influence of weave structure on delamination when milling CFRP. J Mater Process Technol 216:199–205
Islam F, Ramkumar J, Milani AS (2015) A simplified damage prediction framework for milling of unidirectional carbon fiber-reinforced plastics. Adv Manuf Polymer Compos Sci 1(4):175–184
Jenarthanan MP, Jeyapaul R (2013) Optimisation of machining parameters on milling of GFRP composites by desirability function analysis using Taguchi method. Int J Eng Sci Technol 5(4):23–36
Jenarthanan MP, Ramesh Kumar S, Jeyapaul R (2016) Modelling of machining force in end milling of GFRP composites using MRA and ANN. Aus J Mech Eng 14(2):104–114
Jia Z, Fu R, Wang F, Qian B, He C (2018) Temperature effects in end milling carbon fiber reinforced polymer composites. Polym Compos 39(2):437–447
Kiliçkap E, Yardimeden A, Çelik YH (2015) Investigation of experimental study of end milling of CFRP composite. Sci Eng Compos Mater 22(1):89–95
Koplev AA, Lystrup A, Vorm T (1983) The cutting process, chips, and cutting forces in machining CFRP. Composites 14(4):371–376
Mutalib MZA, Jamal MJ (2017) Effect of machining parameters on delamination during milled banana fiber reinforced polyester composites. J Eng Sci Res 1(1):9–13
Palanikumar K, Karunamoorthy L, Karthikeyan R (2006) Assessment of factors influencing surface roughness on the machining of glass fiber-reinforced polymer composites. Mater Des 27(10):862–871
Patel P, Chaudhary V, Patel K, Gohil P (2018) Milling of polymer matrix composites: a review. Int J Appl Eng Res 13(10):7455–7465
Puw HY, Hocheng H (1993) Machinability test of carbon fiber-reinforced plastics in milling. Mater Manuf Process 8(6):717–729
Raj PP, Perumal AE, Ramu P (2012) Prediction of surface roughness and delamination in end milling of GFRP using mathematical model and ANN 19:107–120
Rajmohan T, Vinayagamoorthy R, Mohan K (2019) Review on effect machining parameters on performance of natural fibre-reinforced composites (NFRCs). J Thermoplast Compos Mater 32(9):1282–1302
Razfar MR, Zadeh MZ (2009) Optimum damage and surface roughness prediction in end milling glass fibre-reinforced plastics, using neural network and genetic algorithm. Proc Inst Mech Eng Part B J Eng Manuf 223(6):653–664
Rusinek R (2010) Cutting process of composite materials: an experimental study. Int J Non-Linear Mech 45(4):458–462
Sankar BR, Umamaheswarrao P, Srinivasulu V, Chowdari GK (2015) Optimization of milling process on jute polyester composite using Taguchi based grey relational analysis coupled with principle component analysis. Mater Today Proc 2(4–5):2522–2531

Sorrentino L, Turchetta S (2014) Cutting forces in milling of carbon fibre reinforced plastics. Int J Manuf Eng 8, Article ID 439634

Sreenivasulu R (2013) Optimization of surface roughness and delamination damage of GFRP composite material in end milling using Taguchi design method and artificial neural network. Proc Eng 64:785–794

Wang H, Sun J, Li J, Lu L, Li N (2016a) Evaluation of cutting force and cutting temperature in milling carbon fiber-reinforced polymer composites. Int J Adv Manuf Technol 82(9–12):1517–1525

Wang H, Sun J, Zhang D, Guo K, Li J (2016b) The effect of cutting temperature in milling of carbon fiber reinforced polymer composites. Compos Part A Appl Sci Manuf 91:380–387

Energy Conversion and Partition in Machining Fiber Reinforced Polymer (FRP) Composites

Jamal Sheikh-Ahmad, Fahad Almaskari, and Farrukh Hafeez

Abstract Cutting temperatures during machining are of a particular importance due to its controlling influence on the quality of machined part as well as tool life. Limited understanding and background on this topic for fiber reinforced polymer composites (FRPs) may perhaps be credited to limitless variations of such materials due to layup, fiber orientation, constituent content etc., which are additional to the usual challenges encountered in heat partition evaluation in machining. Direct measurement of the temperatures in this regard is not always possible due to the highly localized temperatures and the unique tribological contact between the tool and the workpiece. Thus, inverse conduction methods are very appealing for studying the heat transfer problem in a wide variety of machining operations. This chapter presents a complete methodology to evaluate the amount of heat dissipated through workpiece, tool and chips during machining of FRPs using an iterative inverse heat conduction technique. Temperatures are measured experimentally and used as boundary conditions. The transient heat conduction problem in the FRP laminate was simulated using the finite element method and the amount of heat flux conducted through the machined surface, cutting tool and chips was determined. Heat partition in two most commonly known FRPs namely CFRP and GFRP subjected to edge trimming operation are presented. It was observed that smaller fraction of the total heat energy is evacuated through the work piece as compared to the tool and chips. CFRP chips carry away more heat as compared to GFRPs. Moreover, cutter type has insignificant effect on the amount of heat dissipated to the work piece.

Keywords Edge trimming · FRPs · Heat partition · Temperature distribution

J. Sheikh-Ahmad (✉)
Department of Mechanical Engineering, Khalifa University of Science and Technology, SAN Campus, Abu Dhabi, United Arab Emirates
e-mail: jamal.ahmad@ku.ac.ae

F. Almaskari
Department of Aerospace Engineering, Khalifa University of Science and Technology, Main Campus, Abu Dhabi, United Arab Emirates

F. Hafeez
DIAC, University of Birmingham Dubai, Block 2, 341799 Dubai, United Arab Emirates

M. T. Hameed Sultan et al. (eds.), *Machining and Machinability of Fiber Reinforced Polymer Composites*, Composites Science and Technology,
https://doi.org/10.1007/978-981-33-4153-1_7

1 Introduction

The heat generated during traditional cutting processes represents a major hindrance to the increase in production rates and the improvement of machined component quality as the cutting temperatures generally increase with the increase in material removal rates. It is widely accepted that the mechanical energy consumed in cutting is converted into heat, which is then dissipated to the workpiece, the cutting tool, the chips and the environment (Shaw 2005). Heat conducted to the machined component causes its temperature to rise. Because fiber reinforced epoxies generally have low temperature tolerances, exposure to high temperatures for long durations might cause irreversible thermodynamic change to the epoxy matrix, which in turn might lead to adverse effects on the mechanical properties of the machined part through the introduction of thermal damage (El-Hofy et al. 2011; Kerrigan and O'Donnell 2016; Delahaigue et al. 2017; Nguten-Dinh et al. 2019). Heat is also a driving factor for various tool wear mechanisms (János and János 2011; Inoue and Hagino 2013; Khairusshima et al. 2013; Halim et al. 2017). Furthermore, heated chips easily adhere to the chip evacuating flutes on the cutter, which prevents proper chip disposal. This leads to more increases in friction and temperatures and further aggravates an already complex situation.

The temperature distribution in the machined component and the cutting tool can be determined by solving classical heat conduction equations for a solid body subjected to known heat flux, initial conditions and boundary conditions. However, the exact amount of heat fluxes conducted to the workpiece and cutting tool are often unknown and cannot be measured by direct methods. Therefore, the solution of the direct heat transfer problem is not possible, and some other means must be followed. Alternatively, many researchers have developed various techniques for determining the temperature distribution, or temperatures at specific points of interest, experimentally. Obviously, the most serious challenge in these attempts is the limited accessibility to areas where the temperature gradients are the highest. Nevertheless, the findings in these studies helped in establishing relationships between process parameters and cutting temperatures, and provided reasonable estimates of the maximum temperatures which might occur during cutting. Among the most widely used techniques for temperature measurement are thermoelectric (thermocouples) and radiation (infrared) methods, but a few attempts have also utilized optical methods (fiber optics). Because the machined component is often stationary, thermocouple wires can be embedded or mounted on the surface of the workpiece close to the cutting zone. Inoue and Hagino (2013), Yashiro et al. (2013), Jia et al. (2018) and An et al. (2018) utilized embedded thermocouples to measure the machined surface temperatures in milling and orthogonal cutting, respectively. To be able to get the measurement very close to the machined surface, the cutting tool was eventually allowed to cut the thermocouple wires. On the other hand, the temperature of the rotating tool was measured by the aid of wireless transmission of the thermocouple signal attached to the tool as demonstrated by El-Hofy et al. (2011), Kerrigan and O'Donnell (2013) and Ghafarizadeh et al. (2016). To avoid the complications of the rotating tool situation, other researchers made the cutting tool stationary and kept the workpiece in motion in order to measure the cutting edge temperature in drilling (Weinert and Kempmann 2004; Sorrentino et al. 2017). Yashiro et al. (2013) utilized the tool-workpiece thermocouple method to measure the cutting temperature in edge milling of carbon fiber reinforced polymer composite (CFRP) by measuring the potential difference between the cutter and the CFRP laminate. Wang et al. (2016) embedded a constantan foil in the CFRP laminate layup and measured the potential difference between the constantan foil and the CFRP. However, thermocouple methods have their known limitations, such as their slow response and rather large measurement tolerances. Thus, information about the dynamic change in cutting temperatures as in the case of interrupted cutting might not be obtained by this method. There are also doubts about the effect of the presence of the thermocouple wire and hole in which it is mounted on the temperature field. Kryzhanivskyy et al. (2018) compared numerically the temperature field in a solid tool to that of a tool with thermocouple holes having a recast layer and filled with thermal paste and showed that a slight error of 2 °C at most was introduced only for the holes closest to the heat source. Nevertheless, extra care must be taken in selecting the location of the drilled holes and in keeping a continuous medium for heat transfer by filling the holes with thermal conductive paste. The tool-workpiece thermocouple method also has its own implementation problems. The assumption is usually made that the measured temperature by this method represents the average temperature at the tool-workpiece interface. Obviously, this assumption would be valid only if the temperature distribution is uniform.

Several studies have demonstrated the use of infrared thermography to measure the cutting temperatures on the stationary workpiece and/or rotating cutting tool. Successful implementation of this technology requires careful estimation of the emissivity of the heat radiating surfaces, and different techniques have been developed to attain this property (Vollmer and Möllmann 2010). Haddad et al. (2014) and Nguyen-Dinh et al. (2019) utilized an infrared camera to measure the maximum temperature on the machined surface in edge trimming of CFRP. Merino-Perez et al. (2015), Giasin and Ayvar-Soberanis (2016) and Fu et al. (2018) utilized IR thermography to measure the temperatures near the edge of the exit hole in drilling CFRP. Sheikh-Ahmad et al. (2019a) measured the temperature of the milling cutter using an infrared camera in edge trimming operation. In addition to radiation techniques, some studies demonstrated the utilization of fiber optic arrays for temperature measurement. Cong et al. (2012) used fiber optic sensor and thermocouple wire to measure the temperatures near the machined hole surface in rotary ultrasonic machining of CFRP. Zitoune et al. (2017) embedded an optical fiber with Bragg Gratings inside a stationary drill to measure the temperatures near the cutting zone. Zhu et al. (2019) measured the two-dimensional temperature profile around the drilled hole using distributed fiber sensing.

A few attempts were also made to solve the heat conduction problem in machining FRPs by analytical and numerical methods. Assumptions were often made regarding the source of heat (e.g. friction), its magnitude and partition, and the nature of tool-workpiece interfaces. These assumptions must be validated by experiment, which is often proven to be very difficult. Furthermore, due to the complexity of practical machining processes, most of the analytical and numerical solutions dealt with simple cutting geometries like orthogonal machining. Santiuste et al. (2014) considered the orthogonal machining of CFRP by a three-dimensional finite element analysis. The coefficient of friction at the tool-chip interface was assumed to be in the range from 0.4 to 0.8 and the heat partitions between the two regions were assumed to be 0.2 and 0.8, respectively. Mkaddem et al. (2017) developed a finite element model to study the heat conduction in a glass fiber reinforced polymer composite (GFRP) laminate during edge trimming by assuming the cutter to act as a moving heat flux of Gaussian distribution. It was not clear in this work how the magnitude of the moving heat flux was determined. Wang et al. (2018) analyzed the heat partition in orthogonal cutting of unidirectional CFRP using the finite element method. Analytical models were used to estimate the contact area between the tool and the workpiece and the heat partition between the two was determined as a function of this contact area. However, the predicted heat partition ratios to the workpiece were shown to be unreasonably high in comparison to what was reported in the literature. Xu and Zhang (2019) studied the heat effects in vibration assisted machining of FRPs using the finite element method. The heat partition ratios were determined by analytical methods and the friction coefficient was assumed to be 0.25.

It can be concluded from the above studies that determination of the heat flux and its partition in machining is a key step in solving the heat transfer problem more accurately and without the need for making assumptions that are difficult to validate. Inverse heat conduction techniques provide a powerful tool for tackling such problems (Colaço et al. 2006). In an inverse heat conduction problem (IHCP), the cause (i.e. heat flux) is estimated by measuring its effect (i.e. temperature) at one or more locations inside the problem boundaries. The solution procedure is usually a two-step process. First, an initial guess is made for the amount of heat flux, the direct heat conduction problem is solved and the temperatures at the measurement points are calculated. Then, a comparison is made between the measured and calculated temperatures. Based on this comparison, another guess for the heat flux is made and the process is repeated until the difference between measured and calculated temperatures is minimized. As such, IHCP is an optimization problem for which different minimization techniques have been developed (Colaço et al. 2006). Furthermore, an important characteristic of an IHCP is it's ill-posedness, and therefore an exact solution does not exit. Several approximation methods are used to search for a solution, such as iterative procedures, regularization, evolutionary and stochastic techniques. Kerrigan and O'Donnell (2013) and Hintze and Klingelhöller (2017) utilized the IHC method to determine the heat partition to the cutter in edge trimming and abrasive cutoff machining, respectively. The direct heat transfer problem in these works was solved numerically. Liu et al. (2014) utilized analytical solution and IHC for determining the heat partition to the workpiece in helical milling of CFRP. Sheikh-Ahmad et al. (2019b) modelled the trimming problem numerically by assuming a moving heat source. The magnitude of the heat source was determined by an iterative IHC technique.

This chapter addresses the heat partition problem in edge trimming of carbon and glass reinforced epoxy composites. The direct transient heat conduction problems in the workpiece and the cutting tool were modeled separately using the finite element method. A uniform moving heat source was applied to the workpiece in order to represent the energy source from machining. A stationary heat source was applied to the cutting tool. Boundary temperatures on the workpiece surface were measured with embedded thermocouples. The average and maximum temperatures of the cutting tool were measured by an infrared camera. Finally, the inverse heat transfer problem for each was solved separately by an iterative procedure. The effects of cutting tool type, workpiece material and cutting conditions on the heat partition ratios and cutting temperatures are discussed.

2 Energy Conversion in Machining

It is widely accepted that most of the mechanical energy spent in machining is converted into heat, which is then partitioned by conduction, convection and radiation into the different regions of the cutting zone and the environment. The determination of the exact partitions of this heat has been the subject of many researches as discussed in the introduction section. However, most of these works did not elaborate on the nature of the original mechanical energy and its conversion from electrical energy, which is the initial source of energy for almost all production machining centers. This section discusses the conversion of electrical energy to mechanical energy in machining.

Figure 1 shows a schematic of edge trimming with a straight-flute cutter with only a single tooth engaged in cutting. This 2-dimensional force and displacement system is chosen for simplicity, and because it represents many of the edge trimming operations with polycrystalline diamond (PCD) cutters as shown in Fig. 2d. The instantaneous cutting forces F_t and F_r in the tangential and radial directions, respectively, are exerted by the cutting edge on the workpiece and eventually lead to material fracture

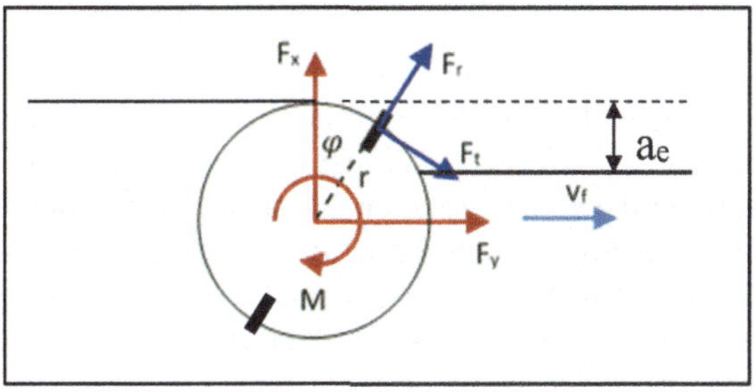

Fig. 1 Schematic of milling with a single tooth showing tangential and radial cutting forces and the equivalent couple-force system

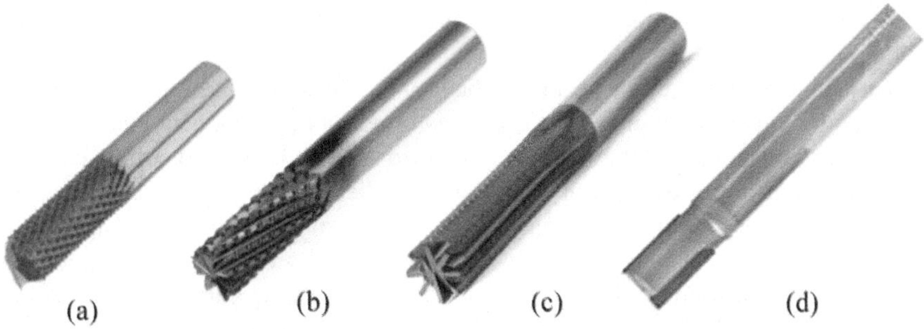

Fig. 2 Various composite cutting tools used for power measurement in edge trimming. **a** TiAlN coated solid carbide burr, **b** diamond coated solid carbide segmented-flute, **c** diamond solid carbide coated segmented straight-flute, **d** two straight-flute PCD

and the formation and removal of the chips. These forces can be transformed into an equivalent force-couple system at the center of rotation as shown, where the couple M is given by

$$M = F_t r \tag{1}$$

where r is the cutter radius. The instantaneous forces in the normal, F_x and feed, F_y directions are given in terms of the radial and tangential forces by the transformation equation,

$$\begin{Bmatrix} F_x \\ F_y \end{Bmatrix} = \begin{bmatrix} \cos\varphi & \sin\varphi \\ -\sin\varphi & \cos\varphi \end{bmatrix} \begin{Bmatrix} F_t \\ F_r \end{Bmatrix} \tag{2}$$

where φ is the rotation angle in degrees, measured from the vertical (cutting edge entry) as shown in Fig. 1.

The instantaneous mechanical power consumed by cutting is then calculated by the expression,

$$P_m = M \cdot \omega + F_y \cdot v_f = \frac{2\pi NM}{60} + F_y \cdot v_f \tag{3}$$

where ω is the spindle speed in rad/s, N is the spindle speed in rev/min and v_f is the feed speed in m/s. The first term in Eq. (3) represents the power consumed by resisting the spindle rotation and the second term represents the power consumed by the feed motion. In most cases, the feed speed is very small when compared to the rotational speed and the second term can be neglected. Thus, the instantaneous mechanical power can be represented in terms of the measured feed and normal forces as,

$$P_m = F_t r \cdot \omega = v_c \cdot \left(F_x \cos\varphi - F_y \sin\varphi\right) \tag{4}$$

where $v_c = 2\pi r N/60$ is the cutting speed in m/s.

Alternatively, if the specific cutting energies (or cutting force coefficients) of the work material are known, the instantaneous machining power can be calculated using the mechanistic modelling approach (He et al. 2020) by the expression,

$$P_m = F_t \cdot v_c = (k_t f_z \sin\varphi + k_{te}) ds \cdot v_c \tag{5}$$

where k_t is the cutting force coefficient in the tangential direction (N/mm^2), k_{te} is the edge force coefficient (N/mm), $f_z sin\varphi$ is the instantaneous uncut chip thickness, $f_z = v_f/NZ$ is the feed per tooth, Z is the number of teeth on the cutting circle and $ds = a_p$ is the axial depth of cut.

The study of heat partition in machining starts with a known magnitude of the mechanical power, which is then converted into heat. This power can be calculated directly from torque measurements as shown by Eq. (3), cutting force measurements as shown by Eq. (4) or known cutting force coefficients and the chip geometry as shown by Eq. (5). For simple cutting geometries where a single tooth is engaged in cutting at any given time, the force signal is a periodic function of time, which can be measured with sufficient accuracy by sampling at a high-enough frequency. To examine the relationship between mechanical and electric power, a number of edge trimming experiments were conducted using various composite cutting tools as shown in Fig. 2. Two quasi-isotropic composite laminates were cut, namely CFRP and GFRP. Table 1 shows the range of cutting parameters used for the different cutting tools. Cutting forces were measured using a Kistler Type 9257B dynamometer. The electric power drawn by the spindle motor was measured by a Load Controls UPC-FR fast response universal power cell.

Figure 3a shows typical cutting force signals obtained by edge trimming quasi-isotropic CFRP laminate with a two-flute PCD cutter. Figure 3b shows a zoom-in segment of the force signals close enough to reveal two force cycles. These force cycles could be approximated by half sine-waves as suggested by Eq. (5). However, considerable distortion to the signals had occurred due to the harmonic vibrations of the machine and fixturing system. Therefore, the implementation of Eq. (4) to calculate machining power using such data is problematic due to wave distortion. The implementation becomes even more complicated when multiple edges are engaged at the same time in cutting and for complex cutter geometries, such as burr tools. In

Table 1 Cutting conditions for power measurement study

Cutter type	N (RPM)	v_f (mm/min)	a_e (mm)	a_p (mm)
Burr (a)	4000, 8000	800, 1000	10	10
Segmented-flute (b, c)	4000, 8000	800, 1000, 1400	10	10
PCD (d)	4000, 8000	400, 800	10	10

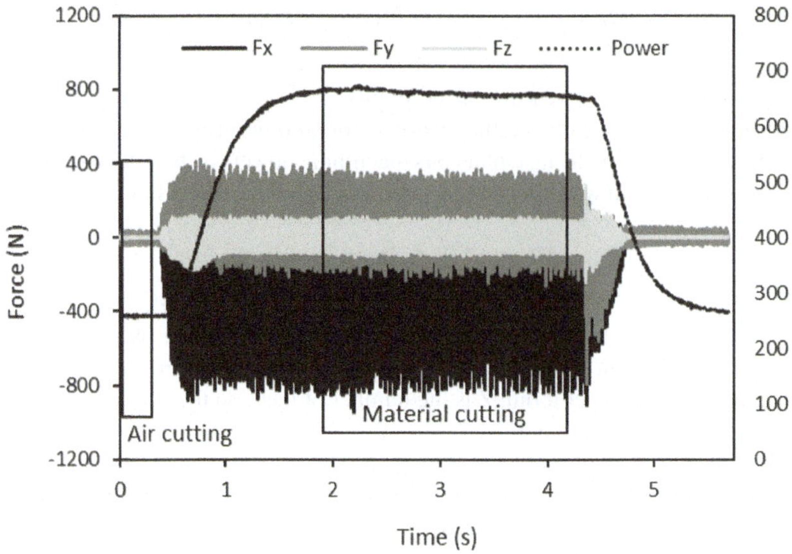

(a)

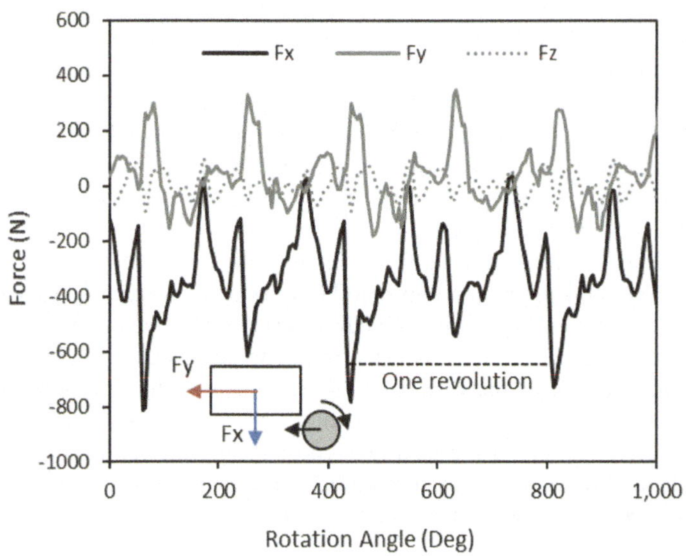

(b)

Fig. 3 Typical cutting force and spindle electric power measurements in edge trimming quasi-isotropic CFRP laminate (**a**), close-up view of the force signals in material cutting region (**b**). Cutter = 10 mm 2FL-PCD, N = 4000 rpm, v_f = 800 mm/min. The arrows on the annotation box indicate dynamometer coordinates

Table 2 Correlation of electric power and functions of cutting force and cutting speed

Function	K	R^2
$F_x \cdot v_c$ (W)	0.6778	0.9548
$R \cdot v_c$ (W)	0.4958	0.9279
$F_{y\_RMS} \cdot v_c$ (W)	0.6450	0.8437
$R_{RMS} \cdot v_c$ (W)	0.4207	0.8251
$F_y \cdot v_c$ (W)	0.7751	0.7450

these cases, it is not possible to determine exactly the shape of the force cycle or the engagement angles for the different teeth of the cutter. Therefore, it was desirable to be able to estimate the machining power from the measured cutting forces without resorting to Eq. (4).

Linear regression was then performed between the net electric cutting power and various functions of the cutting forces and feed speed. The results of these correlations are listed in Table 2, where R^2 represents the correlation coefficient and k is the slope of the regression line. The cutting forces were averaged over a window of time when the tool was engaged in cutting and the process was steady, as shown in Fig. 3a. The resultant cutting force R was evaluated as,

$$R = \sqrt{F_x^2 + F_y^2 + F_y^2} \tag{6}$$

and the feed force RMS was evaluated as,

$$F_{y-RMS} = \sqrt{\frac{\sum_{i=1}^{n} F_y^2}{n}} \tag{7}$$

The net cutting power was calculated as the difference between the average materials cutting power and the average air cutting power. It can be seen that the best correlation was obtained with the products of the average normal force and the average resultant force with the cutting speed. The net electric cutting power could be accurately expressed as a linear function of the product of the average normal force and the cutting speed,

$$P_e = kF_x v_c \tag{8}$$

Figure 4 shows the linear correlation between the product of the average normal force and cutting speed and the net spindle electric power for edge trimming tests conducted at different spindle and feed speeds, using the different cutting tools shown in Fig. 2. The linear correlation appeared to be independent of tool type, work materials and cutting parameters. However, the value of the constant k might be different for different levels of tool wear. This assumption might be further explored in future studies. For the work materials used, the constant k was found to be 0.6778. Thus,

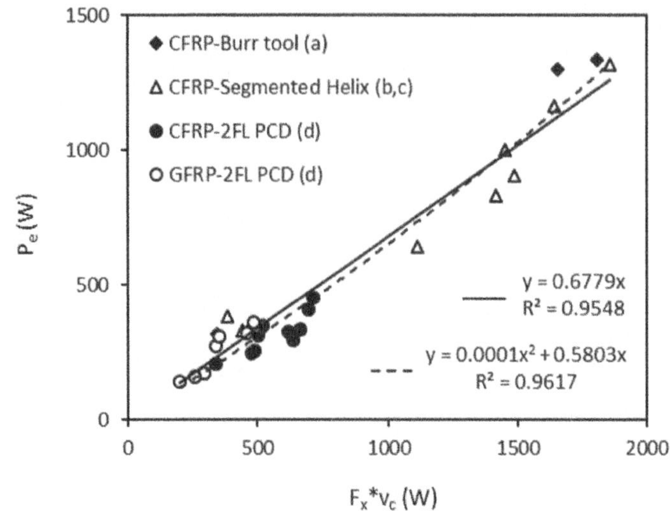

Fig. 4 Relationship between net spindle electric power and the product of average normal force and cutting speed for different tool geometries and FRP materials

Eq. (8) provides a simple empirical relationship which allows the estimation of net electric cutting power from the measured force normal to the feed direction without the need to measure the spindle electric power. A slight improvement in predicting the electric power was obtained when a second order polynomial was used for regression as shown in the figure. Finally, the mechanical machining power can be determined using the relationship $P_m = \varepsilon P_e$, where ε is the spindle efficiency.

3 Energy Partition

Most of the machining power, P_m is eventually converted into heat. The exact efficiency of conversion in machining fiber reinforced polymers is not known. However, in metal cutting literature it is accepted that the efficiency of conversion is greater than 97% (Shaw 2005). The generated heat is then dissipated into four regions as shown in Fig. 5, namely the workpiece ($\dot{Q}_w$), cutter ($\dot{Q}_t$), chips ($\dot{Q}_c$) and the environment ($\dot{Q}_e$). The energy partition equation in machining is expressed as,

$$P_m = \delta\dot{Q} = \dot{Q}_w + \dot{Q}_t + \dot{Q}_c + \dot{Q}_e \tag{9}$$

where δ represents the efficiency of conversion, which is assumed to be 95%. The heat flux $\dot{Q}_w$ is applied to the workpiece over the entire contact area. The distribution function of this flux is not known and most studies have assumed it to be uniform. Kim et al. (2006) proposed different distributions in creep feed grinding of AISI1020,

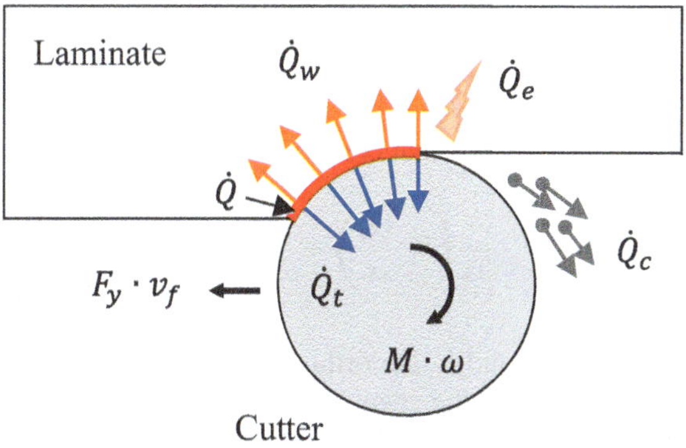

Fig. 5 Heat partition in edge trimming operation (Sheikh-Ahmad et al. 2020)

including uniform and triangular and showed that the temperature distribution is not affected greatly by the shape of the heat flux. Similar findings were also reported by Kryzhanivskyy et al. (2018) in orthogonal cutting of various metals. Sölter and Gulpak (2012), on the other hand, assumed a piece-wise distribution which assumes the shape of the uncut chip in dry milling of steel. Mkaddem et al. (2017) assumed a Gaussian distribution of the heat flux in edge trimming of GFRP. In the present work, the heat flux will be assumed to have a uniform distribution and moving along the machined edge with a constant speed v_f.

4 Estimating the Various Heat Fluxes

The determination of the various heat fluxes in Eq. (9) by direct measurement is not possible. Instead, inverse methods are commonly used. In an inverse method, the effect of heat flux on the system is measured. This effect is typically boundary temperatures at accessible desired locations in the problem domain. The direct heat transfer problem is then solved with an assumed heat flux and the temperatures at the desired locations are determined and compared with the measured ones. An iterative procedure is implemented by adjusting the heat flux until the difference between measured and calculated temperatures is minimum (Sheikh-Ahmad et al. 2019b). In this work, the inverse heat conduction method was applied separately for the workpiece and the cutter and the corresponding heat fluxes were estimated. This separate implementation was useful because it avoided making assumptions about the interactions between the cutter and the workpiece, such as the nature of contact area and the sliding and friction conditions. The direct heat conduction problem for each part of the cutting system was solved by the finite element method as described below.

4.1 Direct Problem for the Workpiece

The heat diffusion problem in the workpiece is 3-D nonlinear and transient as represented by the heat conduction equation:

$$\frac{\partial}{\partial x}\left(k_x\frac{\partial T}{\partial x}\right)+\frac{\partial}{\partial y}\left(k_y\frac{\partial T}{\partial y}\right)+\frac{\partial}{\partial z}\left(k_z\frac{\partial T}{\partial z}\right)=\rho c\frac{\partial T}{\partial t} \tag{10}$$

where k_x, k_y and k_z are orthotropic thermal conductivities of the laminate, ρ is its density and c is the specific heat. Figure 6 shows the heat transfer model used to simulate edge milling of the FRP laminate. The boundary conditions applied to the laminate were as follows:

1. Heat was lost by convection on the top, bottom and machined surfaces, using h_1 *and* h_2 on the respected surfaces as indicated on the figure, where h is the heat transfer coefficient and η is normal to the surface,

$$-k_\eta\frac{\partial T}{\partial \eta}=h(T-T_\infty) \tag{11}$$

2. A moving uniform heat source of magnitude $\dot{q}_w$ represented the heat flux conducted into the workpiece through the machined surface,

$$-k_x\frac{\partial T}{\partial x}(x,y,z,t)=\dot{q}_w \tag{12}$$

3. The remaining surfaces were assumed to be insulated, $\frac{\partial T}{\partial \eta}=0$.

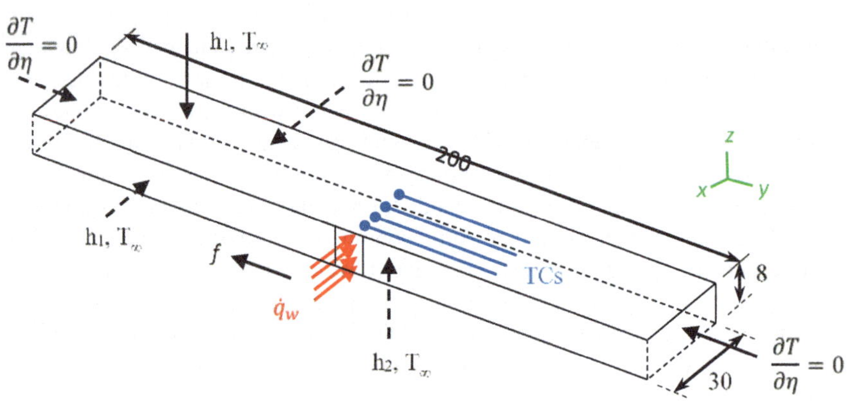

Fig. 6 Model used for simulation of heat conduction in the workpiece. TCs represent the location of thermocouples used for boundary temperature measurement

Table 3 Material properties of CFRP (Joven et al. 2012)

Property	Units	T = 297 °K	T = 473 °K
ρ	g/cm^3	1.550	1.550
k_x, k_y	W/m- °K	2.40	3.00
k_z	W/m °K	0.60	0.66
c	J/g °K	0.820	1.320

Table 4 Thermal properties of the GFRP (Kalogiannakis et al. 2004)

Property	Units	T = 300 °K	T = 423 °K
ρ	g/cm^3	1.570	1.570
k_x, k_y	W/m °K	0.89	0.89
k_z	W/m °K	0.39	0.39
c	J/g °K	0.8882	1.301

4. The problem was subjected to the initial condition $T(x, y, z, t) = T_o$ at t = 0, where t is time.

The heat transfer coefficient on the machined surface and the top surface were set higher than the bottom surface ($h_1 > h_2$) because these two surfaces were subjected to more air movement due to the suction from the dust collector and the rotating tool. The heat transfer coefficients used in the current simulation were $h_1 = 100$ W/°K m^2 and $h_2 = 50$ W/°K m^2. It is worth noting here that during model development it was found that the value of h did not greatly affect the overall solution of the problem. This observation was also reported by Kim et al. (2006). However, it was shown by Kryshanivskyy et al. (2018) that effect of h on the temperature distribution was significant in the simulation of orthogonal cutting of various metals. Thermal properties of the workpiece were assumed to be directional and temperature dependent. Their values are shown in Tables 3 and 4.

The element type used for the workpiece was DC3D8 (thermal analysis), a 3D 8-noded linear heat transfer brick type element, and 60,000 elements and 68,541 nodes were used. Fine mesh with element size 1 × 1 × 0.25 mm^3 was used in the steep thermal gradient region and a coarse mesh of size 1 × 1 × 2.5 mm^3 was used for the remaining regions. Convergence of the numerical solution was determined by adjusting the time increment implemented in Abaqus solver. It was determined that a time increment of 0.01 s guaranteed a converging solution with adequate accuracy (Sheikh-Ahmad et al. 2019b). DFLUX user subroutine was utilized to introduce the moving heat flux in the numerical simulation. The moving heat flux was assumed to be uniformly distributed over the projected area of contact between the cutter and the workpiece, which was 5 mm wide and 10 mm high. During model development, linear heat flux distribution was also tried and found to have negligible effect on the results.

4.2 Inverse Problem for the Workpiece

The magnitude of heat flux $\dot{q}_w$ was determined by minimizing the objective functions $f(\dot{q}_w)$ over time and space domains:

$$f(\dot{q}_w) = \sum_{i=1}^{t}\sum_{j=1}^{m}\left(Y_{ij}(x, y, z, t) - T_{ij}(x, y, z, t, \dot{q}_w)\right)^2 \tag{13}$$

where Y_{ij} is the measured temperature history at location *j* and T_{ij} was simulation temperature history for the same location, m was the number of measurement locations and t was the number of time increments. Under general conditions, the optimization problem in Eq. (13) is ill-posed and the solution might oscillate erratically due to fluctuations in the measured temperatures (Colaço et al. 2006). Regularization techniques are often implemented to keep the solution within acceptable physical limits. However, the heat flux in machining typically does not oscillate with time and can be assumed constant during steady state cutting (Kerrigan and O'Donnel 2013; García et al. 2014). This makes the optimization problem one-dimensional, well-posed and a solution could be found by iteration. Figure 7 shows a flow chart of the iterative procedure used to minimize the objective function in Eq. (13). The trimming experiment was conducted at the required set of cutting parameters and the temperatures were measured at specific locations on the surface of the workpiece. The direct heat conduction problem in Eq. (10) was then solved for an arbitrary moving heat flux, $\dot{q}_w$ and the temperature histories for the measurement locations were determined. A comparison was then made between the measured and simulated temperature histories by evaluating the objective function. The procedure was repeated for a new heat flux $\dot{q}_w + \Delta\dot{q}_w$ and the objective function was calculated. The procedure was stopped when $\partial f(\dot{q}_w)/\partial\dot{q}_w = 0$. The iterative procedure explained

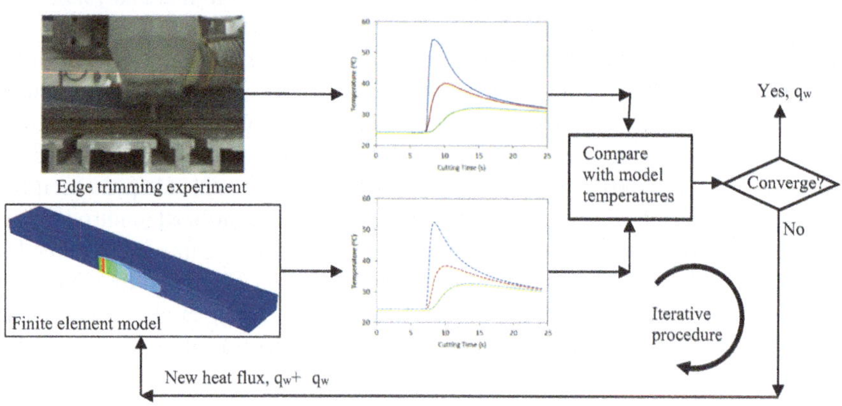

Fig. 7 Iterative procedure for determining the heat flux into the workpiece using inverse method

above was implemented in Matlab, where the initial guess for the heat flux was taken as $\dot{q}_w = 0.5P_m/A_w$, assuming 50% of the total machining power was conducted to the workpiece, A_w was the projected contact area with the workpiece. The heat flux was decremented by an appropriate decrement $\Delta\dot{q}_w$ determined by trial and error.

Figure 8 shows the evolution of the derivative of the objective function $f(\dot{q}_w)$ with iterations. It could be seen that for the cutting conditions of this particular experiment the objective function was minimum at a heat flux of 450 mW/mm^2. Figure 9a shows a comparison of the measured and simulated temperature histories for the estimated heat flux and Fig. 9b shows the difference between the simulated

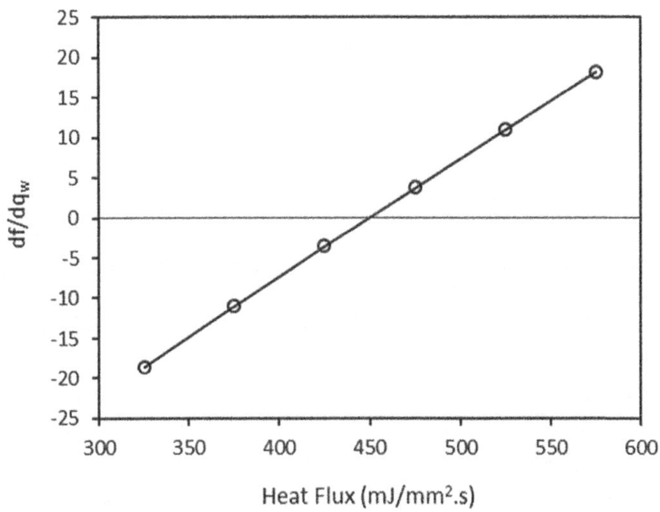

Fig. 8 Evolution of the derivative of the objective function with iterations when edge trimming at 4000 rpm and 800 mm/min

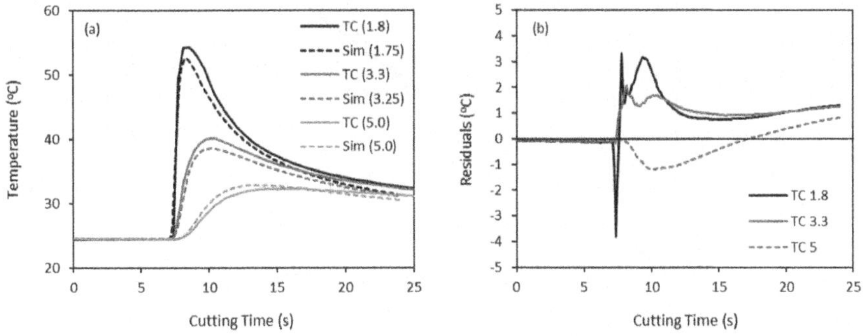

Fig. 9 Comparison between measured and simulated temperature histories at different measurement locations (in mm) from the machined edge for $\dot{q}_w = 450\,\text{mW/mm}^2$ when edge trimming at 4000 rpm and 800 mm/min

and measured temperatures (residuals). It could be seen that difference between the measured and simulated temperatures (residuals) depends on measurement location where the largest residuals occurred for the closest measurement locations.

4.3 Sensitivity Analysis

The inverse problems are difficult because they usually are extremely sensitive to temperature measurement errors. The sensitivity problem was obtained by differentiating the objective function,

$$\frac{\partial f}{\partial \dot{q}_w} = 2\sum_{i=1}^{t}\sum_{j=1}^{m}(Y_{ij} - T_{ij})\frac{\partial T_{ij}}{\partial \dot{q}_w} \tag{14}$$

where $\frac{\partial T_{ij}}{\partial \dot{q}_w}$ is the sensitivity coefficient, which describes the temperature variation ΔT caused by a small perturbation in the heat flux $\Delta \dot{q}_w$. The sensitivity of the solution can be examined by solving the direct problem in Eq. (10) for small increments of the heat flux from the optimum solution. Figure 10 shows the variation of the heat flux due to small variations in the measured temperatures for the case shown in Figs. 8 and 9. It can be seen that small errors in the measured temperatures can lead to small variations in the heat flux. Furthermore, the sensitivity of the solution depends greatly on measurement location. For example, $\pm 1\%$ error in temperature measurement can

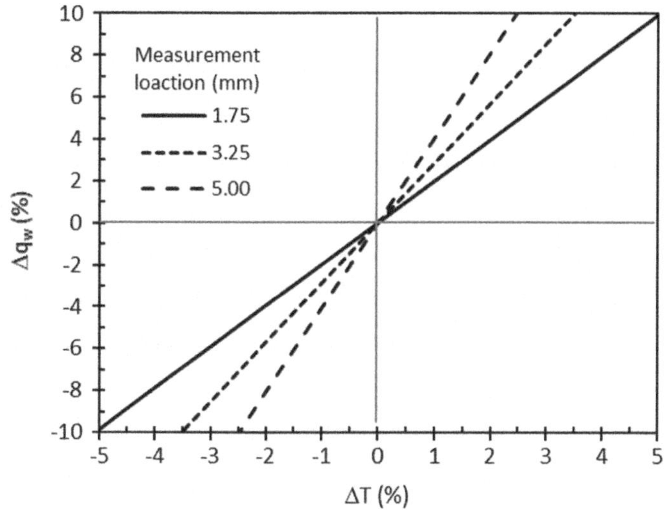

Fig. 10 Sensitivity of the inverse solution to errors in temperature measurement for $\dot{q}_w = 450\,\text{mW/mm}^2$

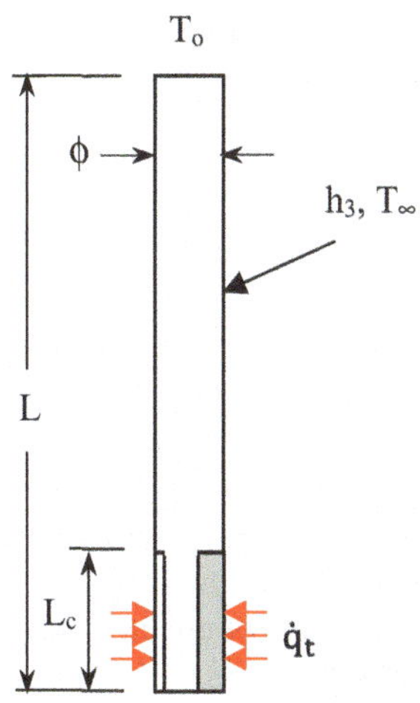

Fig. 11 Numerical model for the PCD cutter

lead to ±2% error in the heat flux for the closest measurement location and 4% error for the farthest location.

Temperature measurement errors also arise from uncertainties in thermocouple location measurement because thermocouples are often placed in high thermal gradient regions. This uncertainty problem was investigated numerically by García et al. (2014) in grinding process for the case of time dependent heat flux. It was shown that ±0.4 mm uncertainty in sensor location might lead to as much as ±8% uncertainty in the heat flux at the start of the process. The magnitudes of the heat flux and uncertainty were shown to stabilize to lower levels as the cutting time was increased.

4.4 Direct Problem for the Cutting Tool

The heat diffusion problem in the cutter was also described by a transient nonlinear expression for each type of tool separately, a model for only one type of tool (PCD) is used for explanation in this section. Same procedure was adopted for each type of the tool. The general expression of the direct problem is of the following form,

$$\nabla \cdot k\nabla T = \rho c \frac{\partial T}{\partial t} \tag{15}$$

Table 5 Material properties of PCD cutter

Property	Units	PCD	WC–Co
ρ	g/cm^3	3.500	15.000
k	W/m °K	2000	100
c	J/g °k	0.518	0.130

subject to the boundary conditions as shown in Fig. 11 and having the initial condition: $T(r, \theta, z, t) = T_o$, at t = 0. The numerical model for the PCD cutter consisted of a cylindrical body 100 mm in length, 10 mm in diameter and with length of blade L_c = 15 mm. Convection heat loss with heat transfer coefficient h_3 = 1000 W/°k m^2 was assumed on all exposed surfaces of the cutter due to its very high speed of rotation. The part of heat from machining, which was conducted to the cutter, $\dot{q}_t$, was applied only at the points of contact between the cutter and the workpiece over a length a_e, equal to the thickness of the laminate. Since the cutter rotated at very high speed, the frequency of application of the heat flux on each of the flutes was extremely high and it was reasonable to assume that the heat source in this case as stationary. Thermal properties at room temperature of both PCD and tungsten carbide used in the cutter model are shown in Table 5. Higher temperature properties were not available.

The total number of elements and nodes used for the numerical model of the cutter were 34,565 and 10,189, respectively. Two element types were used by default, DC3D8 in the cutting region and DC3D4 in the tool shank region. DC3D4 is a 4-node linear heat transfer tetrahedron element.

4.5 Inverse Problem for the Cutting Tool

The magnitude of heat flux $\dot{q}_t$ was determined by minimizing the objective function in Eq. (16):

$$g(\dot{q}_t) = \min\left|\overline{T}_{exp} - \overline{T}_{sim}(\dot{q}_t)\right| \tag{16}$$

where $\overline{T}_{exp}$ is the average steady state cutter temperature determined by IR thermography and $\overline{T}_{sim}$ is the simulated average cutter temperature. The average temperature in each case was evaluated over a 10 × 10 mm^2 area corresponding to the product of cutter diameter and laminate thickness. While this approximation method does not provide a realistic distribution of the tool temperature in the numerical model, it does provide an approximate estimate of the heat flux going into the cutter. The objective function was also evaluated using the maximum temperature of the cutter. However, it was noticed that the maximum cutter temperature determined from the

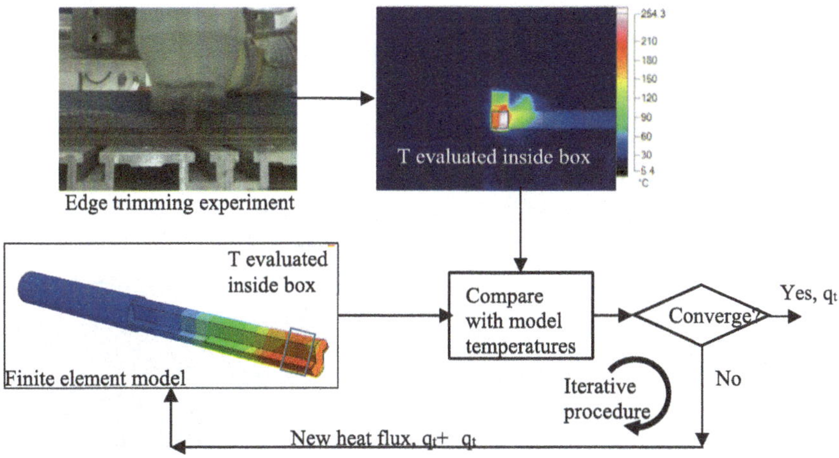

Fig. 12 Iterative procedure used to solve the inverse heat conduction problem for the cutter

thermographic images fluctuated greatly. Figure 12 shows a flow chart of the iterative procedure used to solve the inverse heat conduction problem for the cutter.

5 Experimental Methods

Edge trimming experiments were conducted on a 3-axis CNC router to generate data for machining power and boundary temperatures. Figure 13 shows a schematic of the experimental setup where a 500 mm long fiber reinforced polymer (FRP) laminate was clamped to the machine table so that the long side could be trimmed along the edge in a climb cutting configuration. The spindle speed and feed speed utilized were varied according to the parameters shown in Table 6, while the radial depth of cut was kept constant at 5 mm (half of the cutter diameter). Two types of FRP material were used, namely carbon fiber reinforced polymer composite (CFRP) and glass fiber reinforced polymer composite (GFRP). The CFRP laminate was 10 mm thick and made of twill-weave (2/2) standard modulus carbon fibers with the fiber orientation [0/90°] and fiber volume fraction of approximately 60%. The GFRP laminate was 8.25 mm thick and made of plain-weave E-glass fibers with the fiber orientation [0/90°] and fiber volume fraction of approximately 40%. The cutter used was a two-straight flute PCD (Fig. 2d). The rake and clearance angles of the cutting edge were 0 and 15°, respectively, and the initial edge radius was 12 μm. Electric power required for machining was measured by a Load Control fast response power meter wired in line with the spindle motor. The net electric cutting power was calculated as the difference between the average spindle power during cutting and idling. Measurements of boundary temperatures were made by thermocouples mounted on the top surface of the laminate, approximately halfway as shown in Fig. 13. This allowed for

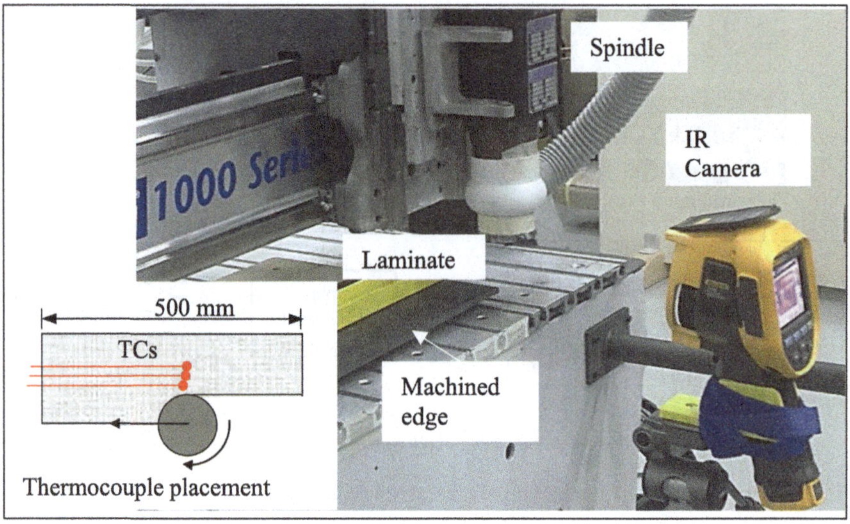

Fig. 13 Experimental setup for the edge trimming operation

Table 6 Electric power consumption in edge trimming of FRPs using PCD cutter

N (rpm)	v_f(mm/min)	f_z(mm)	CFRP P_{el}(W)	GFRP P_{el}(W)
8000	400	0.025	219.0 ± 15.6	157.7 ± 16.4
8000	800	0.05	439.4 ± 17.3	311.1 ± 33.8
4000	800	0.1	373.8 ± 14.4	305.5 ± 56.6

the temperatures to reach steady state by the time the cutter passed by the thermocouples. Gage 36, type *K* sheathed thermocouples (5TC-TT-K-36 by Omega) were used. The tolerance of standard type K thermocouple was reported as ± 2.2 °C or 0.75%, whichever is greater (Omega 2019). Thermocouple beads were placed in 1.0 mm diameter, 1.5 mm deep holes drilled on the top surface at specific locations from the machined edge. The holes were filled with thermal conductive paste before inserting the thermocouple wires. A Fluke Ti400 thermographic camera was used to capture the temperature of the cutter. The camera was placed about 500 mm away from the edge of the laminate and was focused on the cutting zone. A special suction shroud was mounted around the cutter and allowed effective chip evacuation for a clear image capture of the cutter. As the cutting tool passed by the field of view of the camera, thermal images were captured at a speed of one frame every three seconds, which was the maximum sampling speed of the camera. The IR images thus obtained were analyzed using Smart View software. The temperature recorded by the infrared camera was adjusted for the emissivity of the tool and an average tool temperature was estimated over a rectangular area overlapping the exposed surface of the tool as shown in Fig. 15. The emissivity of the tool was determined experimentally using the black body calibration method and was found to be 0.87 (Sheikh-Ahmad et al. 2019a).

5.1 Energy Consumption

Table 6 shows the net electric power consumed in cutting of the two FRP materials at different conditions of spindle and feed speeds. The electric cutting power depended more on feed speed than on spindle speed. As shown, doubling the feed speed caused almost doubling of the consumed power while doubling the spindle speed caused only a slight increase (less than 20%). It is also evident that machining CFRP consumed more power than GFRP due to the high strength of the former. As discussed earlier, most of this power was converted to heat, which was partitioned into the different regions of the cutting zone as shown later.

5.2 Temperatures History of the Laminate

Figure 14 shows typical temperature histories as recorded by surface mounted thermocouples for the CFRP and GFRP laminates during edge trimming at spindle speed of 4000 rpm and feed speed of 800 mm/min. The distance of the hole centers from the cutting edge, where the thermocouples were inserted, are shown between parentheses in the legends. It was noted that the FRP laminate remains at room temperature until the cutter comes in line with the thermocouple array, then the temperature rises abruptly to a maximum before it declines again due to cooling and the cutter moving

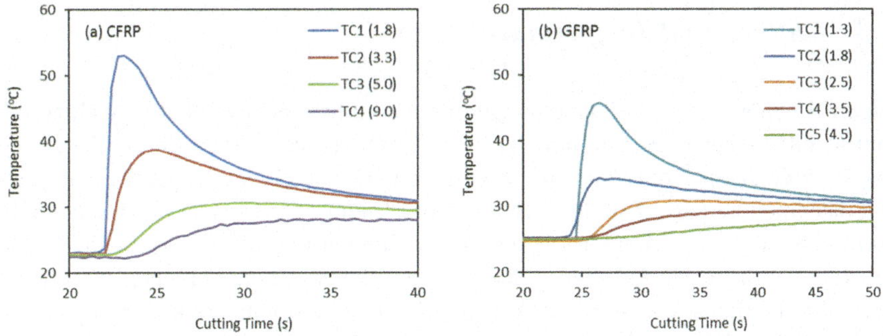

Fig. 14 Temperature histories recorded by surface mounted thermocouples when edge trimming at 4000 rpm and 800 mm/min for **a** CFRP and **b** GFRP laminates. Numbers in parentheses indicate TC locations in mm from the machined edge

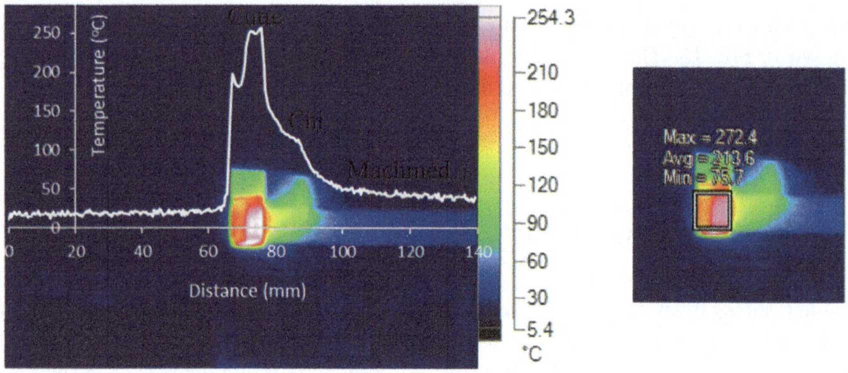

Fig. 15 Thermographic images of the cutting zone when edge trimming at 4000 rpm and 800 mm/min for CFRP laminate

away. The peaks of the different thermocouples were not coinciding due to lags caused by poor thermal conductivity, slow response of the thermocouples and misalignment of the thermocouple locations from the vertical as shown in Fig. 13. The highest temperature of 53 °C was recorded by TC1 at 1.8 mm from the machined edge for the CFRP laminate, while the highest temperature recorded at the same location for the GFRP laminate was 34 °C. This indicates that a higher heat input was received by the CFRP laminate while machining at the same conditions. This was evident from the electric power measurements shown in Table 6. It was also noted that the cooling curve for the CFRP laminate was much steeper than that for the GFRP laminate due to the higher in-plane thermal conductivity as shown in Tables 3 and 4. The temperature peaks and thermal histories obtained from these and similar results were used as boundary temperatures for determining the amount of heat flux conducted to the workpiece as explained in Sect. 4.2.

5.3 Cutting Tool Temperatures

Figure 15 shows thermographic image of the cutting zone taken halfway along the cutting edge when edge trimming CFRP at 4000 rpm and 800 mm/min. The image shows three distinctive regions of the cutting zone, namely the cutter, the chips and the machined surface. The temperature profile along a line extending horizontally from one end of the image to the other and running roughly in the middle of the thickness of the machined edge is shown imposed on the thermal image. The hottest region in the image is that of the cutter, with the highest temperatures being on the trailing side of the cutter leaving the cutting zone. The average cutter temperature was calculated over a square area overlapping the exposed side of the cutter as shown on the insert to the right. This average temperature was used as the boundary temperature for minimizing the objective function in Eq. (16). The average temperature was used

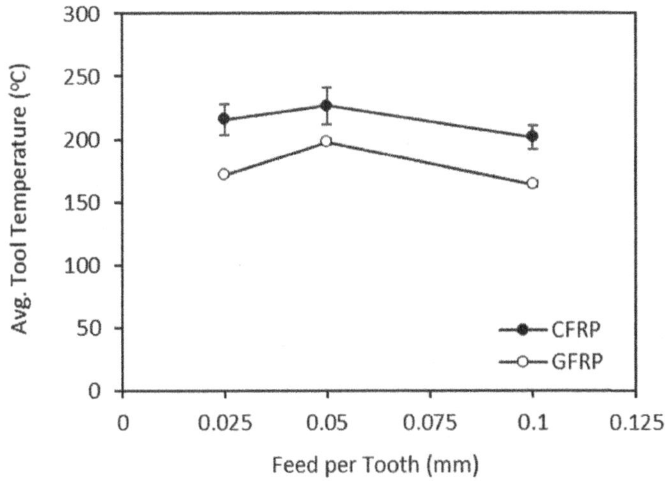

Fig. 16 Effect of cutting conditions on the average PCD tool temperature

in this context instead of the maximum temperature because of the great fluctuations in the recording of the maximum temperature over a very small area as shown in the figure.

Figure 16 shows the average tool temperatures for each laminate as a function of the feed per tooth. It was shown that the average tool temperature in cutting CFRP is approximately 40 °C higher than that in cutting GFRP. This was due to the higher strength of the CFRP laminate, which translates into higher cutting energy, a portion of which was conducted to the tool. The average cutter temperature changes slightly with cutting conditions. It increased slightly with an increase in the feed speed and an increase in the spindle speed because both increases cause an increase in the machining power as indicated by Eq. (3). The increase in feed speed caused an increase in the machining power by increasing the cutting forces due to the larger size of uncut chips. The increase in rotation speed increased the machining power directly. Only a few researches investigated the cutter temperature in edge trimming FRPs due to the complexity imparted by high speed rotation. Kerrigan et al. (2012) reported that increasing the spindle speed and feed speed resulted in increasing the cutter temperature, with the former having the greater effect. Ghafarizadeh (2016) and El-Hofy et al. (2017) also confirmed the effect of cutting speed on cutter temperature. However, the latter reported a decrease in the cutter temperature with an increase in feed per tooth. This was the result of strong interactions between the cutting speed and feed speed (Kerrigan et al. 2012).

6 Estimated Heat Fluxes

6.1 Heat Partition into the Workpiece

The heat conduction problem for the workpiece was solved numerically for different values of the heat flux $\dot{q}_w$ and different feed speeds. The heat flux conducted to the workpiece was increased systematically in increments of 50 mW/mm^2 and the simulated boundary temperatures for each heat flux value were recorded. The correct magnitude of the heat flux $\dot{q}_w$ for each FRP material at each cutting condition was determined by solving the inverse heat conduction problem as described in Sect. 4.2. Once the minimum heat flux was determined, the portion of total heat conducted to the workpiece was determined by multiplying the heat flux by the projected contact area with the workpiece, A_w,

$$\dot{Q}_w = \dot{q}_w \cdot A_w \tag{17}$$

Finally, the energy partition ratio, R_w was determined by dividing the portion of heat conducted to the workpiece by the total power.

$$R_w = \frac{\dot{Q}_w}{\varepsilon \cdot P_e} = \frac{\dot{Q}_w}{P_m} \tag{18}$$

Tables 7 and 8 show the magnitude of heat conducted to the CFRP and GFRP laminates, respectively. The errors reported represent the standard deviation of the data obtained at each specific combination of cutting conditions. The large variation in some of the data sets is attributed mainly to uncertainties in the thermocouple measurements and their locations. The main issue with thermocouple temperature measurements was the response time, which might not be short enough to capture the real dynamic response of thermal behavior of the laminate. Nevertheless, the

Table 7 Energy partition in edge trimming of CFRP with PCD cutter

N (rpm)	v_f (mm/min)	f_z (mm)	P_{el} (W)	$\dot{Q}_w$ (W)	$\dot{Q}_t$ (W)	$\dot{Q}_c$ (W)
8000	400	0.025	219.0 ± 15.6	24.2 ± 5.1	97.9 ± 6.0	53.1 ± 14.7
8000	800	0.05	439.4 ± 17.3	29.6 ± 1.9	103.5 ± 7.5	218.4 ± 15.9
4000	800	0.1	373.8 ± 14.4	23.5 ± 2.6	90.8 ± 4.6	184.7 ± 12.6

Table 8 Energy partition in edge trimming of GFRP with PCD cutter

N (rpm)	v_f (mm/min)	f_z (mm)	P_{el} (W)	$\dot{Q}_w$ (W)	$\dot{Q}_t$ (W)	$\dot{Q}_c$ (W)
8000	400	0.025	157.7 ± 16.4	8.6 ± 1.4	75.9 ± 0.85	41.7 ± 13.2
8000	800	0.05	311.1 ± 33.8	9.8 ± 0.9	89.1 ± 1.43	150.0 ± 27.1
4000	800	0.1	305.5 ± 56.6	6.9 ± 2.4	72.0 ± 1.20	165.6 ± 45.3

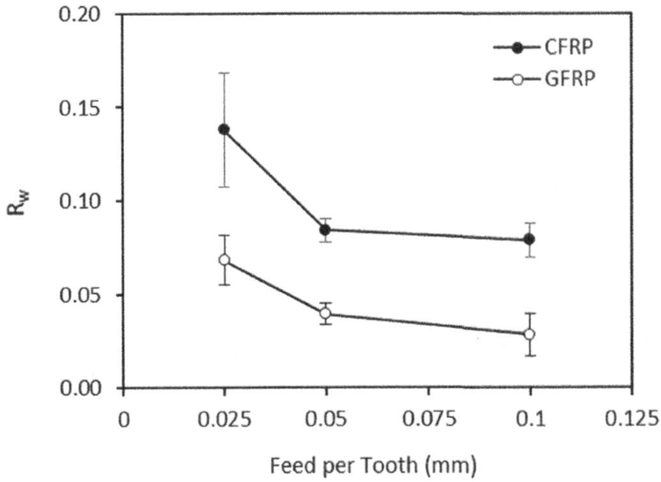

Fig. 17 Variation of the heat partition ratio R_w with cutting conditions

certainty of the results were reasonable for the most part. The heat conducted to the CFRP laminate was much higher in magnitude than that conducted to the GFRP laminate. This was due to the higher heat generated when machining CFRP and the higher heat conductivity of carbon fibers. The heat conducted to the workpiece appeared to increase with an increase in the feed speed and an increase in the spindle speed. Again, this is mainly attributed to the proportional increase in the machining power with the increase in both parameters.

Figure 17 shows the partition ratios determined by Eq. (18) for both FRP laminates. The heat partition to the CFRP is almost twice as much as that to the GFRP laminate. The highest heat partition ratio for CFRP was 0.14 at the smallest feed per tooth and stabilizes at 0.08 for feed per tooth values greater than 0.05 mm. The highest heat partition ratio for GFRP was 0.07 at the smallest feed per tooth and decreased to about 0.03 for higher feeds per tooth. The heat partition ratio decreased significantly with an increase in the feed speed for both FRP laminates and its variation with spindle speed was much less. The main reason for this behavior was exposure time. For slow feed speeds the time available for heat conduction was higher and thus more heat was conducted to the workpiece. Similarly, König and Graß (1989) reported higher heat flux conducted to the CFRP in drilling and that its magnitude decreased with an increase in feed per revolution. Furthermore, the partition ratios obtained for GFRP were comparable to those reported in the current study, but those of CFRP were slightly higher. Liu et al. (2014) determined the heat partition ratios in helical milling to be between 0.18 and 0.21. It is noted here that the heat partition ratios for FRPs are relatively small in comparison to those observed in metal cutting. Sölter and Gulpak (2012) reported that the heat partition into the workpiece in the dry milling of steel varied with undeformed chip thickness from 0.10 to 0.50, with the larger partition ratio associated with the smaller undeformed chip thickness. Luchesi and

Coelho (2012) determined the partition ratio in face milling of 4340 steel to be 0.35. Both of these works utilized an inverse heat conduction method to determine the heat partition ratio.

6.2 Heat Partition into the Cutting Tool

The heat flux applied to the cutter, $\dot{q}_t$, was obtained by minimizing the difference between measured and simulated average cutter temperatures using an iterative inverse problem as described in Sect. 4.5. The portion of heat conducted to the cutter, $\dot{Q}_t$ and the heat partition ratio, R_t were then determined by Eqs. (19) and (20), respectively.

$$\dot{Q}_t = \dot{q}_t \cdot A_t \quad (19)$$

$$R_t = \frac{\dot{Q}_t}{\varepsilon \cdot P_e} \quad (20)$$

Tables 7 and 8 show the magnitude of heat conducted to the cutter in trimming CFRP and GFRP laminates, respectively. Figure 18 shows the variation of the heat partition ratio, as a function of the feed per tooth for both FRP laminates. The behavior of the heat partition into the cutter was similar to that for the workpiece, except that the magnitudes were much higher for the cutter. It was shown that as much as 60% of the heat generated in machining was conducted into the cutter at the smallest feed per tooth. Further increase in the feed per tooth caused the heat partition ratio to

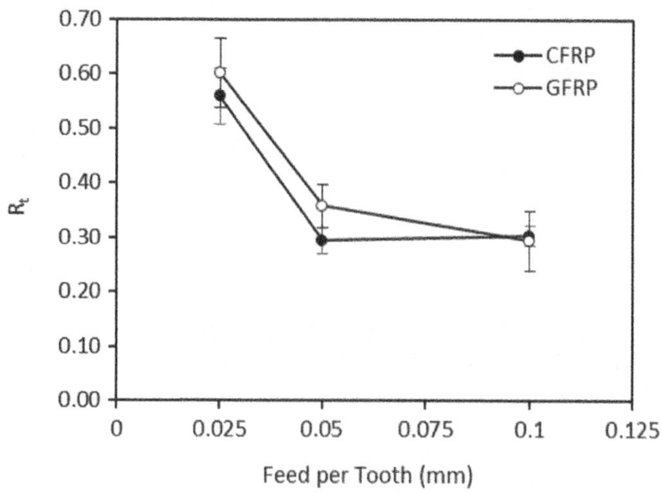

Fig. 18 Variation of the heat partition ratio R_t with cutting conditions

drop to approximately 0.3. Furthermore, the heat partition ratio was slightly higher for the GFRP than the CFRP. Apparently, the lower heat conductivity of the GFRP material allowed less heat to be dissipated through the workpiece and more heat to be dissipated through the cutter. Similarly, König and Graß (1989) reported higher heat partition ratios for GFRP for the case of drilling and the magnitudes of these ratios were between 0.51 and 0.62.

6.3 Heat Partition to the Chips

The complete heat partition at the cutting zone can now be determined from Eq. (9). Neglecting the heat dissipated to the environment, $\dot{Q}_e$ the heat carried away by the chips was determined by subtracting the values for $\dot{Q}_t$ and $\dot{Q}_w$ from the total heat,

$$\dot{Q}_c = \dot{Q} - \dot{Q}_w - \dot{Q}_t \tag{21}$$

Similarly, the heat partition ratio for the chips was determined by Eq. (22).

$$R_c = \frac{\dot{Q}_c}{\varepsilon \cdot P_e} \tag{22}$$

Tables 7 and 8 show the magnitude of heat carried away by the chips in trimming CFRP and GFRP laminates, respectively. Figure 19 shows the variation of the heat partition into the chips as a function of the feed per tooth. It is noted here that $R_t + R_w + R_c = 1$. The heat partition into the chips generally increased with the

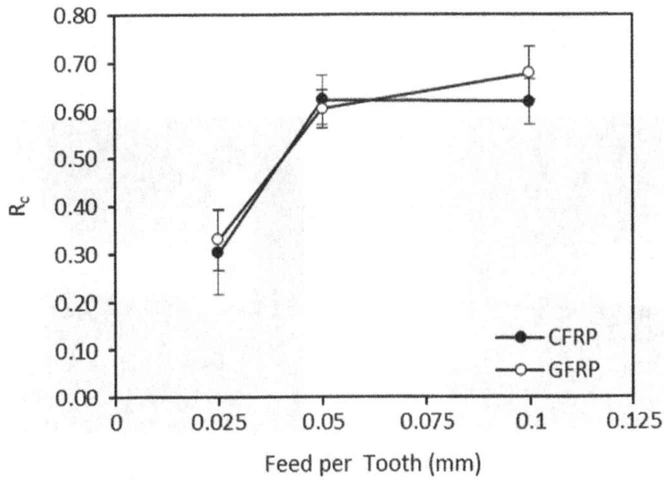

Fig. 19 Variation of the heat partition ratio R_c with cutting conditions

increase in the uncut chip size (feed per tooth). This is expected as larger chips carry more heat than smaller chips. Under the same cutting conditions, the CFRP chips carried more heat than the GFRP chips, even though their heat capacities were almost the same. This is due to the fact that more total energy was used in cutting CFRP than GFRP as shown in Tables 7 and 8. The heat partition ratio for the chips did not appear to depend greatly on the fiber material for the smaller feeds per tooth. For the larger feed per tooth of 0.1 mm, the heat partition to the GFRP chips was slightly larger than that for the CFRP chips. König and Graß (1989) also reported higher heat partition to the GFRP chips in the case of drilling, with the partition ratio in the range from 0.35 to 0.37.

The results for heat partition to the chips might be roughly verified by the calorimetric method and considering the chip temperature field from the thermographic images. Using the values of $\dot{Q}_c$ from Tables 7 and 8, the average temperature rise of the chips due to machining might be calculated using the expression,

$$\Delta T_c = \frac{\dot{Q}_c}{R_m \rho c} \tag{23}$$

where $R_m = v_f r a_p$ is the material removal rate, v_f is the feed speed, r is the tool radius and a_p is thickness of the laminate. For the cutting condition N = 4000 rpm and v_f = 800 mm/min the average chips temperatures for CFRP and GFRP were calculated as 165.1 °C and 146.1 °C, respectively. The chips temperatures can also be estimated from the infrared images as shown in Fig. 20. In this case, the temperatures of the chips were evaluated in a small region immediately as they exit the cutting zone. The emissivity of CFRP and GFRP materials was considered as 0.85 and 0.95, respectively. Since the chips cool rapidly due to their small size and high velocity, it would be more reasonable to consider the maximum temperature in this region as the representative temperature of the chips before exiting the cutting zone. Furuki et al. (2014) also measured the chips temperatures in edge trimming CFRP with PCD cutter

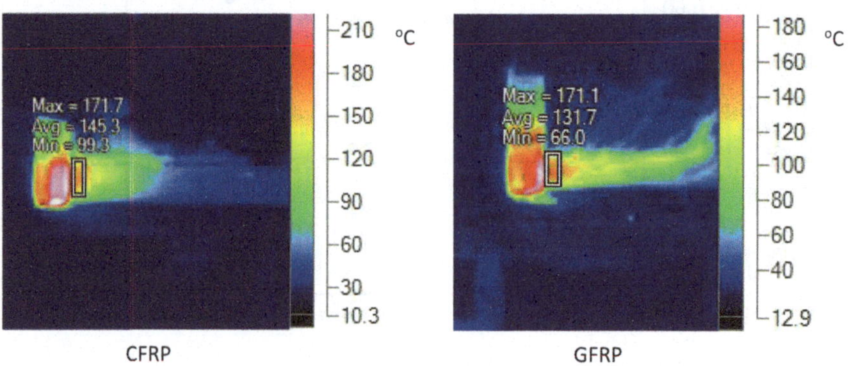

Fig. 20 Estimation of chips temperatures for the cutting condition N = 4000 rpm and v_f = 800 mm/min

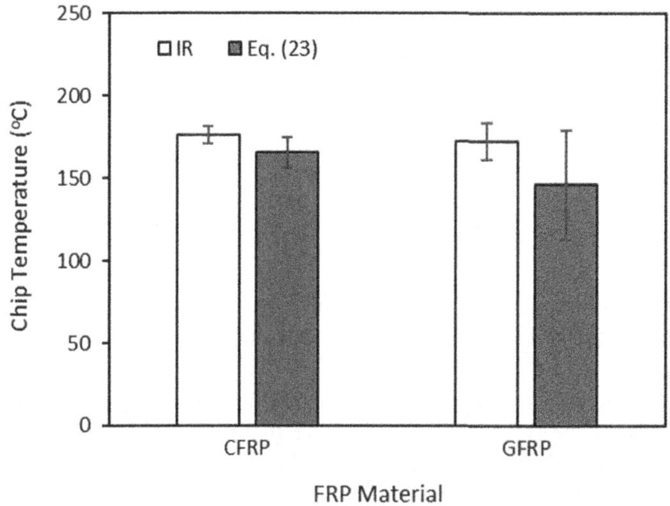

Fig. 21 Comparison between calculated and IR estimated chips temperatures for the cutting condition N = 4000 rpm and v_f = 800 mm/min

and reported a maximum temperature of 170 °C for $f_z = 0.05$. This is in agreement with the temperatures shown in Fig. 20. Figure 21 shows a comparison between the chips temperatures obtained from Eq. (23) and the thermographic images. It can be seen that the estimated temperatures are in close agreement for CFRP (7.2% difference) and in less agreement for GFRP (18% difference). These large differences arise from the accumulated uncertainty in the calculation of $\dot{Q}_c$ by Eq. (21) and the rough estimation of the chips temperatures from the thermographic images. Nevertheless, the closeness of chips temperatures obtained by the two methods support the methodology used to calculate the heat fluxes for the workpiece and the cutter.

Examining the results in Tables 7 and 8, it could be concluded that most of the thermal energy in edge trimming FRPs was dissipated through the cutting tool and chips and that only a small portion of this energy was conducted to the workpiece. A clear change of roles took place as the feed speed was increased. For the low feed speed, more heat was dissipated through the tool than by the chips (approx. 60% vs. 30%). These ratios were reversed when the feed speed was increased due to the increased volume of the chip and its ability to carry more heat. The effect of spindle speed on the heat partition seemed to be insignificant.

6.4 *Effect of Cutter Type on Heat Partition*

To investigate the effect of cutter type on heat partition, three different types of cutters were used for edge trimming the CFRP laminate as shown in Table 9 and Fig. 2. These included a TiAlN coated burr, diamond coated segmented flute and a

Table 9 Heat partition ratios for different cutters

Cutter type	N (rpm)	F (mm/min)	f_z(mm/tooth)	T_{avg}(°C)	P_e(W)	R_w	R_t	R_c
Burr (a)	4000	800	0.02	203	305.0 ± 7.30	0.080	0.455	0.466
Segmented-flute (b)	4000	800	0.02	218	366.8 ± 6.30	0.082	0.410	0.509
PCD (d)	4000	800	0.10	202	373.8 ± 14.4	0.079	0.304	0.618

polycrystalline diamond tool, all with 10 mm diameter. The burr tool consisted of 13 cutting and 15 intersecting flutes. The segmented flute cutter had 12 cutting flutes and 10 intersecting flutes. The PCD milling tool had two straight flutes. The emissivity of the tools were determined experimentally using the black body calibration method and was found to be 0.88, 0.68 and 0.87 for the burr, segmented flute and PCD tools, respectively (Sheikh-Ahmad et al. 2019a). Table 9 shows the average cutter temperatures and the net cutting power for the three different cutters. The table also shows the heat partition ratios for the complete cutting system. It could be seen that regardless of cutter type, the heat partition into the workpiece was very small (approx. 8%) as compared to that into the cutter and the chips. Among the three cutters, the PCD received the smallest portion of heat (30%) and its chips carried away the largest portion (62%). This was largely due to the large uncut chip thickness (0.1 mm/tooth) as compared to the other two cutters (0.02 mm/tooth). The two coated tools (a and b) conducted more heat than the PCD cutter due to the small chip size. Inoue and Hagino (2013) also showed that a PCD cutter registered the lowest cutting temperatures when compared to uncoated and TiAlN coated carbide tools. In terms of cutting effectiveness, it can be seen that the PCD cutter outperforms the other two because of its ability to remove larger chips while maintaining the cutter and workpiece temperatures at lower levels.

7 Conclusions

An inverse heat conduction method was used to determine the heat partition and temperature distribution in edge trimming of fiber reinforced composites. Three-dimensional transient heat conduction problem was modeled and solved independently in Abaqus for the workpiece and cutting tools. The heat flux applied to the workpiece and cutter was estimated by minimizing the difference between measured and calculated boundary temperatures. The heat carried away by the chips was determined from the energy balance and verified by simple calorimetric calculations. Temperatures were measured experimentally by installing thermocouples on the workpiece as well as with insitu infrared thermography. Total electric power consumed during machining operation was measured with a powercell and cutting forces were measured by a force-dynamometer. Correlations were made between the cutting forces and consumed electric power. Conclusions drawn from this work are:

1. Consumed electric power in machining FRPs can be accurately modeled as a linear function of the product of the average normal force and the cutting speed. The same function holds for both CFRP and GFRP materials, as well as different types of cutting tools.
2. Increase in feed speed proportionally increased the power consumption. For example, doubling the feed speed caused almost doubling of power consumption. However, the effect of spindle speed was relatively insignificant. Doubling the spindle speed caused slight increase in power consumption.
3. Inverse heat conduction method is an effective and efficient technique for determining the heat partition during machining FRPs. It provides valuable insight into the effect of machining parameters including spindle speed, feed speed and chip size onto the temperature fields and conversion of the heat energy.
4. Highest temperature was recorded for CFRP at almost same location in both types of materials suggesting higher heat input was received by CFRP under same cutting conditions. Moreover average tool temperature in cutting CFRP was approximately 40 °C higher than that in cutting GFRP.
5. Relatively small portion of heat dissipated through both types of workpiece at any given chip size i.e. 13.8% and 6.8% for CFRP and GFRP respectively at 0.025 mm feed per tooth. Remaining heat was dissipated through the tool and chips.
6. Increase in chip size reduced the amount of dissipation through the workpiece with as little as 3% through GFRP at 0.1 mm feed per tooth. Smaller amount of heat dissipating through GFRP was attributed to its lower heat conductivity as compared to CFRP.
7. Heat carried away by the chips was determined from the total energy balance and verified by calorimetric method. Reasonable agreement was shown between the two methods.
8. Cutter type had a very little effect on the amount of heat evacuated through the workpiece. It primarily effected the heat dissipation through the chips and cutter. Among the different types of cutters, the PCD received the smallest portion of heat (30%) and its chips carried away the largest portion (62%).

References

An Q, Chen J, Cai X, Peng T, Chen M (2018) Thermal characteristics of unidirectional carbon fiber reinforced polymer laminates during orthogonal cutting. J Reinf Plast Compos 37(13):905–916. https://doi.org/10.1177/0731684418768892

Colaço MJ, Orlande HRB, Dulikravich GS (2006) Inverse and optimization problems in heat transfer. J Braz Soc Mech Sci Eng 28(1):1–24

Cong W, Zou X, Deines T, Wu N, Wang X, Pei Z (2012) Rotary ultrasonic machining of carbon fiber reinforced plastic composites: An experimental study on cutting temperature. J Reinf Plast Compos 31(22):1516–1525. https://doi.org/10.1177/0731684412464913

Delahaigue J, Chatelain J-F, Lebrun G (2017) Influence of cutting temperature on the tensile strength of a carbon fiber-reinforced polymer. Fibers 5(4):46. https://doi.org/10.3390/fib5040046

El-Hofy HH, Soo SL, Aspinwall DK, Sim WM, Pearson D, M'Saoudi R, Harden P (2017) Tool temperature in slotting of CFRP composites. Procedia Manuf 10:371–381

El-Hofy MH, Soo SL, Aspinwall DK, Sim WM, Pearson D, Harden P (2011) Factors affecting workpiece surface integrity in slotting of CFRP. Procedia Eng 19:94–99

Fu R, Jia Z, Wang F, Jin Y, Sun D, Yang L, Cheng D (2018) Drill-exit temperature characteristics in drilling of UD and MD CFRP composites based on infrared thermography. Int J Mach Tool Manuf 135:24–37

Furuki T, Hirogaki T, Aoyama E, Kodama H, Ogawa K (2014) Influence of tool shape and coating type on machining surface quality in face milling of CFRP. Adv Mater Res 1017:310–315

Ghafarizadeh S, Lebrun G, Chatelain J-F (2016) Experimental investigation of the cutting temperature and surface quality during milling of unidirectional carbon fiber reinforced plastic. J Compos Mater 50(8):1059–1071

Giasin K, Ayvar-Soberanis S (2016) Evaluation of workpiece temperature during drilling of GLARE fiber metal laminates using infrared techniques: effect of cutting parameters, fiber orientation and spray mist application. Materials 9:622. https://doi.org/10.3390/ma9080622

García E, Méresse D, Pombo I, Harmand S, Sánchez JA (2014) Identification of heat partition in grinding related to process parameters, using the inverse heat flux conduction model. Appl Therm Eng 66(1–2):122–130

Haddad M, Zitoune R, Eyma F, Castine B (2014) Study of the surface and dust generated during trimming of CFRP: Influence of tool geometry, machining parameters and cutting speed range. Compos 66:142–154

Halim NFH, Ascroft H, Barnes S (2017) Analysis of tool wear, cutting force, surface roughness and machining temperature during finishing operation of Ultrasonic Assisted Milling (UAM) of Carbon Fibre Reinforced Plastic (CFRP). Procedia Eng 184:185–191

He Y, Sheikh-Ahmad J, Zhu S, Zhao C (2020) Cutting force analysis considering edge effects in the milling of carbon fiber reinforced polymer composite. J Mater Process Technol 279:116541. https://doi.org/10.1016/j.jmatprotec.2019.116541

Hintze W, Klingelhöller C (2017) Analysis and modeling of heat flux into the tool in abrasive circular cutting of unidirectional CFRP. Procedia CIRP 66:210–214

Inoue T, Hagino M (2013) Cutting characteristics of CFRP materials with carbon fiber distribution. Int J Autom Technol 7(3):285–291

János L, János K (2011) Tool wear and cutting temperature at machining composites. Adv Mater Res 325:381–386

Jia Z, Fu R, Wang F, Qian B, He C (2018) Temperature effects in end milling carbon fiber reinforced polymer composites. Polym Compos 39(2):437–447

Joven R, Das R, Ahmed A, Roozbehjavan P, Minaie B (2012) Thermal properties of carbon fiber-epoxy composites with different fabric weaves. Sampe 2012–Charleston, SC.

Kalogiannakis G, Van Hemelrijck D, Van Assche G (2004) Measurements of thermal properties of carbon/epoxy and glass/epoxy using modulated temperature differential scanning calorimetry. J Compos Mater 38(2):163–175. https://doi.org/10.1177/0021998304038647

Kerrigan K, O'Donnell G (2016) On the relationship between cutting temperature and workpiece polymer degradation during CFRP edge trimming. Procedia CIRP 55:170–175

Kerrigan K, O'Donnell GE (2013) Temperature measurement in CFRP milling using a wireless tool-integrated process monitoring sensor. Int J Autom Technol 7(6):742–750

Kerrigan K, Thil J, Hewison R, O'Donnel GE (2012) An integral telemetric thermocouple sensor for process monitoring of CFRP milling operations. Procedie CIRP 1:449–454

Khairusshima MKN, Hassan CH, Jaharah AG, Amin AKM, Idriss AN (2013) Effect of chilled air on tool wear and workpiece quality during milling of carbon fibre-reinforced plastic. Wear 302:1113–1123

Kim H-J, Kim N-K, Kwak J-S (2006) Heat flux distribution model by sequential algorithm of inverse heat transfer for determining workpiece temperature in creep feed grinding. Int J Mach Tool Manuf 46:2086–2093

König W, Graß P (1989) Quality definition and assessment in drilling of fibre reinforced thermosets. Annals CIRP 38(1):119–124
Kryzhanivskyy V, Bushlya V, Gutnichenko O, M'Saoubi R, Ståhl J-E (2018) Heat flux in metal cutting: experiment, model, and comparative analysis. Int J Mach Tools Manuf 134:81–97
Liu J, Chen,G, Ji C, Qin X, Li H, Ren C (2014) An investigation of workpiece temperature variation of helical milling for carbon fiber reinforced plastics (CFRP). Int J Mach Tool Manuf 86:89–103
Luchesi VM, Coelho RT (2012) An inverse method to estimate the moving heat source in machining process. Appl Therm Eng 45–46:64–78. https://doi.org/10.1016/j.applthermaleng.2012.04.014
Merino-Perez JL, Royer R, Ayvar-Soberanis S, Merson E, Hodzic A (2015) On the temperatures developed in CFRP drilling using uncoated wc-co tools part I: workpiece constituents, cutting speed and heat dissipation. Compos Struct 123:161–168. https://doi.org/10.1016/j.compstruct.2014.12.033
Mkaddem A, Zain-ul-abdein M, Mezlini S, Bin Mahfouz AS, Jarraya A (2017) Sensitivity of GFRP composite integrity to machining-induced heat : a numerical approach. In: Fakhfakh T, Chaari F, Walha L, Abdennadher M, Abbes M, Haddar M (eds) Advances in acoustics and vibration. Applied condition monitoring, vol 5. Springer, Switzerland
Nguten-Dinh N, Bouvet C, Zitoune R (2019) Influence of machining damage generated during trimming of CFRP composite on the compressive strength. J Compos Mater. https://doi.org/10.1177/0021998319883335
OMEGA (2019) Thermocouple types. https://www.omega.com/en-us/resources/thermocouple-types. Accessed 17 Dec 2019
Santiuste C, Díaz-Álvarez J, Soldani X, Miguélez H (2014) Modelling thermal effects in machining of carbon fiber reinforced polymer composites. J Reinf Plast Compos 33(8):758–766. https://doi.org/10.1177/0731684413515956
Shaw MC (2005) Metal cutting principles, 2nd edn. Oxford Universiy Press, Oxford, UK
Sheikh-Ahmad JY, Almaskari F, Hafeez F (2019a) Thermal aspects in machining CFRPs: effect of cutter type and cutting parameters. Int J Adv Manuf Technol 100(9–12):2569–2582
Sheikh-Ahmad JY, Almaskari F, Hafeez F, Meng F (2019b) Evaluation of heat partition in machining CFRP using inverse method. Mach Sci Technol 23(4):530–546. https://doi.org/10.1080/10910344.2019.1575401
Sheikh-Ahmad JY, Almaskari F, Hafeez F (2020) Heat partition in edge trimming of fiber reinforced polymer (FRP) composites. J Composite Materials. https://doi.org/10.1177/0021998320904129
Sölter J, Gulpak M (2012) Heat partitioning in dry milling of steel. CIRP Annals—Manuf Technol 61(1):87–90. https://doi.org/10.1016/j.cirp.2012.03.046
Sorrentino L, Turchetta S, Bellini C (2017) In process monitoring of cutting temperature during the drilling of FRP laminate. Compos Struct 168:549–561
Vollmer M, Möllmann K-P (2010) Infrared thermal imaging: fundamentals. Wiley-VCH Verlag GmbH & Co, Research and Applications
Wang F-J, Yin J-W, Ma J-W, Niu B (2018) Heat partition in dry orthogonal cutting of unidirectional CFRP composite laminates. Compos Struct 197:28–38
Wang H, Sun J, Zhang D, Guo K, Li J (2016) The effect of cutting temperature in milling of carbon fiber reinforced polymer composites. Compos Appl Sci Manuf 91:380–387. https://doi.org/10.1016/j.compositesa.2016.10.025
Weinert K, Kempmann C (2004) Cutting temperatures and their effects on the machining behaviour in drilling reinforced plastic composites. Adv Eng Mater 6(8):684–689
Xu W, Zhang L (2019) Heat effect on the material removal in the machining of fibre-reinforced polymer composites. Int J Mach Tools Manuf 140:1–11
Yashiro T, Ogawa T, Sasahara H (2013) Temperature measurement of cutting tool and machined surface layer in milling of CFRP. Int J Mach Tools Manuf 70:63–69. https://doi.org/10.1016/j.ijmachtools.2013.03.009
Zhu P, Wang Y, Wang S, Soto MA (2019) Measuring the two-dimensional temperature profile of carbon fiber reinforced polymers during drilling using distributed fiber sensing. J Lightw Technol 37(18):4687–4696

Zitoune R, Cadorin N, Collombet F, Šíma M (2017) Temperature and wear analysis in function of the cutting tool coating when drilling of composite structure: In situ measurement by optical fiber. Wear 376–377:1849–1858. https://doi.org/10.1016/j.wear.2016.12.015

Influence of Different Tool Materials on the Machining Performance in μED-Milling of CFRP Composites

K. Debnath, H. Dutta, and D. K. Sarma

Abstract This work investigates the micro-electrical-discharge-milling (μED-milling) of carbon fiber-reinforced plastic (CFRP) using assisting-electrode and rotating tool. Micro-channels of a length-to-depth ratio of 7.5 (depth: 400 μm and length: 3000 μm) were fabricated in CFRP using copper and brass tools of 960 μm in diameter. An assisting-electrode in the form of a copper sheet of thickness of 70 μm was used to initiate the sparking between the inter-electrodes gap. The input factors investigated in this experimental work were (i) input energy, (ii) feed rate, and (iii) tool speed. The machining performance was evaluated in terms of machining time, deviation in channel width, and morphology of the machined surface. The full factorial experimental design was applied to perform the experiments. The statistical analysis was carried out to find the relative significance of the input factors. The surface morphology of the micro-channel was studied using field emission scanning electron microscope (FE-SEM). The different mode of failures such as fiber breakage, spalling, peripheral surface damage, and debris accumulation were evident on the microscopic images of the machined surface obtained through FE-SEM.

Keywords CFRP · μED-Milling · Machining time · Channel width · FE-SEM

1 Introduction

Miniaturization of components used in electronics, aerospace, and biomedical applications is a major challenge to the technological advancement (Mehfuz and Ali 2009). μEDM is a nontraditional machining method which is frequently used for making miniature components. It is an important machining method that can be applied to the materials that are difficult to machine by traditional machining methods (Prakash et al. 2019). This method can be applied for manufacturing of intricate shape micron-size features coupled with superior surface quality to any electrically conductive

K. Debnath (✉) · H. Dutta · D. K. Sarma
Department of Mechanical Engineering, National Institute of Technology Meghalaya, Shillong 793 003, India
e-mail: debnath.iitr@gmail.com

M. T. Hameed Sultan et al. (eds.), *Machining and Machinability of Fiber Reinforced Polymer Composites*, Composites Science and Technology,
https://doi.org/10.1007/978-981-33-4153-1_8

material irrespective of the hardness of the material to be machined (Singh et al. 2016; Singh et al. 2017). The working principle of μEDM is similar to that of EDM except that the size of the tool, the energy involved, and the resolutions of the axes movement are at micron level. The low pulse energy required is supplied by the resistance-capacitance type generator (Yang et al. 2018). The addition of servo-controlled axes and computer numerical control into the μEDM system led to the development of a new variant of μEDM method namely μED-milling. This new variant of μEDM is frequently used for fabricating complex three-dimensional micro-components for various applications (Saito et al. 1986; Kaneko and Tsuchiya 1988; Zhang et al. 2015).

CFRP is a composite material that consists of carbon fiber as reinforcing and polymer as binding materials. CFRP is used in various applications due to its superior properties namely (i) high mechanical strength, (ii) high modulus, and (iii) low weight to strength ratio, to name a few. It is used for fabricating a variety of structural parts in aircraft such as cowling, keel beam, J-nose, and wing (Wang et al. 2017). The secondary processing techniques such as (i) shaping, (ii) milling, (iii) drilling, etc. are indispensible machining operations for making of different components based on CFRP. But it is quite difficult to machine this material due to its anisotropic and inhomogeneous structure coupled with abrasive nature of the carbon fiber. The machining of CFRP by traditional methods results in severe tool wear and surface defects such as fiber pullouts, delamination, spalling, etc. (Madhavan et al. 2015). Therefore, there is an imminent need to develop a cost-effective method that can offer defect-free features in CFRP. The nontraditional machining processes namely (i) water jet machining (WJM), (ii) ultrasonic machining (USM), (iii) plasma arc machining (PAM), and (iv) EDM were applied for machining of CFRP as the performance of the traditional machining methods are inadequate in achieving required surface quality and shape (Gil et al. 2014).

2 μED-Milling of Different Materials

A few works have been reported on μED-milling of various materials. Karthikeyan et al. (2012) studied the effect of feed rate, energy, aspect ratio, and tool speed on the tool wear rate (TWR) and material removal rate (MRR) during μED-milling of EN 24 using tungsten tool. It was observed that the TWR and MRR increased with an increase in the energy from 500 to 2000 μJ and tool speed from 100 to 800 RPM. Both TWR and MRR decreased as the aspect ratio and feed rate exceed 1.5 and 45 μm/s, respectively. Karthikeyan and coworkers (2012) concluded that the input energy and tool speed were the significant factors for MRR and TWR. Mehfuz and Ali (2009) studied the intrinsic influence of voltage and capacitance on MRR, TWR, maximum peak-to-valley roughness height (R_y), and average surface roughness (R_a) during μED-milling of beryllium-copper (Be-Cu) using tungsten tool. Chiou et al. (2015) studied the μED-milling of high-speed steel alloy using silver (Ag) and copper (Cu) coated tungsten carbide (WC) tool. It was found that the coating improved the MRR and surface roughness (SR). The highest MRR and the lowest SR were achieved by Cu coated WC and Ag coated WC tools, respectively. Kuriachen and Mathew (2015) concluded that the factors that influence the machining performance of μED-milling of Ti-6Al-4 V by WC tool most are feed rate and capacitance. The increase in MRR is not much due to unproductive spark generation as the tool speed exceeds 1000 RPM. The formation of recast layer was observed on the machined surface which was formed due to resolidification of the melted material. Kuriachen and Mathew (2016) further investigated the machining behavior of Ti-6Al-4 V during powder-mixed μED-milling using WC tool. High MRR and low TWR were achieved at a powder concentration of 5 g/L (lowest), voltage of 115 V (medium), and a capacitance of 0.1 μF (medium). The formation of microcracks and microvoids were observed at the two ends of the micro slot which were formed due to frequent temperature change. Jafferson et al. (2016) performed μED-milling of stainless steel and studied the effect of tool speed, feed rate, and layer thickness on MRR and TWR using tungsten tool. The highest MRR and the lowest TWR were achieved at a feed rate of 100 mm/min and tool speed of 2500 RPM. Unune and Mali (2017) observed a reduction in frontal tool wear and an improvement in MRR when the workpiece is subjected to a low-frequency vibration during μED-milling of Inconel 718. D'Urso et al. (2018) suggested a method for the evaluation of peaks and valleys shape and distribution in machining steel and ceramic materials by μED-milling. The melting point and thermal conductivity were found as the vital parameters in erosion by discrete sparks during μEDM (Tsai and Masuzawa 2004). Feng et al. (2018) performed μED-milling of Ti-6Al-4 V, SUS304, and SKH59 and found that the thermal conductivity and specific heat play important role in extending the depth of the weld pool. The surface morphology of Ti-6Al-4 V was superior to SUS306 and SKH59 owing to formation of shallow craters in Ti-6Al-4 V. The low thermal conductivity and high specific heat capacity were accountable for the formation of shallow craters in Ti-6Al-4 V.

3 EDM/WEDM/μEDM of CFRP

The research work carried out on EDM, WEDM, and μEDM of CFRP is scanty. Lau et al. (1990) experimentally investigated the feasibility of using EDM method for machining of CFRP. It was found that the copper tool performed better than the graphite tool in terms of TWR. Lau and Lee (1991) carried out a comparative analysis between WEDM and laser machining in cutting of CFRP. The machined surface produced by WEDM showed less damage, better surface finish, and less extent of heat-affected zone than the surface produced by laser cutting. Teicher et al. (2013) investigated the μEDM method in fabricating holes in CFRP. The study suggested that a layer of epoxy resin from the top surface of the CFRP laminate needs to be removed to expose the carbon fiber so that the sparking can be initiated between the inter-electrodes gap. Korlos et al. (2014) investigated and confirmed the feasibility of machining CFRP by EDM. Lodhi et al. (2014) studied the effect of pulse duration, gap voltage, and discharge current on surface roughness during EDM of CFRP. It was found that the discharge current was the most influencing parameter. Park et al. (2015) compared the performance of μEDM method on CFRP and metal workpiece. It was observed that the damage at the entrance of the hole was more in μEDM of CFRP. Sheikh-Ahmad and Shinde (2016) fabricated holes in CFRP composites by EDM method using copper and graphite tools. The highest MRR was achieved for the graphite tool at intermediate pulse on time and high current. The same input condition also resulted in the highest TWR, delamination, and hole deviation. Sheikh-Ahmad (2016) investigated the surface morphology of CFRP in making holes by EDM method and found more delamination at the entrance of the hole. The delamination was predominantly affected by the gap current. Habib and Okada (2016) concluded that the MRR increased with an increase in the pulse on time, pulse off time, current, voltage, and rotation of the tool in EDM of CFRP. Islam et al. (2017) applied the EDM method for burr removal from the holes drilled in CFRP components. The performance of the copper tool was better than the aluminium, brass, and steel tools in terms of deburring time and MRR. Kurniawan et al. (2017) made a comparison between the performance of ultrasonically-assisted-EDM and dry-EDM by removing the burr from the CFRP component. The capacitance was the most affecting parameter and the performance of the copper tool was better than the aluminum and brass tools in terms of burr removal. Yue et al. (2018) correlated the material removal mechanism with the mechanical, chemical, and thermal aspects during EDM of CFRP. The study showed that the material was removed by thermal decomposition, vaporization, oxidation, and sublimation. Kumar et al. (2018) studied the electrical-discharge drilling (EDD) of CFRP for fabricating micro-holes. The drilling of CFRP was performed by concentrating the discharge on the conductive carbon fiber exposed by mechanical process. Kumar et al. (2018) further studied the EDD method in making blind micro-holes in CFRP. Dutta et al. (2019) established the material removal mechanism during μEDM of CFRP using rotating tool and assisting-electrode. It was reported that the voltage was the most significant factor for machining time. The highest MRR was achieved at a voltage of 170 V, pulse duration of 10 μs, and the tool speed of 800 RPM. Dutta et al. (2020) also investigated the feasibility of cutting a thin CFRP plate by WEDM method using sandwich assisting-electrodes. It was found that the machining time decreased with an increase in both input current and voltage and increased with the pulse off time. Dutta and co-workers (2020) further studied the μED-milling of CFRP while fabricating micro-channels in CFRP using the copper tool.

From the literature survey, it is clear that work reported on μED-milling of CFRP is quite rare. The study on μED-milling of CFRP using different tool materials was also not explored. Thus, in the present work, the performance of μED-milling of CFRP was investigated using two different tool materials namely copper and brass. A full factorial experimental design was applied considering three different levels of each input factor (input energy, feed rate, and tool speed). The effect of the input factors on the machining time and deviation in channel width were extensively studied both experimentally and statistically.

4 Experimental Details

The composite was fabricated by hand-layup process using carbon fiber (3 k plain weave) and epoxy resin (LY 556/HY 951). The fiber mat was stacked on the lower mold plate and the resin mixture was applied on it. The process was repeated to incorporate desired number of fiber mats. A hand roller was used to press the stack so that the resin is uniformly distributed. Finally, the upper mold plate was placed over the stack and both plates were tightened using mechanical fasteners. The whole setup was kept for 24 h at room temperature for curing.

The HYPER 15 micro-machining setup was used (Fig. 1) to perform the μED-milling of CFRP. The travel limits of the work table are 130, 75, and 80 mm in X, Y, and Z directions. It has a repeatability of $\pm$ 1 μm for all the axes. SPO-A EDM oil was used as dielectric medium. A copper sheet of a thickness of 70 μm was used as an assisting-electrode to initiate the sparking phenomenon. The solid copper and brass rods of diameter of 960 μm were used as the tool (Fig. 2). The tools were fabricated by the WEDM process. The depth and length of the channel fabricated in

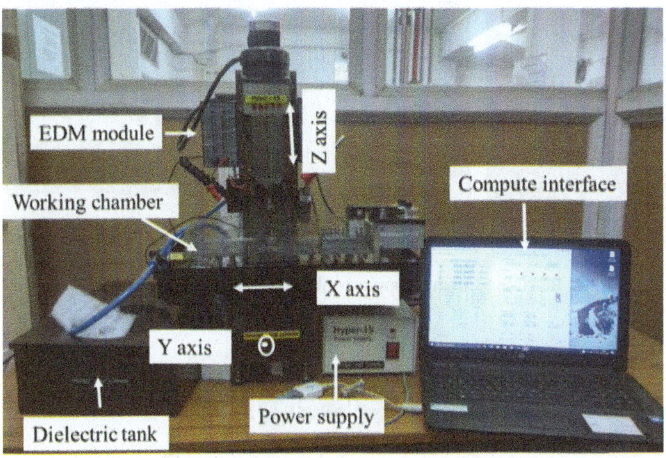

Fig. 1 HYPER 15 micro-machining setup

Fig. 2 Copper and brass tool

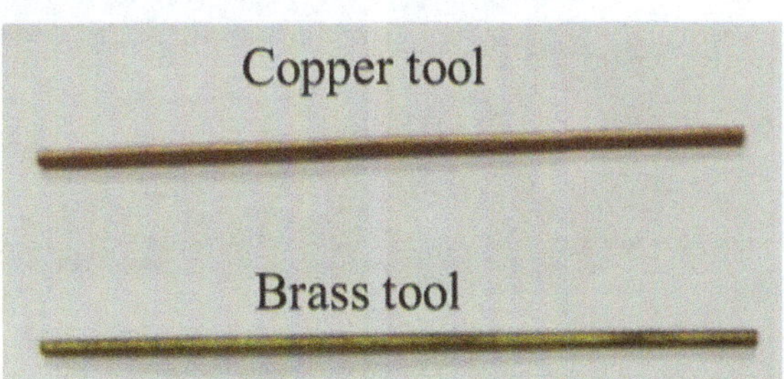

CFRP were 400 and 3000 μm, respectively. The depth was achieved in two steps i.e., 200 μm in each step. Figures 3 and 4 show the micro-channel fabricated in CFRP using copper and brass tool, respectively.

The different levels and values of the input factors are listed in Table 1. Each set of experiment was repeated three times to obtain the better results. The regression analysis was carried out using Minitab-17 to find the relative significance and contribution of the process parameters on the machining time and $W_{(dev.)}$. The width of the micro-channel was measured using an optical microscope (Make: Olympus and Model: BX51).

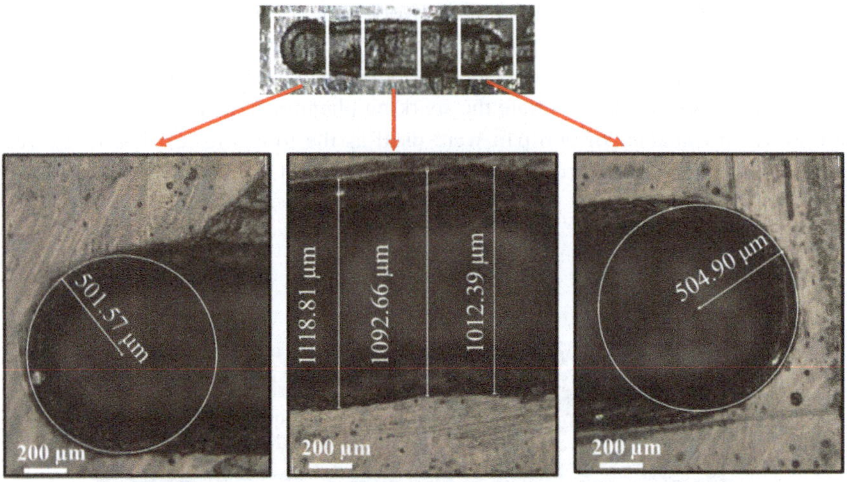

Fig. 3 Micro-channel fabricated using copper tool

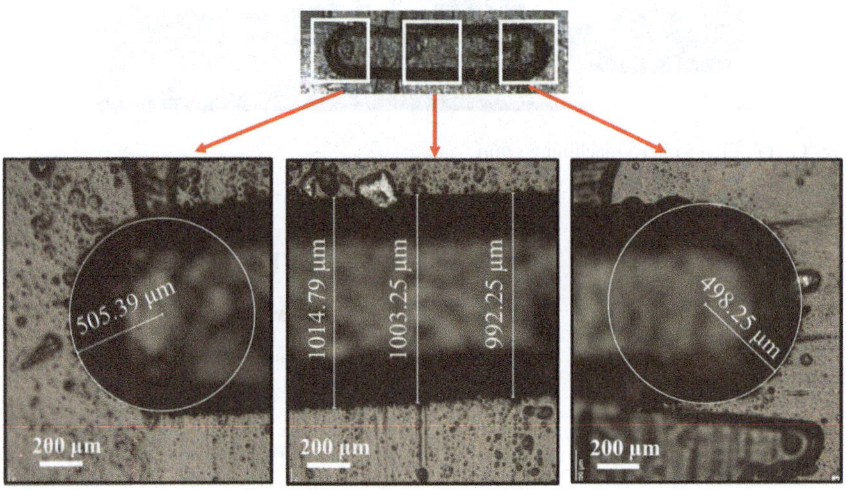

Fig. 4 Micro-channels fabricated using brass tool

Table 1 Input factors and levels

Factors	Coding	Level 1	Level 2	Level 3	Unit
Input energy	E	1.805	18.05	180.35	μJ
Feed rate	F	3	4	5	μm/s
Tool speed	S	400	500	600	RPM

5 Results and Discussion

5.1 Statistical Analysis of Experimental Results

The settings of input parameters and experimental results are shown in Table 2. The ANOVA for machining time obtained using a copper tool ($MT_{(copper)}$) is shown in Table 3. It can be observed from the table that input energy is the most significant factor showing a contribution of 70.66% followed by feed rate (8.58%) and tool speed (1.13%). The primary mechanism of material removal in μEDM of CFRP is melting and vaporization (Dutta et al. 2019). The temperature generates to melt and vaporize the constituents of CFRP is significantly affected by the input energy. Therefore, the input energy is the most significant factor among all the factors chosen for the purpose of investigation. The R-sq. value for $MT_{(copper)}$ is 80.68%, as shown in Table 4. This confirms that the model is well fitted with the experimental results. The optimum condition of input factors was found to be E3/F3/S600 i.e., input energy of 180.35 μJ, the feed rate of 5 μm/s, and tool speed of 600 RPM. The percentage difference in the optimum value of MT $_{(copper)}$ was found by calculating the $MT_{(copper)}$ using the regression equation (Eq. 1) and comparing it with the optimum value obtained from the experiment. The percentage difference for $MT_{(copper)}$ was as low as 5.64% which shows that the experimental results are adequately fitted with the regression model. The ANOVA for machining time obtained using the brass tool ($MT_{(brass)}$) is shown in Table 5. The results show a similar trend as that of copper tool. The input energy has the highest percentage contribution (69.23%) followed by the feed rate (10.00%) and tool speed (1.11%). The R-sq. value for $MT_{(brass)}$ is 80.35% (Table 6). The optimum level of input factors for $MT_{(brass)}$ was found to be E1/F1/S1. The regression relation between $MT_{(brass)}$ and the input factors is shown by Eq. 2. The percentage difference between the experimental and regression value of optimum $MT_{(brass)}$ is 4.61%.

The regression analysis was also performed for the variation in channel width for both copper ($W_{dev.\ (copper)}$) and brass tool ($W_{dev.\ (brass)}$). Table 7 shows the results of ANOVA for $W_{dev.\ (copper)}$. The machining time was found to be the most significant factor with a contribution of 76.36% followed by feed rate (7.67%), and tool speed (1.36%). The input energy determines the amount of heat produced in the machining zone that causes removal of a significant amount of materials through melting and vaporization. The material removal increases with the input energy which results in more variation in the channel width. The R-sq. value for $W_{dev.\ (copper)}$ is as high as 85.39% (Table 8). Thus, it can be inferred that the model is well fitted with the

Table 2 Settings of input parameters and experimental results

Exp. run	Input energy (μJ)	Feed rate (μm/s)	Tool speed (RPM)	$MT_{(copper)}$ (s)	$MT_{(brass)}$ (s)	$W_{dev. (copper)}$ μm	$W_{dev. (brass)}$ μm
1	1.805	3	400	5083	3191	114.62	43.43
2	1.805	3	500	4871	3133	118.98	46.74
3	1.805	3	600	4761	3083	121.67	49.76
4	1.805	4	400	4602	3078	113.18	42.54
5	1.805	4	500	4452	2969	115.56	44.12
6	1.805	4	600	4354	2789	117.45	47.58
7	1.805	5	400	4225	2705	112.53	40.92
8	1.805	5	500	4091	2597	114.23	43.46
9	1.805	5	600	3912	2553	116.15	46.72
10	18.05	3	400	3694	2477	137.16	65.81
11	18.05	3	500	3502	2429	138.98	67.77
12	18.05	3	600	3371	2293	140.57	70.80
13	18.05	4	400	3193	2209	129.19	57.33
14	18.05	4	500	3085	2123	130.45	59.42
15	18.05	4	600	2976	2051	133.69	63.61
16	18.05	5	400	2879	1981	123.11	51.26
17	18.05	5	500	2722	1932	126.23	54.58
18	18.05	5	600	2626	1829	127.42	56.71
19	180.5	3	400	2512	1741	154.61	89.11
20	180.5	3	500	2312	1702	157.25	92.33
21	180.5	3	600	2175	1659	159.26	95.56
22	180.5	4	400	2089	1571	148.98	81.32
23	180.5	4	500	1871	1501	151.23	85.44
24	180.5	4	600	1787	1462	152.56	86.34
25	180.5	5	400	1672	1392	142.17	73.98
26	180.5	5	500	1512	1271	144.68	76.12
27	180.5	5	600	1402	1212	145.23	78.51

Table 3 ANOVA for machining time obtained using copper tool ($MT_{(copper)}$)

Source	DF	Adj. SS	Adj. MS	F-value	P-value	% Contribution
Regression	3	26453897	8817966	32.01	0.000	
Input energy	1	23170574	23170574	84.11	0.000	70.66
Feed rate	1	2912089	2912089	10.57	0.004	8.58
Tool speed	1	371235	371235	1.35	0.258	1.13
Error	23	6335959	275476			19.32
Total	26	32789857				

Table 4 Model summary for MT $_{(copper)}$

R-sq.	R-sq. (adj.)	R-sq. (pred.)
80.68%	78.16%	74.38%

Table 5 ANOVA for machining time obtained using brass tool (MT $_{(brass)}$)

Source	DF	Adj. SS	Adj. MS	F-value	P-value	% Contribution
Regression	3	8007465	2669155	31.34	0.000	
Input energy	1	6899516	6899516	81.01	0.000	69.23
Feed rate	1	996872	996872	11.70	0.002	10.00
Tool speed	1	111078	111078	1.30	0.265	1.11
Error	23	1958883	95169			
Total	26	9966348				

Table 6 Model summary for MT $_{(brass)}$

R-sq.	R-sq. (adj.)	R-sq. (pred.)
80.35%	77.78%	74.05%

Table 7 ANOVA for channel width obtained using copper tool ($W_{dev. (copper)}$)

1	DF	Adj. SS	Adj. MS	F-value	P-value	% Contribution
Regression	3	5164.14	1721.38	44.82	0.000	
Input energy	1	4618.07	4618.07	120.24	0.000	76.36
Feed rate	1	463.83	463.83	12.08	0.002	7.67
Tool speed	1	82.23	82.23	2.14	0.157	1.36
Error	23	883.34	38.41			
Total	26	6047.48				

experimental results. Equation 3 shows the regression relation between the input factors and $W_{dev. (copper)}$. The values of input energy, feed rate, and tool speed for obtaining the optimum $W_{dev. (copper)}$ was found to be 1.805 μJ, 5 μm/s, and 400 RPM. The percentage difference between the experimental and regression value of the optimum $W_{dev. (copper)}$ is quite low (2.23%). Table 9 shows the ANOVA for $W_{dev. (brass)}$. The results show a similar trend as that of $W_{dev. (copper)}$ where the input energy has the highest contribution of 80.02% followed by feed rate (6.93%) and tool speed (1.76%). The fitness of the model can be deduced from the value of R-sq. which is

Table 8 Model summary for $W_{dev. (copper)}$

R-sq.	R-sq. (adj.)	R-sq. (pred.)
85.39%	83.49%	80.42%

Table 9 ANOVA for channel width obtained using brass tool ($W_{dev. (brass)}$)

Source	DF	Adj. SS	Adj. MS	F-value	P-value	% Contribution
Regression	3	6981.2	2327.07	60.19	0.000	
Input energy	1	6297.88	6297.88	162.90	0.000	80.02
Feed rate	1	545.05	545.05	14.10	0.001	6.93
Tool speed	1	138.28	138.28	3.58	0.071	1.76
Error	23	889.20	38.66			
Total	26	7870.40				

Table 10 Model summary for $W_{dev. (brass)}$

R-sq.	R-sq. (adj.)	R-sq. (pred.)
88.70%	87.23%	84.74%

88.70% (Table 10). Equation 4 depicts the regression relation between the $W_{dev. (brass)}$ and the input factors. The optimum levels of input factors for $W_{dev. (brass)}$ are the same as obtained for $W_{dev. (copper)}$. The percentage difference in the experimental and regression value of optimum $W_{dev. (brass)}$ is 4.62%.

$$MT_{(copper)} = 6269 - 11.48 - 402F - 1.44S \quad (1)$$

$$MT_{(brass)} = 3935 - 6.265 - 235.3F - 0.786S \quad (2)$$

$$W_{dev. (copper)} = 131.65 + 0.1621E - 5.08F + 0.0214S \quad (3)$$

$$W_{dev. (brass)} = 58.89 + 0.1893E - 5.50F + 0.0277S \quad (4)$$

5.2 *Effect of Input Parameters on Machining Time*

The variation of machining time with different input factors for both copper and brass tools are depicted in Fig. 5 and Fig. 6, respectively. It can be seen that the machining time decreases with an increase in the input energy, feed rate, and tool speed for both copper and brass tools. The heat that is generated during machining increases with the input energy which leads to more amount of material removal through melting and vaporization of constituents of CFRP. Thus, the increase in input energy from 1.805 to 180.35 μJ resulted in decreased machining time irrespective of variation in feed rate and tool speed. The increase in feed rate also resulted in faster machining. The machining time decreases with an increase in the feed rate from 3 to 5 μm/s for both copper and brass tools. The effective flushing of debris from the machining

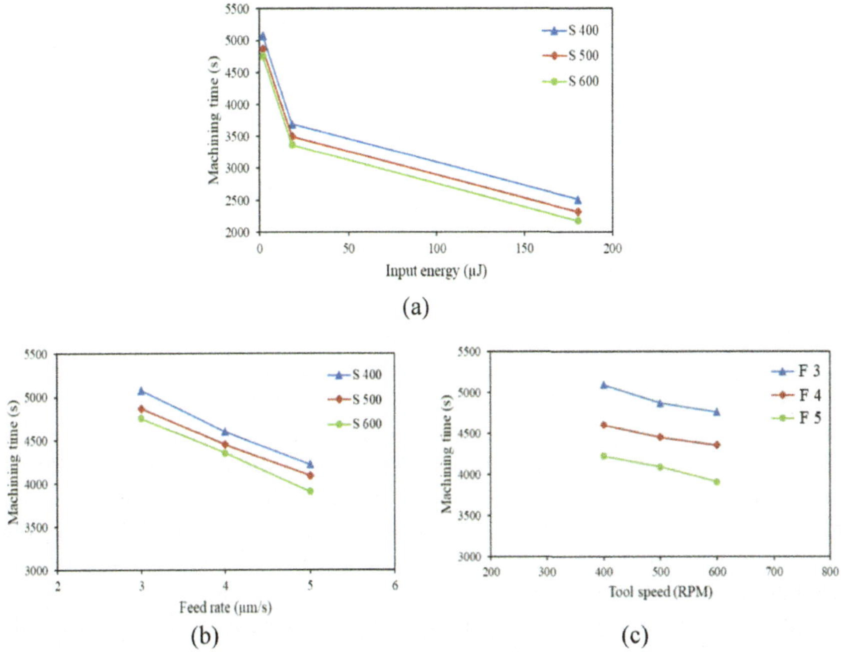

Fig. 5 Variation of MT $_{\text{(copper)}}$ with **a** input energy, **b** feed rate, and **c** tool speed

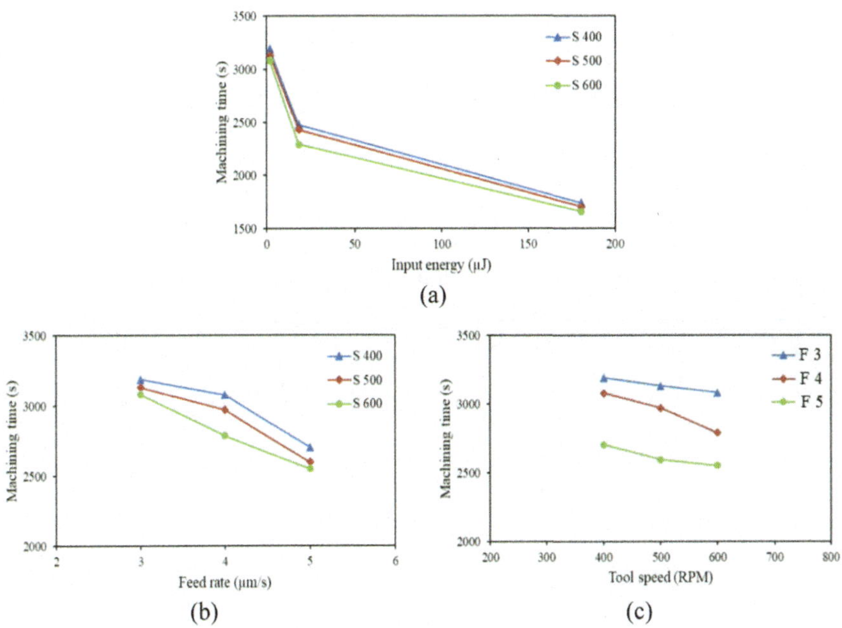

Fig. 6 Variation of MT $_{\text{(brass)}}$ with **a** input energy, **b** feed rate, and **c** tool speed

zone as the tool speed is increased from 400 to 600 RPM resulted in better machining performance in terms of increased MRR (Dutta et al. 2019). Therefore, the machining time decreases with an increase in the speed of the tool.

5.3 *Effect of Input Parameters on the Deviation of the Channel Width*

The effects of input energy, feed rate, and tool speed on the deviation of the channel width ($W_{(dev.)}$) for copper and brass tools are depicted in Fig. 7 and Fig. 8, respectively. It can be seen from the figures that $W_{(dev.)}$ increases with the input energy and tool speed and decreases with the feed rate for both copper and brass tools. The thermal decomposition of the composite constituents increases due to the higher heat generation with an increase in the input energy. The extent of heat-affected zone is increases due to this increased heat which results in more deviation in the channel width. The formed debris is efficiently flushed from the machining gap due to increased centrifugal force exerted on the dielectric by the tool as the rotational speed of the tool is increased from 400 to 600 RPM. This subsequently resulted in more deviation in channel width as more amount of material is removed from the work surface. The increase in the feed rate resulted in less deviation in the channel

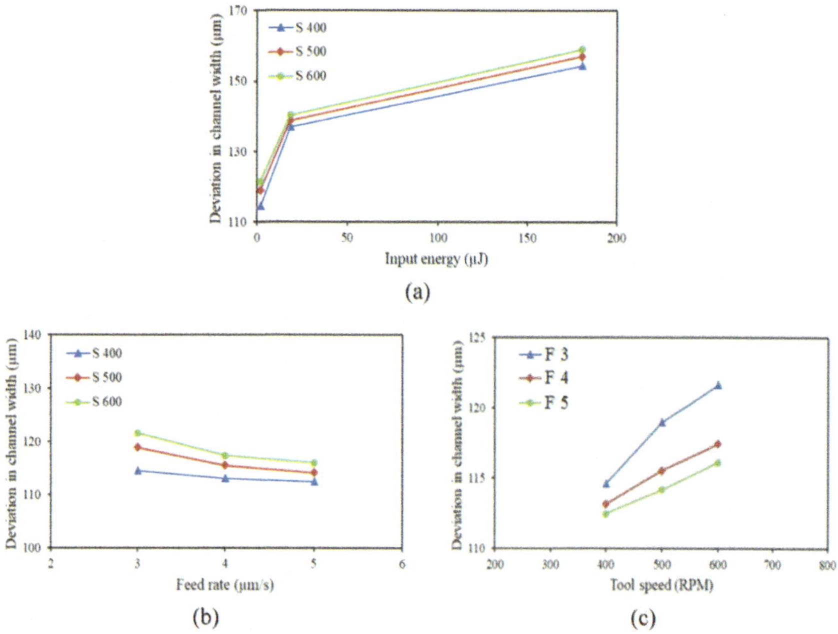

Fig. 7 Variation of $W_{dev. (copper)}$ with **a** input energy, **b** feed rate, and **c** tool speed

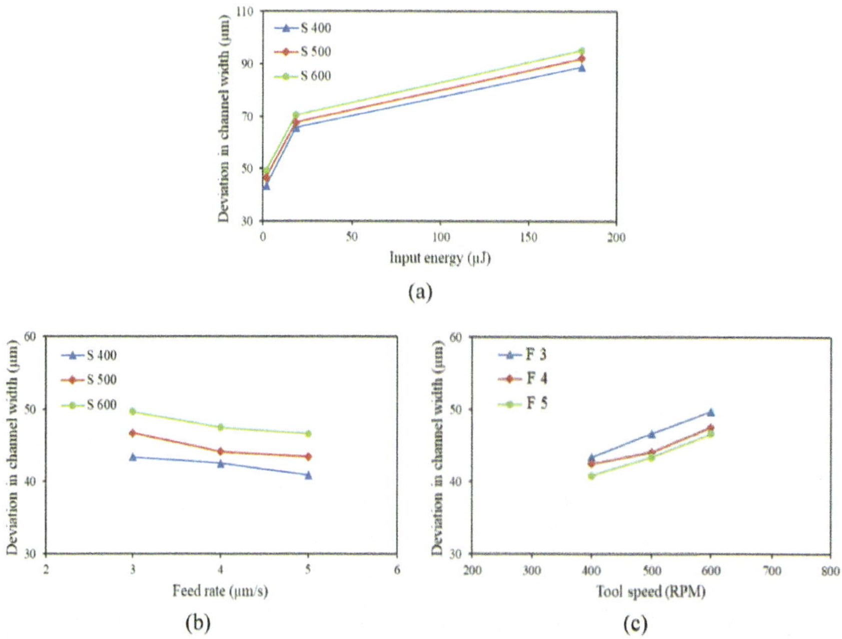

Fig. 8 Variation of $W_{dev.\ (brass)}$ with **a** input energy, **b** feed rate, and **c** tool speed

width though more amount of material is removed as the feed rate is increased from 3 to 5 μm/s.

5.4 Comparative Analysis of the Copper and Brass Tools

The variation of machining time with the input factors at different levels for copper and brass tools is shown in Fig. 9a. It can be observed from the figure that the

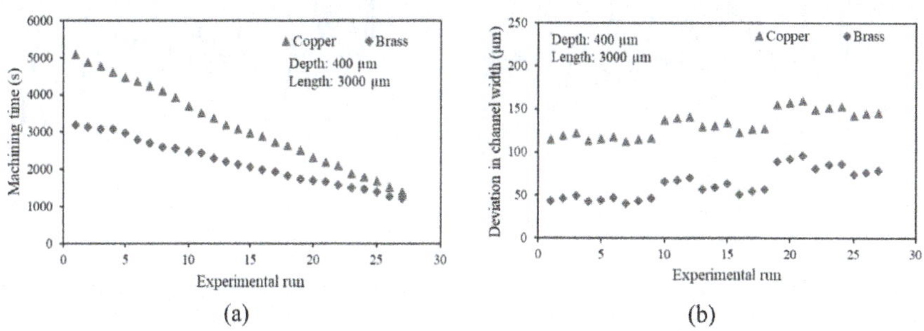

Fig. 9 Effect of copper and brass tool on the output factors **a** machining time and **b** $W_{(dev.)}$

machining time for brass tool is relatively less when compared one-on-one with copper tool. Copper has higher thermal conductivity (401 W/mK) than that of brass (159 W/mK). Thus the heat dissipation through conduction is faster for copper tool due to its high thermal conductivity. Subsequently, spark ignition is continued for less time when copper is used as tool material as compared to brass. Thus a lesser amount of heat is being transferred to the workpiece which eventually results in lower amount of material removal (Bhaumik and Maity 2018). Hence, the machining time for copper is more as compared to brass in μED-milling of CFRP.

Figure 9b shows the varaiation of $W_{(dev.)}$ with the different settings of input factors. It can be seen that the $W_{(dev.)}$ for brass tool is less than that of the copper tool. The sparking is more at the inter-electrodes gap due to the higher electrical conductivity of copper. Also, the dissipation of heat energy in copper tool is more than the brass tool due to the high thermal conductivity of copper. The extent of the heat-affected zone is more for the copper tool as compared to the brass tool. Thus more amount of material is removed from the works surface which leads to higher deviation in the width of the channel obtained using copper tool than that of brass tool. The roughness profiles of the micro-channels fabricated by copper and brass tools are shown in Fig. 10. The average surface roughness of the fabricated micro-channel obtained using copper and brass tools are 1.4 and 1.1 μm, respectively.

The morphology of the machined surface was studied through FE-SEM analysis. Figures 11 and 12 show the FE-SEM images of the micro-channel fabricated using copper and brass tools, respectively. The breakage of fibers, peripheral surface damage, and debris accumulation were detected in the machined surface. The peripheral damage to the machined surface was due to the high heat generation during μED-milling of CFRP. Moreover, spalling was also detected in the microscopic image of the machined surface obtained using brass tool. The failure of the composite constituents due to the thermal gradient resulted in spalling of the machined surface. The machined surfaces obtained using copper and brass tools showed relatively rough texture due to nonuniform sparking during μED-milling of CFRP as CFRP is composed of conductive carbon fiber and nonconductive polymer.

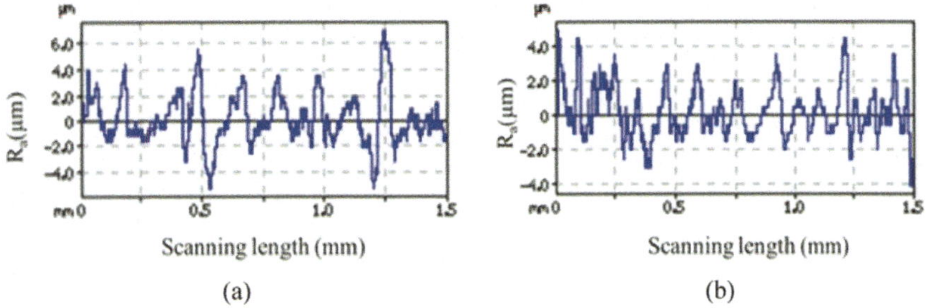

Fig. 10 Surface roughness profiles of the machined surface obtained using **a** copper and **b** brass tools

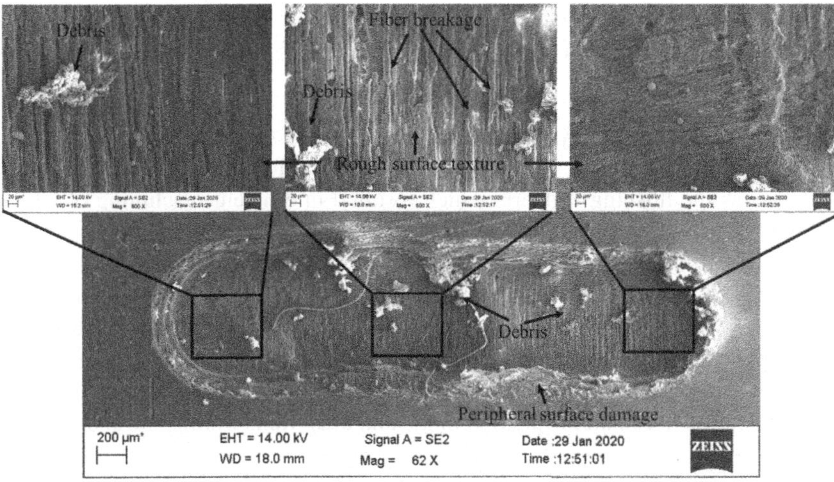

Fig. 11 Micro-channel fabricated using copper tool

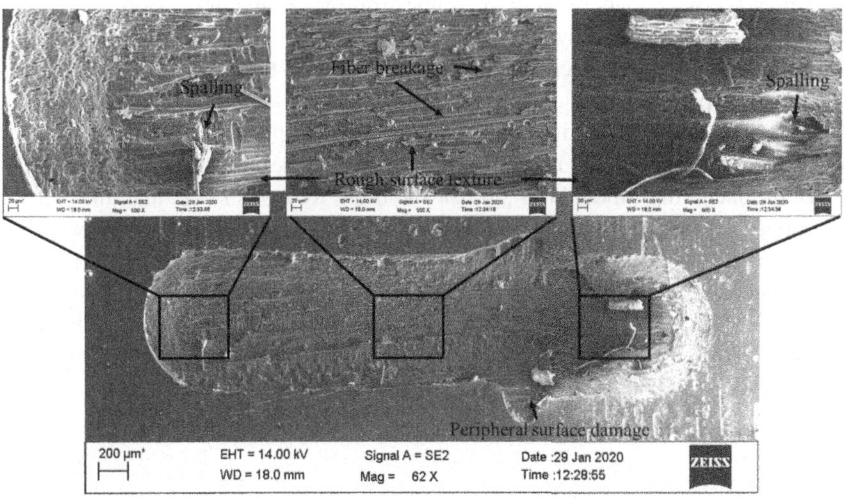

Fig. 12 Micro-channel fabricated using brass tool

6 Concluding Remarks

µED-milling of CFRP composite was performed to fabricate micro-channels of length-to-depth ratio of 7.5 using copper and brass tools of 960 µm in diameter. The intrinsic effect of different process parameters namely discharge energy, feed rate, and tool speed on the machining time and the deviation in the channel width was experimentally and statistically analyzed. The machined surface morphology was also studied to investigate the mechanism of material removal. The following salient observations can be made from the current experimental investigation and statistical analysis:

1. μED-milling of CFRP was performed using the assisting-electrode and rotating tool. The assisting-electrode facilitates the spark initiation and the rotating tool improves the flushing if debris which resulted in efficient machining of CFRP by means of μED-milling.
2. The regression analysis showed that input energy is the most significant parameter followed by feed rate and tool speed for both machining time and deviation in channel width obtained using copper and brass tools. The percentage contribution of input energy on the machining time for copper and brass tools were 70.66 and 69.23%, respectively. Similarly, the percentage contribution of input energy on the deviation in channel width for copper and brass tools were 76.36 and 80.02%, respectively.
3. R-sq. value for machining time for copper and brass tools were found to be 80.68 and 80.35%, respectively. The high value of R-sq. indicates adequate fitness of the adopted model with the experimental results. The similar conclusion can be drawn for $W_{(dev.)}$ where the R-sq. values for copper and brass tools were 85.39 and 88.70%, respectively.
4. The machining time decreased with an increase in the input energy, feed rate, and tool speed for both copper and brass tools. Whereas $W_{(dev.)}$ was increased with an increase in the input energy and the tool speed and decreased with an increase in the feed rate in case of both the tool materials.
5. Both machining time and $W_{(dev.)}$ for the brass tool was lower than that for the copper tool as the thermal and electrical conductivity of the brass is lower than the copper. Thus it can be inferred that the brass tool perform better than the copper tool in machining CFRP by means of μED-milling.
6. FE-SEM analysis revealed that fiber breakage and spalling were the major modes of failure of the composite constituents. The peripheral surface damage was also observed in the microscopic images of the machined surface.

References

Bhaumik M, Maity K (2018) Effect of different tool materials during EDM performance of titanium grade 6 alloy. Eng Sci Technol Int J 21(3):507–516

Chiou AH, Tsao CC, Hsu CY (2015) A study of the machining characteristics of micro EDM milling and its improvement by electrode coating. Int J Adv Manuf Technol 78:1857–1864

Dutta H, Debnath K, Sarma DK (2019) A study of material removal and surface characteristics in micro-electrical discharge machining of carbon fiber reinforced plastics. Polym Compos 40(10):4033–4041

Dutta H, Debnath K, Sarma DK (2020) A study of wire electrical discharge machining of carbon fibre reinforced plastic. In: Shunmugam M, Kanthababu M (eds) Advances in unconventional machining and composites. Lecture notes on multidisciplinary industrial engineering, Springer, Singapore, pp 451–460. (Proceedings of the AIMTDR 2018)

Dutta H, Debnath K, Sarma DK (2020). Improving the performance of μED-milling using assisting electrode for fabricating micro-channels in CFRP composites. Mater Today Proc. https://doi.org/10.1016/j.matpr.2019.12.293

D'Urso G, Giardini C, Quarto M (2018) Characterization of surfaces obtained bymicro-EDM milling on steel and ceramic components. Int J Adv Manuf Technol 97:1–9

Feng W, Chu X, Hong Y, Zhang L (2018) Studies on the surface of high-performance alloys machined by micro-EDM. Mater Manuf Process 33:616–625

Gil R, Sanchez JA, Plaza S, Ortega N, Izquierdo B, Pombo I (2014) Modeling recast layerand surface finish in the manufacturing of high—aspect ratio micro-tools usingthe inverse slab electrical discharge milling process. Proc Inst Mech Eng B J Eng Manuf 228:553–562

Habib S, Okada A (2016) Influence of electrical discharge machining parameters on cutting parameters of carbon fiber-reinforced plastic. Mach Sci Technol 20(1):99–114

Islam MM, Li CP, Won SJ, Ko TJ (2017) A deburring strategy in drilled hole of CFRP composites using EDM process. J Alloys Compd 703:477–485

Jafferson JM, Hariharan P, Ram Kumar J (2016) Effect of nonelectrical parameters in μED milling: an experimental investigation. Int J Adv Manuf Technol 85:2037–2047

Kaneko T, Tsuchiya M (1988) Three-dimensional numerically controlled contouring by electric discharge machining with compensation for the deformation of cylindrical tool electrodes. Prec Eng 10(3):157–163

Karthikeyan G, Garg AK, Ramkumar J, Dhamodaran S (2012a) A microscopic investigation of machining behavior in μED-milling process. J Manuf Process 14:297–306

Karthikeyan G, Ramkumar J, Aravindan S (2012b) Performance analysis of μED-milling process using various statistical techniques. Int J Mach Mach Mater 11:183–203

Korlos A, Tzetzis D, Mansour G (2014) Investigation of the electro-discharge open-hole machining on the structural behavior of carbon fiber reinforced polymers. Appl Mech Mater 657:321–325

Kumar R, Agrawal PK, Singh I (2018a) Fabrication of micro holes in CFRP laminates using EDM. J Manuf Processes 31:859–866

Kumar R, Kumar A, Singh I (2018b) Electric discharge drilling of micro holes in CFRP laminates. J Mater Process Technol 259:150–158

Kuriachen B, Mathew J (2015) Experimental investigations into the effects of microelectric-discharge milling process parameters on processing Ti–6Al–4 V. Mater Manuf Process 30:983–990

Kuriachen B, Mathew J (2016) Effect of powder mixed dielectric on material removal and surface modification in microelectric discharge machining of Ti–6Al–4V. Mater Manuf Process 31:439–446

Kurniawan R, Kumaran ST, Prabu VA, Zhen Y, Park KM, Kwak YI, Islam MM, Ko TJ (2017) Measurement of burr removal rate and analysis of machining parameters in ultrasonic assisted dry EDM (US-EDM) for deburring drilled holes in CFRP composite. Measurement 110:98–115

Lau WS, Lee WB (1991) A comparison between EDM wire-cut and laser cutting of carbon fiber composite materials. Mater Manuf Process 6(2):331–342

Lau WS, Wang M, Lee WB (1990) Electrical discharge machining of carbon fiber composite materials. Int J Mach Tools Manuf 30(2):297–308

Lodhi BK, Verma D, Shukla R (2014) Optimization of machining parameters in EDM of CFRP composite using Taguchi technique. International Journal of Mechanical Engineering and Technology 5(10):70–77

Madhavan V, Lipczynski G, Lane B, Whitenton E (2015) Fiber orientation angle effects in machining of unidirectional CFRP laminated composites. J Manuf Process 20:431–442

Mehfuz R, Ali MY (2009) Investigation of machining parameters for the multiple-response optimization of micro electrodischarge milling. Int J Adv Manuf Technol 43:264–275

Park SH, Kim G, Lee W, Min BK, Lee SW, Kim TG (2015) Microhole machining on precision CFRP components using electrical discharging machining. In: 20th international conference on composite materials, Copenhagen, Denmark

Prakash V, Kumar P, Singh P, Hussain M, Das A, Chattopadhyaya S (2019) Micro-electrical discharge machining of difficult-to-machine materials: a review. Proc Inst Mech Eng B J Eng Manuf 233:339–370

Saito K, Kishinami T, Konno H, Sato M, Takeyama H (1986) Development of numerical contouring control electrical discharge machining (NCC-EDM). CIRP Annals—Manufac Tech 35(1):117–120

Sheikh-Ahmad JY (2016) Hole quality and damage in drilling carbon/epoxy composites by electrical discharge machining. Mater Manuf Processes 31(7):941–950

Sheikh-Ahmad JY, Shinde SR (2016) Machinability of carbon/epoxy composites by electrical discharge machining. Int J Mach Mach Mater 18(1–2):3–17

Singh AK, Patowari PK, Deshpande NV (2016) Experimental analysis of reverse micro-EDM for machining microtool. Mater Manuf Process 31:530–540

Singh AK, Patowari PK, Deshpande NV (2017) Effect of tool wear on microrods fabrication using reverse μEDM. Mater Manuf Process 32:1–8

Teicher U, Muller S, Munzner J, Nestler A (2013) Micro-EDM of carbon fibre-reinforced plastics. Procedia CIRP 6:320–325

Tsai YY, Masuzawa T (2004) An index to evaluate the wear resistance of the electrode in micro-EDM. J Mater Process Technol 149:304–309

Unune DR, Mali HS (2017) Experimental investigation on lowfrequency vibration-assisted μ-ED milling of Inconel 718. Mater Manuf Process 33:964–976

Wang F, Yin J, Ma JW, Jia ZY, Yang Y, Niu B (2017) Effects of cutting edge radius and fiber cutting angle on the cutting-induced surface damage in machining of unidirectional CFRP composite laminates. Int J Adv Manuf Technol 91(9–12):3107–3120

Yang F, Qian J, Wang J, Reynaerts D (2018) Simulation and experimental analysis of alternating-current phenomenon in micro-EDM with a RC-type generator. J Mater Process Technol 255:865–875

Yue X, Yang X, Tian J, He Z, Fan Y (2018) Thermal, mechanical and chemical material removal mechanism of carbon fiber reinforced polymers in electrical discharge machining. Int J Mach Tools Manuf 133:4–17

Zhang L, Du J, Zhuang X, Wang Z, Pei J (2015) Geometric prediction of conic tool in micro EDM milling with fix-length compensation using simulation. Int J Mach Tools Manuf 89:86–94

Controlled Depth Milling of Hybrid Aerospace Grade Materials Using Abrasive Water Jet – Critical Review and Analysis

X. Sourd, R. Zitoune, L. Crouzeix, and D. Lamouche

Abstract Due to the expensive price of their structures, both in terms of material and manufacturing costs, aerospace manufacturers prefer to perform repair operations as much as possible in case of damage. In this context, the damaged area is removed by conventional or non-conventional milling processes, depending on the material constitutive of the structure to repair. However, all these techniques present issues in case of aerospace grade materials (e.g. Carbon Fibers Reinforced Plastics—CFRPs—or titanium alloys) machining such as excessive tool wear, thermal damage or use of corrosive products. In this context, Abrasive Water Jet (AWJ) machining has proven to be a good alternative to these processes. However, some issues have still to be faced when machining with AWJ. This review presents the influence of the machining parameters on the material removal features (viz. depth of cut and Material Removal Rate) and on the types and expanse of the machine-induced defects and damage when using AWJ for machining of CFRP and metallic materials (mainly aerospace grade titanium alloy Ti6Al4V). Moreover, the link between the machining process and the modifications in the mechanical behaviour are also discussed.

Keywords Abrasive water jet machining · Surface quality · Material integrity

1 Introduction

Composite materials are widely employed for their high strength to weight ratio in many fields such as sporting goods, robotics or transportation (train and aviation). In the specific field of aeronautics, this increasing use is emphasized by the presence of aircrafts' outer parts made of composite materials (monolithic and/or hybrid structures). However, aircrafts are subjected to several types of damage during flight or

X. Sourd · R. Zitoune (✉) · L. Crouzeix
Institut Clément Ader, UMR 5312, CNRS, 3 Rue Caroline Aigle. 31400, Toulouse, France
e-mail: redouane.zitoune@iut-tlse3.fr

X. Sourd · D. Lamouche
Safran Aircraft Engines (Villaroche). Rond-Point René Ravaud, 77550 Moissy-Cramayel, France

M. T. Hameed Sultan et al. (eds.), *Machining and Machinability of Fiber Reinforced Polymer Composites*, Composites Science and Technology,
https://doi.org/10.1007/978-981-33-4153-1_9

on ground such as erosion or impact, which implies maintenance in order to keep the airplanes in service. Due to the expensive price of some parts (material cost and manufacturing process), complete replacement is avoided as much as possible and the option of repair is strongly recommended. This operation is performed by removing the damaged area. Depending on the type of material and structure needing repairs, various processes of machining are already employed by aircraft manufacturers. In case of monolithic or sandwich parts (metallic or composite materials), the main process of machining used by the aircraft manufacturers is conventional milling (Crouzeix et al. 2012). However, conventional machining of composite materials presents challenging tasks. Indeed, the contact of the tool with the workpiece leads to two main problems when dealing with composite materials. On one hand, the high temperature generated during machining can damage heat-sensitive materials such as Carbon Fibers Reinforced Plastics (CFRP) (Nguyen-Dinh et al. 2019). On the other hand, the heterogeneity and abrasive character of the carbon fibers are responsible of the premature wear of the cutting tool. Moreover, the physical interaction between the rigid tool and the workpiece induces cutting forces, needing the machined part to be clamped which can be difficult in case of complex shapes. In addition, removing a constant depth of cut on composite structure characterized by a complex form with conventional process of machining requires a Computer Numerical Control (CNC) machine with 5 axes. In this case the machining cost is drastically increased. From these considerations, non-conventional processes have been developed, such as Electro-Discharge Machining (EDM), Laser Beam Machining (LBM) or chemical machining. However, electro-discharge machining efficiency is highly dependent on the conductivity of the materials, making this method mainly dedicated for metallic materials. Laser beam machining generates heat-affected zones, which can be problematic for heat sensitive materials such as composites. Besides, LBM is expensive because of its low Material Removal Rate (MRR), defined as the volume (or the mass) of machined material per second. When thick workpieces are considered, this process of machining is not recommended (Hashish 1989). In the case of hybrid materials such as CFRP/Titanium, chemical machining process can be employed in the aerospace industry. Nonetheless, the MRR is small compared to other process. Moreover, chemical machining is based on the use of highly corrosive products, affecting both the operator's health and the environment.

Based on these observations, Abrasive Water Jet (AWJ) machining process seems to be an efficient machining method thanks to its unique abilities. With this process, the material removal is performed thanks to a coupling between high velocity impact and erosion. Indeed, by analysing the kerf profile generated by AWJ machining in Plexiglas, Hashish (Hashish 1984), observed the presence of two main material removal mechanisms, which are "cutting mode" (Finnie 1958) and "deformation mode" (Bitter 1963a, b). Based on the work of Finnie (Finnie 1958), the cutting mode of erosion is due to the particle impact on the material surface at small angles of attack. In this case, a shear load exceeding the shear strength of the material is generated. In fact, this type of cutting occurs at a material portion close to the surface which exhibits a texture characterized by high quality (low roughness and waviness). However, when the impact of abrasive particles at large angle occurs deeper in the workpiece, the material removal occurs by plastic deformation and creation of microchips (Bitter 1963a,b). The area subjected to the deformation mode of erosion can be highlighted by a wavy surface texture and grooming. In fact, with the AWJ process, the material removal occurring by high speed impact of the abrasive particles and erosion is suitable to machine a wide range of materials whatever their machinability or mechanical properties (ceramics, metals, and composites) (Hashish 1984, 1989; Ramulu and Arola 1994; Axinte et al. 2010; Zitoune et al. 2013). In addition, with this process of machining, the heat-affected zones, generated by the impact of the abrasive grits on the work-piece material, are considerably reduced due to the cooling effect of the water. This feature favours the machining of heat-sensitive materials like composites. Moreover, with AWJ machining process, no high clamping forces are required to maintain the workpiece compared to conventional process. This allows to machine easily low stiffness materials without out of plane deflection (van Luttervelt 1989). In addition, AWJ is able to perform different machining operations, such as cutting (Hashish 1984), milling (Hashish 1989), trimming (Arola and Ramulu 1994), drilling (Phapale et al. 2016), turning (Hloch et al. 2014) or peening (Arola and McCain 2000). In the work of Hashish (Hashish 1989), various non-conventional 3D machining methods have been conducted on metals. It was shown that AWJ gives the best compromise between high MRR and low surface roughness. Moreover, contrary to conventional milling which removes material to a distance from a reference plane, AWJ milling machines a constant depth from the upper surface of the workpiece. This makes AWJ milling a suitable technique for bended parts or performing ply-per-ply machining of composite materials (Cénac et al. 2008). Finally, from the environmental point of view, waterjet machining can be considered as an environmentally friendly solution compared to other processes of machining. Waterjet machine generally possesses a water tank below its machining table, permitting to collect almost all the water contaminated both by remaining abrasive grits and material chips (dust particles). With a proper filtration system, it is possible to clean and reuse the water.

Though waterjet machining presents undeniable advantages over conventional or other non-conventional methods, some issues persist such as difficulty to control the depth of cut or variability of the machining quality due to the heterogeneity of the workpiece material. Indeed, the time dependence of the process can produce non-uniform erosion and induce various quality, particularly in case of decelerations or accelerations at jet direction changes. The tool path has thus to be studied in order to minimize the variations of traverse speed and scan step during the machining while trying to obtain the wanted geometry. Moreover, as explained previously, the footprint generated by waterjet machining is greatly influenced by the process parameters. Which can be considered as flexibility becomes an issue when the parameters evolve in a stochastic way over the machining time, also leading to heterogeneous erosion (Lozano Torrubia et al. 2015).

It is important to mention that every machining technique generates its own types of damage depending on the process parameters and the materials constitutive of the workpiece. Indeed, mechanisms of material removal affect the geometry and quality of the surface, inducing external and internal defects and damage which can reduce the mechanical performances of the machined part (Saleem et al. 2013). In addition, in a context of repair application by adhesive bonding method the AWJ process can be considered for texturing the surfaces before bonding.

The goal of this review is to sum up and analyse studies dealing with controlled-depth abrasive water jet machining (i.e. milling) of aerospace grade materials such as metals (and more especially titanium which is extensively used in the aeronautical field) and composite materials with organic matrix (mainly CFRP). This review is organized into three main topics.

The first part focuses on milling efficiency of the abrasive water jet technique both in terms of depth of cut (and/or MRR) and generated surface quality. The influence of the main process parameters (e.g. jet pressure, traverse speed and abrasive size) on the depth of cut and the surface quality is discussed in a first section. At first glance, the geometrical characteristics (i.e. depth of cut and surface quality) of the machined specimens using waterjet are influenced by complex combinations of process parameters. However, after deeper analysis, it is confirmed that some of these parameters are particularly influent compared to other of secondary importance. Then, for industrial concerns, the prediction of the depth of cut as a function of the machining parameters is necessary. Different models have been proposed over the years and can be classified into four categories viz. analytical/geometric modelling, artificial intelligence, numerical simulation and statistical approaches. Some examples amongst each of these models are presented in a second section. Moreover, in case of milling with water jet, the selection of a proper set of machining parameters is necessary but not sufficient to guaranty a controlled depth over the machined area. Indeed, the path followed by the nozzle has to be correctly designed depending on the shape of the workpiece to machine. Hence, some tool path strategies used for milling with waterjet are presented in the last section.

The second part of the review covers the geometrical and material changes consecutive to AWJ milling of both titanium and composite parts. First, the main machine-induced defects are presented. It has been shown that the main defects on metals are in form of micropits located around the grain boundaries and cracks. The addition of abrasive grits also introduces various texturing of the machined surface, mainly depending on the nozzle's traverse speed and inclination (Fowler et al. 2005a). In addition, waterjet milling of a metallic material hardens its microstructure through few tens of microns below the machined surface and induces a bi-axial compressive residual stress state (Arola and McCain 2000; Arola et al. 2002). Water jet machining (without abrasive grits) of composite laminates seems to be prohibited because of high probability of delamination. However, AWJ machining produces several kinds of defects and damage such as broken fibers or craters of various sizes which are linked to the selected set of process parameters. In addition, grit embedment within the milled surface is observed. Depending on the machining parameters and the milled material, grits can be trapped at various depths, acting like stress concentration sites and contaminating the machined surface. These issues can be problematic respectively for the mechanical behaviour of the milled part and its adhesion properties in case of repairs by bonding. In this context, several authors (Hashish 1998; Arola and McCain 2000; Fowler et al. 2005a; Kong et al. 2011; Liu et al. 2012; Huang et al. 2013) proposed some solutions in order to reduce grit embedment such as tilting the nozzle when machining, milling backwards or adding a post-machining cleaning operation with Plain Water Jet (PWJ—i.e. without abrasive particles). However, these different solutions can induce several modifications of the surface and material properties.

The last part of the review focuses on the effect of the waterjet milling process on the mechanical behaviour during static and fatigue loading. In the case of composite laminates, Hejjaji et al. (2017; Hejjaji 2018) have studied the influence of the craters size, induced by the machining parameters of the AWJ process, on the mechanical properties such as endurance limit and tensile strength. It was clearly observed that the roughness criteria (Ra and Sa) usually employed for metallic material cannot be used for composite materials. This conducts the authors to propose an innovative criterion called "crater volume" which presents a good correlation between the surface quality and the material integrity of the specimens consecutive to AWJ machining. However, in case of metallic materials, two phenomena compete. On one hand, compressive residual stresses have a beneficial effect on the endurance limit of the machined specimens. This can be explained by the fact that compressive residual stresses favour the closing of cracks induced by the milling operation (Tönshoff et al. 1997; Ramulu et al. 2002; Arola et al. 2006). On the other hand, when the material removal rate becomes too high, the number of defects greatly increases, leading to a degradation of the fatigue properties (Arola et al. 2002; Azhari et al. 2016; Lieblich et al. 2016). The set of machining parameters has then to be carefully chosen in order to find a compromise between the amounts of MRR and generated residual stresses.

2 Machining Efficiency and Surface Quality

From the industrial point of view, deadlines are of primary concern. In this context, when dealing with machining, the goal is to obtain the final shape of the structures in a minimum amount of time. The machining performances are then characterized through the Material Removal Rate (MRR). In addition, geometrical tolerances are also important. Hence, the depth of cut as well as the post-machining surface quality has to be mastered for each machining process. Several studies on metals (Hashish 1989; Kong et al. 2011) and composites materials (Srinivasu and Axinte 2014a, b) showed that these features are material dependent. However, different materials belonging to the same category presented similar behaviours. Hence, it is interesting to differentiate homogeneous materials (such as metals) from heterogeneous ones (composites). The following part details the effect of the AWJ machining process parameters on the machining efficiency (i.e. the MRR and the depth of cut) and the resulting surface quality of both metals and composite materials.

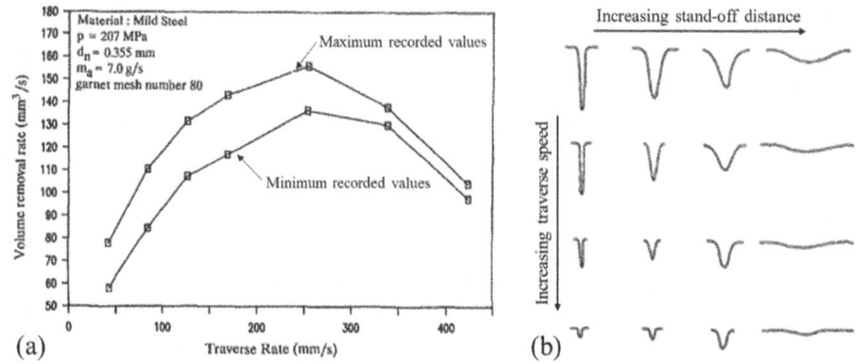

Fig. 1 **a** Evolution of the MRR vs. the traverse speed when milling mild steel and **b** schematic view showing the effect of stand-off distance and traverse speed on kerf's shape (from (Hashish 1989))

2.1 *Influence of the Process Parameters*

2.1.1 On Material Removal

Metals

The first work of trench milling with AWJ was conducted by Hashish (1989) on metallic materials in the late 80s. In order to study the influence of the processing parameters on the main features describing the material removal performances (viz. MRR and depth of cut), the author has investigated several materials (mild steel, aluminium, titanium). The analysis of the evolution of the MRR over large ranges of traverse speed (from 42 to 420 mm/s) and stand-off distance—SOD— (from 2 to 76 mm) have shown that there are optimum values, depending on the other processing parameters and the machined material, leading to a maximum MRR. For example, in order to maximise the MRR when milling mild steel, the optimal traverse speed used is around 250 mm/min for a given set of machining parameters (cf. Fig. 1a). This can be explained by the fact that, although the traverse speed and the SOD influence the depth and width of the kerf in opposite ways (cf. Fig. 1b), it is possible to find a trench which profile has the greatest area in each row (fixed traverse speed) or column (fixed SOD). It was also mentioned that the optimal value of the SOD should be kept between 2 and 5 mm (avoiding the loss of jet focus) to ensure a maximal MRR.

By varying the size and hardness of abrasives particles, Hashish (1989) and Fowler et al. (2009) have observed that both of these parameters influence the MRR. For soft and small grits, a low MRR is recorded and for hard and greater particles a higher MRR is produced (e.g. respectively 37 and 92.6 mm^3/s for grits of 100# and 60#, the other machining parameters being set). In another work, Shipway et al. (2005) have focused on milling of titanium alloy (Ti6Al4V) with AWJ process and similarly

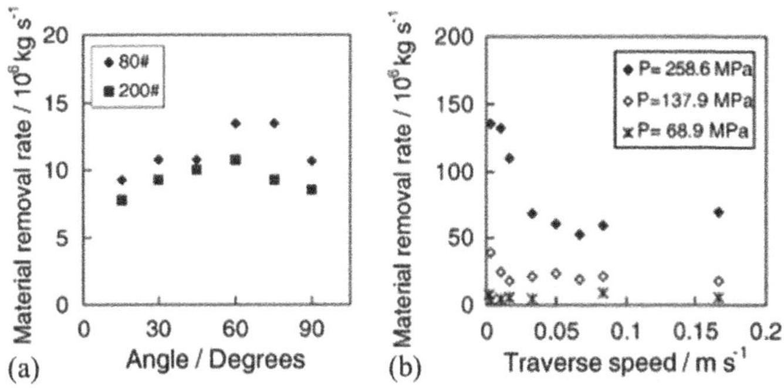

Fig. 2 Influence of the machining parameters on the MRR with: **a** impingement angle and abrasive size, **b** traverse speed and jet pressure. Parameters other than listed are: grit size = 180 μm (80#), SOD = 3 mm, impingement angle = 90°, jet pressure = 137.9 MPa, traverse speed = 5 m/s (Shipway et al. 2005)

noticed that the MRR increases with the abrasive particle size (cf. Fig. 2a) as well as the jet pressure (cf. Fig. 2b). Indeed, an increase in the jet pressure from 68.9 to 258.6 MPa leads to an increase of the MRR in a factor up to 10. According to Pal and Choudhury (2014), who performed AWJ pocket milling in 5 mm thick plates of titanium alloy (Ti6Al4V), an increase in jet pressure and/or grit size favours the increase in MRR. This result has been explained by the fact that the increase in jet pressure and the grit size leads to an increase in kinetic energy given to the abrasive particles and the water droplets to erode the material. This statement is in good agreement with the work conducted by Chillman et al. (2010a, b, 2007) on high-pressure (600 MPa) PWJ machining of titanium alloy Ti6Al4V. Besides, for a given set of parameters, Shipway et al. (2005) showed that an optimal value of impingement angle (cf. Fig. 2a) gives the highest MRR. This optimum is coherent with the conclusions drawn by Kong et al. (2012) who observed that the maximal erosion rate of ductile materials (as Ti6Al4V) occurs for impingement angles around 70°. Indeed, if milling is conducted with an optimized angle (cf. Fig. 2a), the recorded MRR is around 30% superior to the case of milling with normal angle (e.g. with 80# particles, the MRR are respectively 14.10^{-6} kg/s and 11.10^{-6} kg/s for impingement angles of 70° and 90°).

Contrarily to Hashish (1989), Shipway et al. (2005) found that the MRR decreases when the traverse speed increases over a range of 3 to 166 mm/s. This trend is supported by the work of Fowler et al. (2005b) who also studied the effect of the traverse speed on MRR but over a larger range of values (from 3 to 5000 mm/min). Indeed, they noticed that a significant drop of MRR is observed when traverse speed increases till 0.05 m/s (cf. Fig. 2b). However, when milling is conducted with a higher speed, the MRR remains stable (around 50.10^{-6} kg/s for a jet pressure of 258.6 MPa). These results can be explained by the change in material removal mechanisms, strongly affected by the transverse speed, the nature of the material, as well as the impingement angle which can be modified with the depth of cut. Indeed, micro-cutting mechanism occurring at low impact angle leads to a high erosion rate

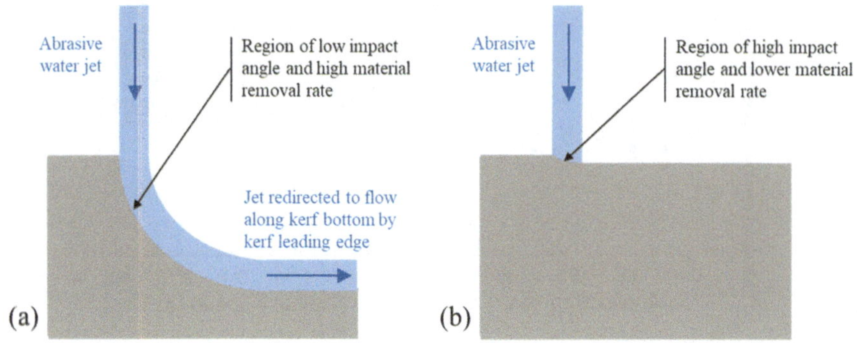

Fig. 3 Schematic diagram of flow patterns in AWJ milling. **a** Low traverse speed developing a large kerf leading edge; **b** high traverse speed developing a small kerf leading edge (from (Fowler et al. 2005b))

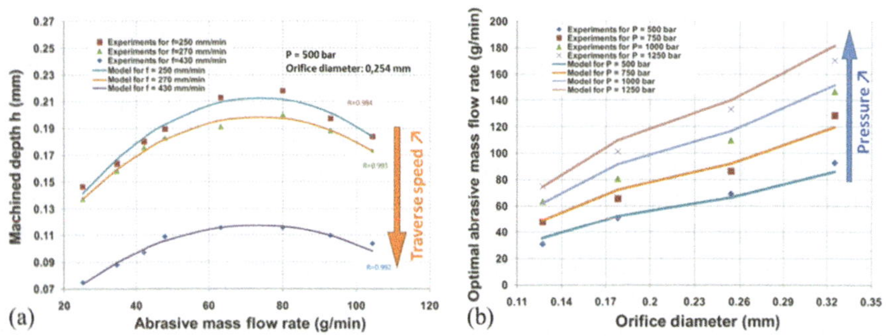

Fig. 4 **a** Machined depth versus the AFR for different jet pressures and **b** evolution of the optimum AFR with the jet pressure and the orifice diameter (Cénac et al. 2013)

according to Bitter theory of erosion (Bitter 1963a, b). In fact, as it can be seen from Fig. 3, a deep kerf is formed at low traverse speeds which results to a material removal mainly occurring at low impact angle. On the contrary, at high traverse speeds, a shallower kerf is formed, and the region of high impact angle is greatly reduced. The same evolution of MRR with an increase of the traverse speed was found by Gupta et al. (2013) when performing pocket milling of stainless steel (SS304).

In another work, Cénac et al. (2013) performed AWJ milling of aluminium alloy 2024-T3 and clearly showed the existence of an optimum value of Abrasive Flow Rate (AFR). This optimum value, which depends on the other process parameters, is responsible of the deepest machining. From the Fig. 4a, which represents the evolution of the milling depth (measured and predicted values) as a function of the AFR, it can be observed that the milling depth increases for an AFR inferior or equal to 63 g/min then decreases for AFR values superior to 80 g/min. The increasing phenomenon of the milling depth can be related to the important quantity of abrasive particles taking part in the material removal mechanisms. However, for AFR values

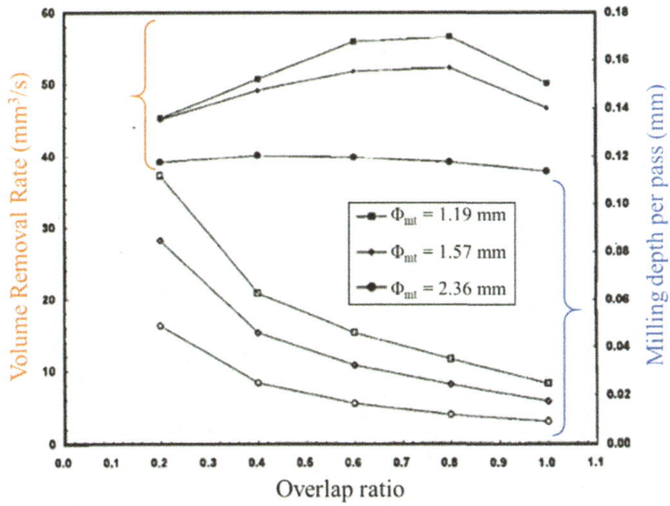

Fig. 5 Evolution of the MRR and the depth of cut per pass with respect to the overlap ratio for different mixing tube diameters (Φ_{mt}) (Hashish 1995)

superior to 80 g/min, the milling depth decreases due to the conservation of total momentum within the jet provided by fluid mechanics. In this case, the velocity of the particles for a given jet pressure decreases because the particles are more numerous, reducing their individual impact energy by hitting each other. It can also be observed that the traverse speed has a minor effect on the available energy within the jet, thus on the optimum value of AFR. Indeed, the optimum AFR remains at around 70 g/min when milling is conducted with traverse speeds of 250, 270 and 430 mm/min. In the same work, Cénac et al. (2013) have shown that the augmentation of the optimum value of the AFR can be performed when milling is conducted with higher values of jet pressure and orifice diameter (cf. Fig. 4b). This can be related to the increase of the available energy for the particles, mainly piloted by the jet pressure.

Hashish (Hashish 1995) studied the degree of overlap (overlap ratio), defined as the scan step (i.e. the distance between two consecutive milled trenches) over the diameter of the mixing tube, when milling aluminium. He concluded that the overlap ratio is a crucial parameter influencing the material removal performances. Indeed, as seen from Fig. 5, the MRR reaches its maximum value for an overlap ratio of around 75% and the milling depth per pass decreases as the overlap ratio increases (a drop of 70% was recorded when varying the overlap ratio from 20 to 80%). The same trend was observed by Kanthababu et al. (2016) concerning the depth of cut. Indeed, several works concerning metals (Alberdi et al. 2011; Anwar et al. 2013a, b; Billingham et al. 2013) proved the influence of the overlap ratio on erosion regime (cf. Fig. 6). Billingham et al. (2013) distinguished two ways of modelling overlapped trenches profile depending on the regime of erosion. Indeed, for two consecutive trenches and in the case of the linear regime (shallow kerfs), there is no influence

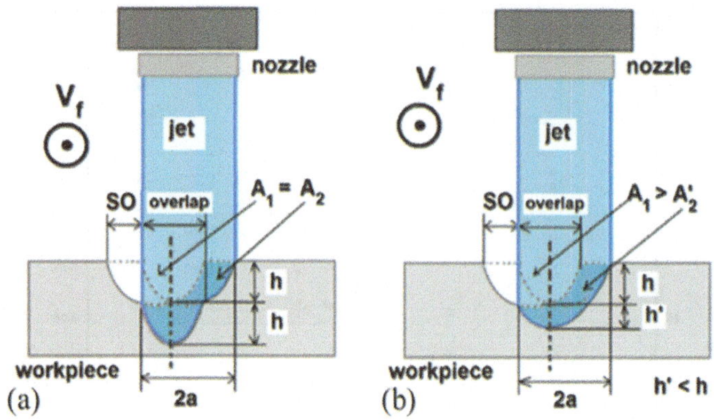

Fig. 6 Differences in overlapped footprints in case of **a** linear and **b** non-linear regimes of erosion (Billingham et al. 2013)

of a trench on the next one. Whatever the degree of overlap, the final profile is the algebraic sum of the N single trenches separated from each other by the scan step. It was supposed that the final profile is symmetric, and each pass removes the same amount of material (cf. Fig. 6a). In the case of non-linear regime (deeper kerfs), each trench influences the initial conditions of the next one so that the initial profile of the *k*th kerf is the profile after the k-1 previous kerfs. Thereby, the final profile is not symmetric anymore because of the difference in impingement on the inner and outer sides of the trench edge and each pass removes less material than the previous one (cf. Fig. 6b). When increasing the scan step, the non-linearity decreases until disappearing when the scan step is greater than the nozzle diameter.

Composites

In order to analyse the effect of machining parameters (jet pressure, AFR, traverse speed, SOD and scan step) on the material removal when milling composite materials, Hocheng et al. (1997) have conducted an experimental work during milling of thick CFRP laminate (8 mm of thickness, 64 plies with a stacking sequence of $[0/90]_{32}$). As seen from Fig. 7, when machining composite materials, the MRR also depends on the abrasive flow rate, the jet pressure and the traverse speed, increasing when one or more of these parameters increase. Indeed, when the abrasive flow rate varies from 100 to 200 g/min, the MRR increases from 120 to 155 mm^3/s. This can be explained by the fact that, when the AFR increases, more abrasive particles take part in the material removal mechanisms. However, when AWJ milling CFRP (HexPLY M21/35%/268/T700GC), Zitoune et al. (2013) showed the same tendency (parabolic evolution) of the depth of cut when increasing the AFR that Cénac et al. (2013) with aluminium.

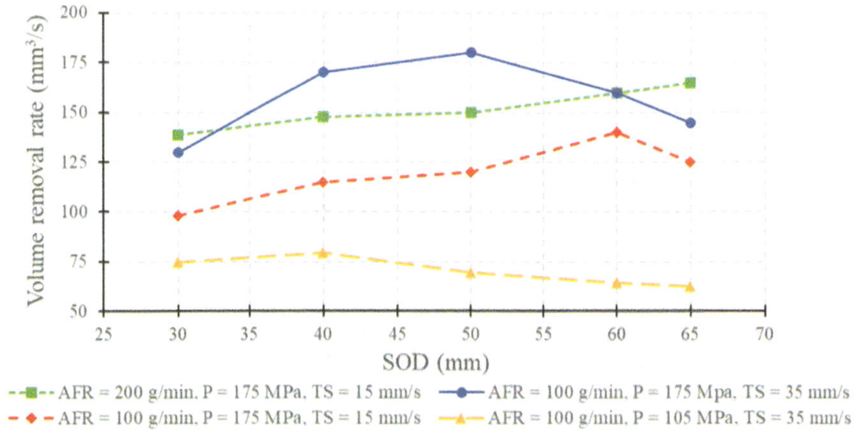

Fig. 7 Influence of the stand-off distance (SOD), abrasive flow rate (AFR), pressure (P) and traverse speed (TS) on MRR (from results of (Hocheng et al. 1997))

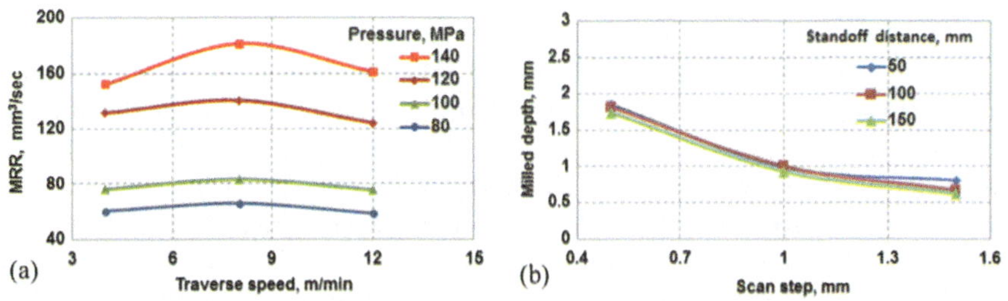

Fig. 8 Influence of **a** the pressure and traverse speed on MRR and **b** the stand-off distance and the scan step on the depth of cut (Hejjaji et al. 2017)

Moreover, Hocheng et al. (1997) observed that an increase in jet pressure from 105 to 175 MPa induces an augmentation of MRR by a factor two (cf. Fig. 7). The same trend was observed by Hejjaji et al. (Hejjaji et al. 2017) when performing AWJ pocket milling on unidirectional (UD) laminates of CFRP (Hexply T700-M21.A). As seen from Fig. 8a, when the jet pressure rises from 80to 140 MPa, the MRR is greater by a factor 2.7. It has to be highlighted that for all the tested pressures, a traverse speed of around 8 m/min leads to an optimum value of MRR. This is similar to the conclusions drawn by Hashish (1989) on mild steel presented in Sect. 2.1.1.1. The results of Hocheng et al. (1997) are also coherent with the previous works (Cénac et al. 2013; Zitoune et al. 2013). Indeed, the authors recorded an increase of 30% of the MRR when the jet traverse speed is raised from 15 to 35 mm/s (cf. Fig. 7). This augmentation is in good agreement with other works (Hashish 1989; Hejjaji et al. 2017), where the low values of traverse speed correspond to the left part of the concave curves. The significant influence of the jet pressure and the traverse speed on the material removal is consistent with the conclusions of numerous studies

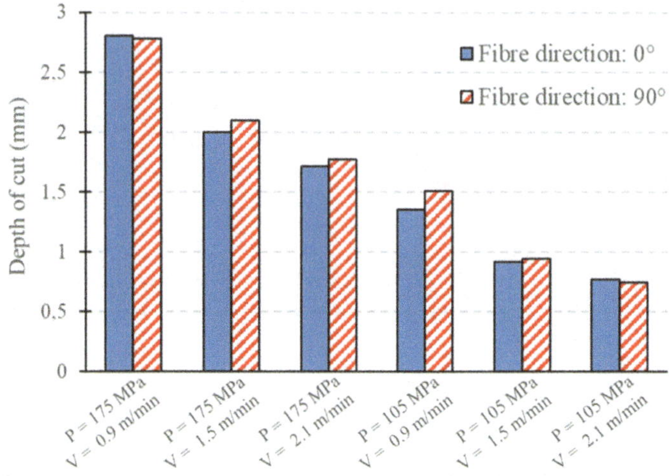

Fig. 9 Influence of the fiber orientation on the depth of cut AFR = 100 g/min—SOD = 60 mm—Garnet 80# (from results of (Hocheng et al. 1997))

(Chillman et al. 2007, 2010a, b; Azhari et al. 2012; Gupta et al. 2013) conducted on metals. However, the impact of the stand-off distance on the MRR is totally different compared to the effect of the jet pressure, the transverse speed as well as the AFR. In fact, in the works of Hejjaji et al. (2017) and Cénac et al. (2008, 2009) which focus on the milling of CFRP made of unidirectional laminate and GFRP made of 2D woven fabric with large ranges of stand-off distances (respectively from 50 to 150 mm and from 32 to 73 mm), it was noticed that this machining parameter neither has significant influence on the MRR, nor on the depth of cut (cf. Fig. 8b) when milling with high traverse speeds (> 5 m/min). For this reason, Cénac et al. (2009) concluded that AWJ milling is hence a suitable technique to perform ply-per-ply machining in composite laminates even in the case of slightly bended parts like fuselage panels. In the same way, Hocheng et al. (1997) have concluded that the orientation of the plies has no significant influence on the depth of cut, since the same mechanism of material removal (impact fracture) occurs in the composite. Indeed, as seen from Fig. 9, the maximum difference in depth of cut between 0° and 90° plies orientations for a given set of machining parameters is around 10%.

Based on the analysis conducted by Hejjaji et al. (2017) on several milled CFRP laminate coupons, the features related to material removal (depth of cut and MRR) greatly depend on the scan step. As a matter of fact, from Fig. 8b, when the distance between two consecutive trenches increases from 0.5 to 1.5 mm, the depth of cut is divided by a factor of 2.3 (from 1.75 to 0.75 mm). The reasons of this trend are the same than the ones described in Sect. 2.1.1.1., which brought out the importance of the degree of overlap in the erosion process (Alberdi et al. 2011; Anwar et al. 2013a, b; Billingham et al. 2013). Though multi-layer milling was not part of the study, Hocheng et al. (1997) presumed that the depth of cut obtained by milling a second pass might be greater than the first (up to 25%). Two main reasons could explain

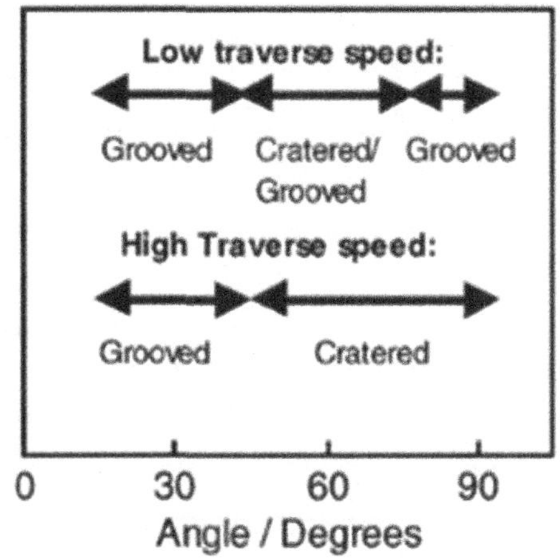

Fig. 10 Morphologies developed in milling of titanium alloy (Fowler et al. 2005a)

this assumption. Firstly, the potential micro-cracks and the texturing generated during milling the first layer would weaken the material and improve the removal. Secondly, the depth of the first milled layer permits to reduce the jet rebounds and thus to use a greater part of its energy in the wear process.

2.1.2 On Surface Quality

Metals

Fowler et al. (2005a) analysed the effect of abrasive particle size, traverse speed and impingement angle on the machined surface morphology by performing AWJ milling tests on titanium alloy (Ti6Al4V). As shown from Fig. 10, the authors observed that a mix of traverse speed and impingement angle conditions influences the surface morphology. For low attack angles (below 45°), the surface is groovy whatever the traverse speed is. The differences in morphology between surfaces machined with low and high speeds appear for impingement angles above 45°. At low speeds, a transitive grooved/cratered surface zone is noticed for impingement angles between 45° and 75°, before the reappearance of grooves when the angle keeps increasing. At higher speeds, the surface is cratered since the impingement angle is above 45°. Similarly, Shipway et al. (2005) found that the surface roughness increases with the traverse speed and impingement angle (below 60°). It decreases for impingement angles moving towards the normal. The authors observed that surface waviness decreases as the traverse speed increases. Variation in impingement angle has particular effects on the surface waviness: for low angles (below around 60°) the increase in impingement angle leads to an increase in surface waviness but for higher angles (towards the normal) it decreases. These trends correspond to the different surface morphologies presented in the work of Fowler et al. (2005a) and seem to link craters and grooves to surface roughness and waviness respectively. This assumption is supported by the

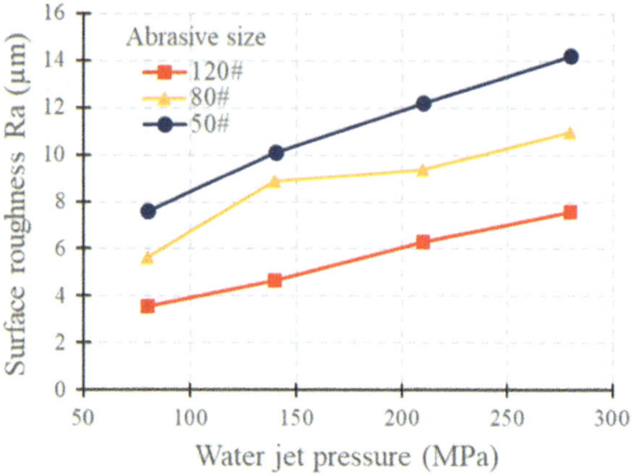

Fig. 11 Influence of the abrasive size and the waterjet pressure on the surface roughness (from results of (Arola et al. 2002))

results of other works on normal milling of titanium (Fowler et al. 2009) and stainless steel (Gupta et al. 2013) explaining that an increase in the traverse speed leads to a rougher but flatter surface.

Arola et al. (2002) performed AWJ peening where the jet is normal to the titanium alloy (Ti6Al4V) specimen to investigate the influence of water jet pressure and abrasive particle size on surface roughness. The stand-off distance (150 mm) and the traverse speed (3.81 m/min) are set for all the tests. The authors have mentioned that surface roughness increases with an increase in jet pressure and/or in abrasive size (cf. Fig. 11). Indeed, when a small grit is used (e.g. 120# instead of 50#), the surface roughness is divided by a factor two. Similarly, when machining with a pressure of 80 MPa instead of 280 MPa, the surface is also twice smoother. These observations were later on supported by other authors testing different metals (Arola et al. 2006; Azhari et al. 2012; Lieblich et al. 2016). Several works studying peening (Azhari et al. 2012; Lieblich et al. 2016) or milling (Shipway et al. 2005; Pal and Choudhury 2014) operations draw the same conclusions concerning the machined surface waviness. Azhari et al. (2012) explained that increasing the kinetic energy of the particles permits to remove bigger parts of the material because of a more severe erosion, then leading to a poorer quality. A compromise has to be made between the machining time and the wanted surface quality.

Gupta et al. (2013) and Vasanth et al. (2016) have performed AWJ milling of stainless steel (SS304) and titanium alloy (Ti6Al4V) respectively. They found out that the SOD has a particular influence on the surface roughness. If the nozzle is too close to the machined surface, the abrasive particles are dampened by the workpiece. On the other hand, if the distance between the nozzle and the machined surface is too important, the jet is scattered. In both of these cases, the surface has an important roughness. Thus, the goal is to find on compromise between these two situations. As for SOD, surface roughness is sensitive to the abrasive flow rate. If the AFR is too low, the material will be mainly deformed rather than cut. On the other hand, if the number of abrasive particles in the jet is too important, the particles collided, and the jet became turbulent. In both these cases, it results in a high roughness

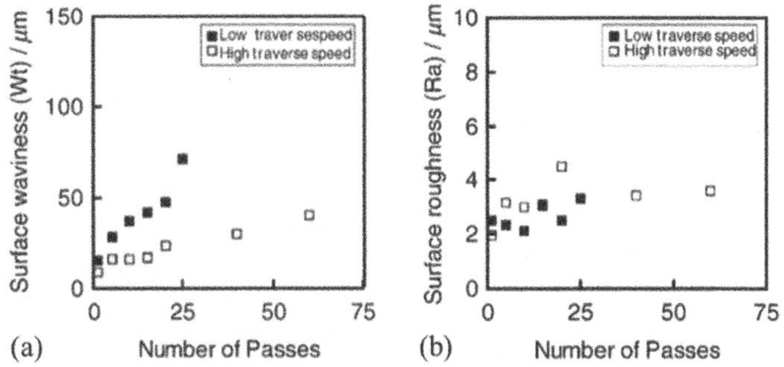

Fig. 12 Surface waviness **a** and roughness **b** developed during linear AWJ-CDM of Ti6Al4V as a function of number of jet passes at high (0.166 ms^{-1}) and low (0.05 ms^{-1}) jet traverse speeds with 200# garnet grit (Fowler et al. 2005b)

surface. The authors noticed that a high abrasive flow rate combined with a high stand-off distance lead to a large and random energy distribution, thus to a high surface roughness. Nevertheless, the limited ranges of SOD and AFR studied in both the works and the small variations in surface roughness (between 4.7 and 6.1 μm) does not seem to promote a clear influence of these two machining parameters.

Fowler et al. (2005b) studied the effect of the number of passes on surface roughness and waviness parallel to the jet path. The surface waviness, as a macro-scale measurement, is not reduced by multiple jet passes but rather exaggerated. The same trend was observed by Hashish (Hashish 1989) who proposed to take into account the problem of finished surface quality since the beginning of the machining process. The lower the traverse speed, the faster the waviness increases with the number of passes. Indeed, as seen from Fig. 12a, the slope of the curve is greater for a low traverse speed than for a high one. On the contrary, the surface roughness is a micro-scale measurement: hence, it is independent on the number of passes (cf. Fig. 12b). Shipway et al. (2005) also noticed the same evolutions of longitudinal waviness and roughness with the number of passes. They explained that the surface roughness is a local measure of the surface profile mainly depending on impact conditions.

Composites

Hejjaji et al. (2017) studied the influence of the machining parameters on the AWJ milled surface quality of unidirectional CFRP (HexplyT700-M21.A). They found out that the surface roughness in both longitudinal (parallel to the milled trenches) and transverse directions depends on the jet pressure, the traverse speed and the SOD. As seen from Fig. 13a, an increase in jet pressure and/or traverse speed leads to an increase in transverse surface roughness (from around 6.5 μm at 4 m/min with a pressure of 80 MPa to almost 10 μm at 12 m/min with a pressure of 140 MPa). This is due to the fact that the jet pressure and traverse speed define the jet energy transmitted to the workpiece, thus the degree of erosion. Hence, the more eroded the

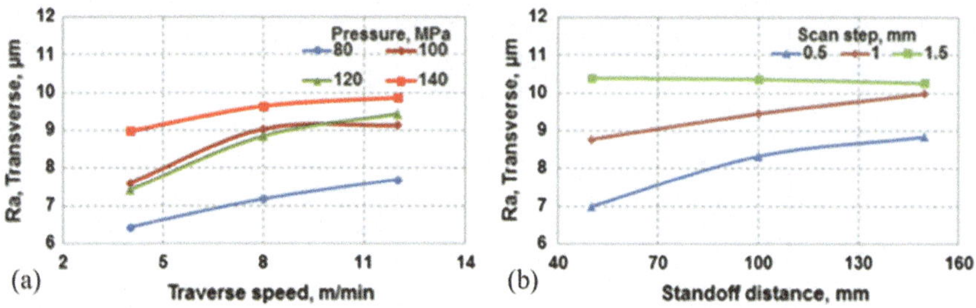

Fig. 13 Influence of **a** the jet pressure ad traverse speed and **b** the scan step and the SOD on the transverse surface roughness (Hejjaji et al. 2017)

material with a single jet pass, the rougher the surface (Azhari et al. 2012). The same evolution is observed for longitudinal roughness. The role of the SOD on transverse surface roughness (cf. Fig. 13b) is consistent with the work of Vasanth et al. (2016) on titanium. Indeed, an increase in SOD from 50 to 150 mm leads to a rougher surface (respectively 7 and almost 9 µm) when the scan step is equal to 0.5 mm. In case of high values of stand-off distance, the jet has a large span, leading to a large and random energy distribution, thus to a high surface roughness. However, it has to be noticed that when the distance between two consecutive trenches is superior to 1 mm, corresponding to the focusing nozzle diameter, the effect of the SOD becomes negligible. It then seems to prove that the overlap ratio modifies the influence of the SOD on surface roughness. The work of Hocheng et al. (1997) supports this assumption. They concluded that the width-to-depth (WTD) ratio of the kerf, in association with the scan step, influence the roughness of the milled pocket bottom. Indeed, according to the authors, milling multiple kerfs with a WTD ratio over two and a scan step around 40% of the kerf's width produced surfaces with low roughness values (around 6 µm) as shown from Fig. 14.

According to Hejjaji et al. (2017), longitudinal and transverse waviness are mainly influenced by the SOD and the scan step respectively. Though it seems that longitudinal waviness is not clearly influenced by the scan step (cf. Fig. 15a); there exists an optimum value of scan step inducing a minimal value of transverse waviness (cf. Fig. 15b). This value, equal to 1 mm in the study, corresponds to the diameter of the focusing nozzle. Hence, the authors concluded that overlapped or separated trenches create surface waviness over the workpiece.

Contrarily to metals, no study focusing on multi-pass milling has been performed on composites to our knowledge.

2.2 *Tool Path Strategies*

The problem of over erosion linked to the variations in jet speed during machining, mainly due to start and stop of the jet but also when the jet direction changes, has

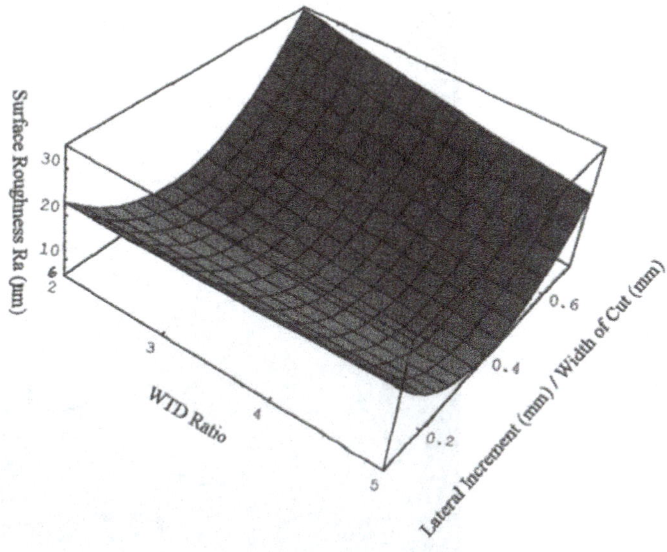

Fig. 14 Influence of the WTD ratio and the lateral feed on the obtained surface roughness (Hocheng et al. 1997)

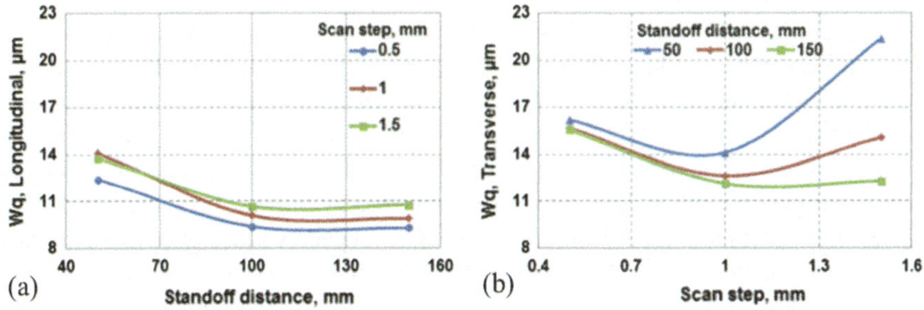

Fig. 15 Influence of the scan step and the SOD on the **a** longitudinal and **b** transverse waviness (Hejjaji et al. 2017)

been of great concern from the beginning of studies involving water jet milling operations. To minimise this issue, the first proposed solution was to use masks made of materials with low machinability. In this case, the jet accelerations and decelerations are made over the mask and the machining of the workpiece is done at constant speed. Figure 16 shows the result of AWJ milling with masks for elliptical pockets on CFRP composite laminate. However, Fowler (2003) demonstrated the limitations of masks: when the jet hits the mask, it is deflected inside the pocket and further erodes the already machined surface. Moreover, the preparation of the masks adds significant machining time, which can be incompatible with the industrial

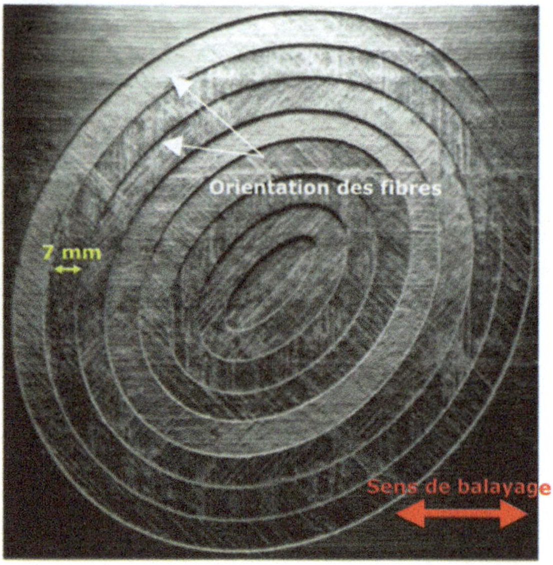

Fig. 16 Elliptical pockets milled using AWJ with masks (Zitoune et al. 2013)

requirements. The studies presented below proposed different path strategies in order to perform mask-less pocket milling.

Kong et al. (2011) developed an innovative jet path to perform multi-pass PWJ pocket milling which reduces the speed variations of the cutting head. The machining conditions are fixed as follow: jet pressure 345 MPa, traverse speed 500 mm/min, SOD 3 mm and scan step 0.5 mm. The goal of this study was to prove the capability of this milling method to perform a geometrically and dimensionally accurate pocket with good surface quality. The tests were conducted by milling a rectangular pocket with radial corners in titanium alloy Ti6Al4V. The jet path was designed to maintain a constant cutting head speed during the entire machining time (i.e. no stops between the layers). Figure 17 shows the jet path pattern used by the authors. The cutting path of each layer is perpendicular to the previous one and the same number of identical jet paths must be performed in order to ensure a uniform removal and then to enhance the finished surface quality. Five cycles, hence 20 layers, were machined. The selected jet path has proven to produce a pocket with good surface quality ($Ra < 4$ μm and $Wa < 11$ μm) and geometrical tolerance (straightness < 200 μm), without over-eroded zone. However, it seems that this method is effective only for machining with low traverse speeds, thus for low machinability materials. Indeed, in case of high-speed machining, the milling direction changes would lead to over erosion due to local decelerations.

Alberdi et al. (2011) studied the influence of the tool path on the depth of triangular pockets in case of single-pass normal milling with overlapped footprints in 7075-T651 aluminium alloy. They considered four different tool paths with both spiral and zigzag strategies, with different starting points (SP). However, the scan step and the machining parameters are set for all the specimens. The differences in pocket depth

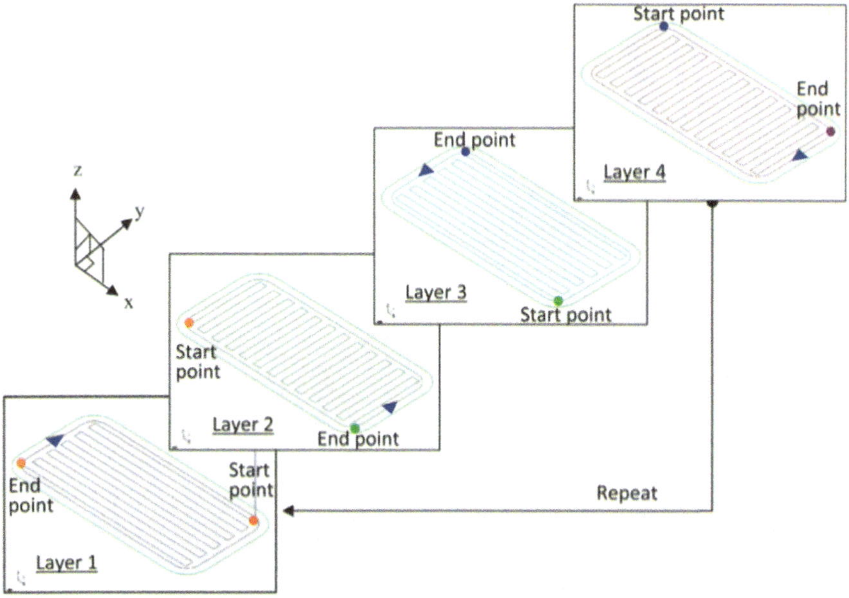

Fig. 17 Innovative jet path pattern of one cycle for milling a rectangular pocket with radial corners proposed by Kong et al. (Kong et al. 2011)

for all the specimens tend to prove that, in addition to the operating parameters, the interaction between the trenches (the shortest distance to any previous slots should be considered as shown from Fig. 18) and the cutting head kinematics have to be taken into account in AWJ milling. However, the first additional parameters seemed to have a more pronounced influence on the pocket's depth than the second one. This can be concluded from the comparison between the total depths generated by a

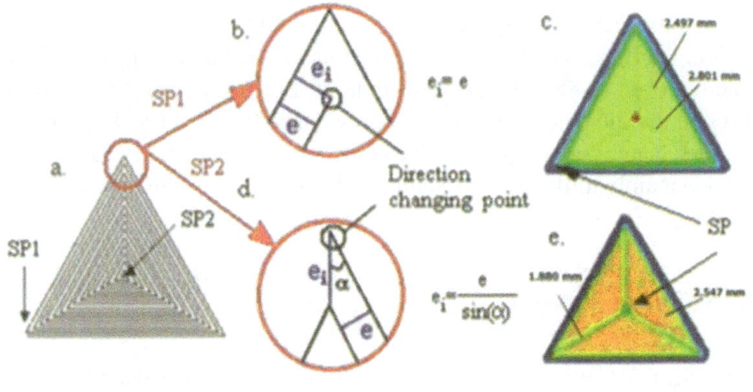

Fig. 18 **a** Spiral pattern; **b** lateral feed calculation and **c** actual surface with SP1; **d** lateral feed calculation and **e** actual surface with SP2 (Alberdi et al. 2011)

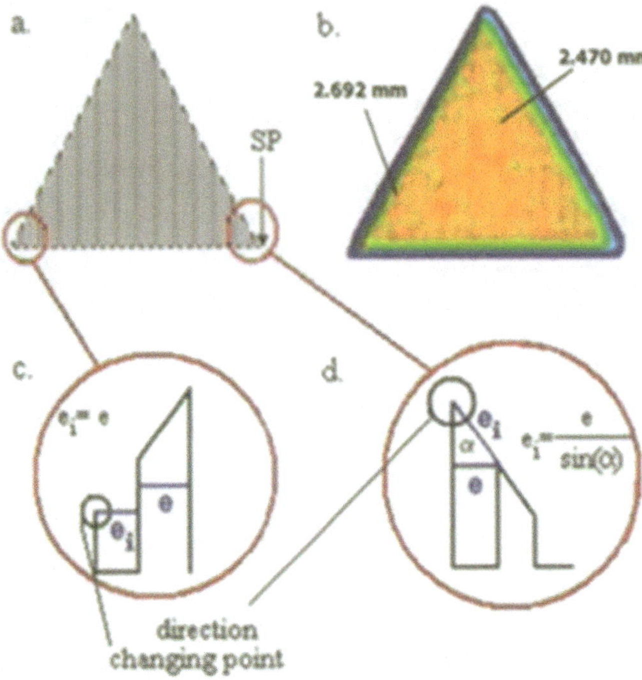

Fig. 19 **a** Zigzag pattern; **b** actual surface; lateral feed calculation **c** on the left and **d** right sides of the triangle (Alberdi et al. 2011)

spiral pattern with two jet path directions (cf. Fig. 18). In case of starting at SP1, the direction changing points are deeper than the rest of the bottom surface (2.801 mm versus 2.497 mm, cf. Fig. 18c) due to the deceleration of the cutting head leading to a longer exposure time, the instantaneous pitch being kept constant ($e_i = e$, cf. Fig. 18b). On the other hand (case SP2), the increase in lateral feed at the direction changing points ($e_i > e$, cf. Fig. 18d) originate a shallower depth of cut than the rest of the bottom surface. This can be explained by the fact that more "fresh" material has to be removed (1.880 versus 2.547 mm, cf. Fig. 18e) even if the cutting head goes through the same deceleration than previously. In case of a zigzag pattern (cf. Fig. 19), the changes in depths of the generated surface are greatly reduced because the pocket is mainly milled with straight parallel trenches (mean depth of 2.47 mm).

2.3 Prediction Models

As seen in Sect. 2.1, both geometry and quality generated by PWJ/AWJ machining depend on various process parameters. For a better control of these features, several works have been conducted in order to develop predictive models. The different methods allowing to predict the jet footprint profiles generated by PWJ/AWJ machining can be sorted into four categories: analytical/geometric modelling (Axinte et al. 2010; Kong et al. 2012; Billingham et al. 2013; Rabani et al. 2016), artificial intelligence (Artificial Neural Networks (Gupta et al. 2015), Genetic Algorithms (Zain et al. 2011)), simulation (FEM) (Anwar et al. 2013a, b; Lozano Torrubia et al. 2015) and statistical approaches (interpolation, regression analysis) (Alberdi et al. 2010, 2011; Cénac et al. 2013; Zitoune et al. 2013; Bui et al. 2017).

Axinte et al. (2010), Kong et al. (2012) and Billingham et al. (2013) used the same analytical/geometric approach in order to predict the depth of kerfs generated by AWJ milling in straight (Axinte et al. 2010) or variable (Kong et al. 2012) directions and of overlapped trenches created by the same machining process (Billingham et al. 2013). They described the kerf profile thanks to nonlinear partial differential equations in which the influence of a given configuration of parameters (material, pressure, abrasive size and flow rate, etc.) on the material erosion is given by an empirically determined function called the "etching rate". The number of input variables increased gradually starting from the traverse speed (Axinte et al. 2010), then adding the impingement angle, the direction of milling (Kong et al. 2012) and the overlap ratio (Billingham et al. 2013). The models accurately predicted the depth of cut from single kerf (error < 4%) to pockets of numerous overlapped trenches with up to 3 layers (error < 8%). Nonetheless, the non-linear increase of errors with the number of layers and overlap percentage indicates that the model will not be valid for a high number of milled layers or in the case of multiple 100% overlapped trenches. Billingham et al. (2013) considered that their model could precisely predict the depth of milled pockets up to 2 mm deep in Ti6Al4V. The model is also able to spot complex phenomena such as the difference in erosion between backward and forward milling, supporting the conclusions drawn by Fowler et al. (2005a). However, some restrictions and considerations have to be noticed about these models. Only average values of the footprint profile have been taken into account, which means that the variation of the profile along the footprint generated by the fluctuations of the process parameters during milling are considered. Moreover, the empirically determined etching rate is valid for a given configuration of parameters (material, jet pressure, abrasive size and flow rate, etc.) which implies that any modification in one or more of these parameters leads to additional tests for new calibration.

Gupta et al. (2015) utilized Artificial Neural Network (ANN) in order to forecast the shape and surface quality of single-pass micro-channels generated by AWJ normal milling of stainless steel SS304. The input process parameters were abrasive particles size, abrasive flow rate, traverse speed and SOD, each of them having three levels, leading to a full factorial set of 81 (3^4) tests. The innovation of this study was to also consider jet vibrations (influencing the trench's width and measured with an accelerometer fixed on the mixing chamber) and impact force (linked to the trench's depth and gauged with a dynamometer located under the workpiece) as additional input parameters. The modification of both theses parameters leads to changes in surface roughness and kerf taper. The output parameters were the trench's depth, width, taper and surface roughness. The 81 input sets were split into 70% (57 sets) for training the artificial network, 20% (16 sets) for validating and 10% (8

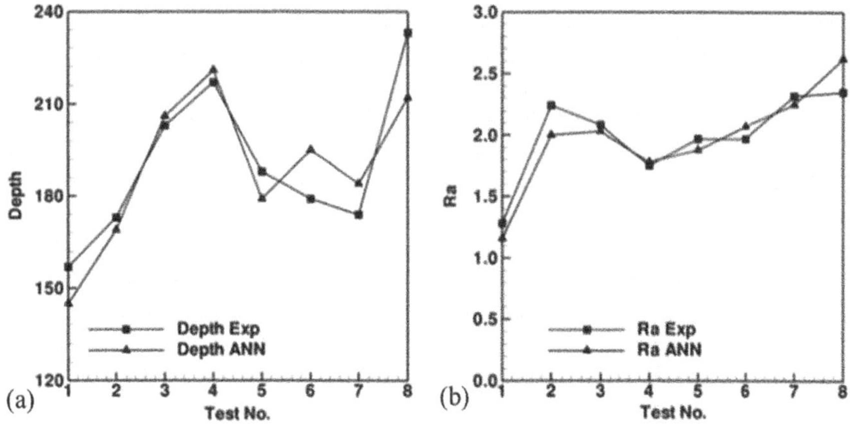

Fig. 20 Comparison between experimental and ANN predicted values of **a** the depth of cut (μm) and **b** the surface roughness (μm) (Gupta et al. 2015)

sets) for testing. Several architectures were tested and the 6-12–10-4 architecture (6 input parameters, 2 hidden layers of respectively 12 and 10 synaptic weights, 4 output parameters) gave the most accurate outputs: the mean prediction errors for depth of cut and surface roughness were respectively 3 and 6% as shown from Fig. 20. These results are all the more acceptable that numerous machining parameters were influencing the process. The main limitation of artificial intelligence prediction methods is the need of a huge amount of data in order to perform the training process.

Anwar et al. (2013a, b) developed a finite element model in order to describe the footprint generated by AWJ normal milling of Ti6Al4V at different jet pressures and traverse speeds, in case of a single trench and single-layered overlapped footprints with variable scan step. The jet, with a diameter equal to the nozzle exit, is modelled by a succession of equally spaced (50 μm) layers of particles with various shapes. Each layer is composed of different particle shapes and sizes whose number and position are arranged to fit a Gaussian spatial distribution around the jet axis (cf. Fig. 21). Two features have been compared between experimental and simulation data: the erosion rate, defined as the ratio of the target weight loss over the mass of impacting particles, and the mean 2D footprint profile. The authors concluded that their model of single pass milled trenches gave consistent results compared to the experimental values for both footprint profiles and erosion rates (error < 10%). Nonetheless, in the case of overlapping footprints, the comparison between experimental and simulated footprints revealed the dependence of the footprint's shape on the mass distribution within the jet plume. Indeed, the Gaussian spatial distribution underestimated the mass distribution inside a radius of 0.3 mm around the jet axis. With this modification, the simulation provided more accurate footprint profiles with an error below 15%, which is acceptable considering the fluctuations of the water pressure and the particle fragmentation over the machining time. The relatively good estimation of the erosion rate (error < 8%) proved that neglecting the effect of water in the erosion

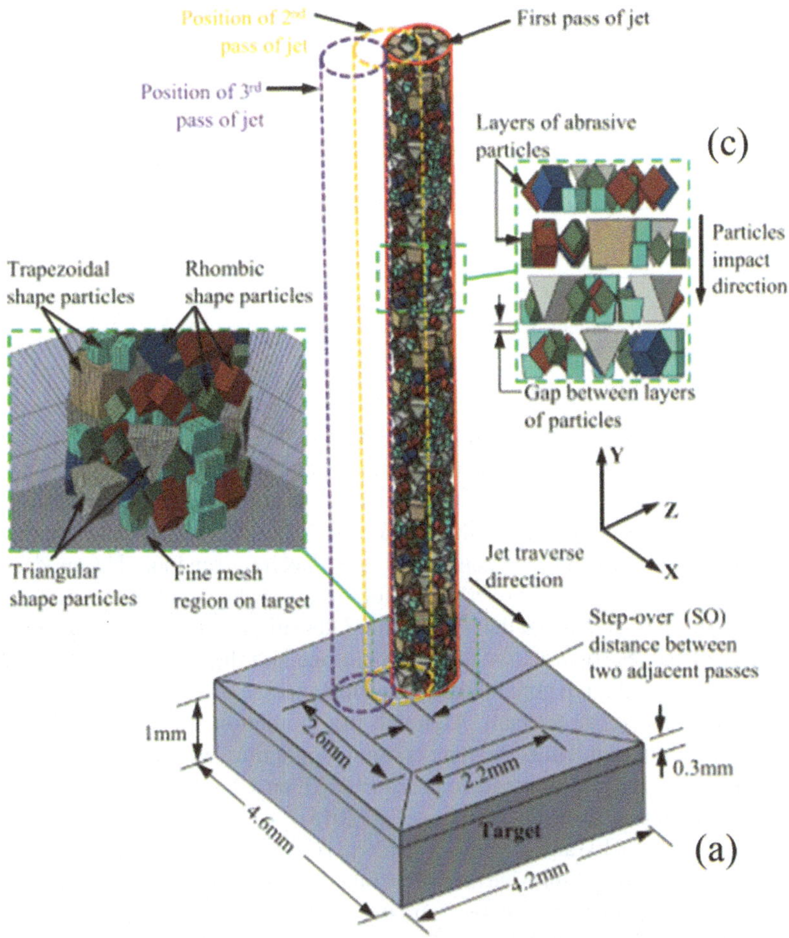

Fig. 21 **a** 3D view of the overlapped footprint FE model; **b** Meshing on the target and the particles; **c** Zoomed in view of layers of abrasive particles (Anwar et al. 2013b)

process was accurate for the range of pressures tested. The finite element model also well described the non-linearity of the erosion process when performing overlapped trenches—as explained in (Billingham et al. 2013)—by generating non-symmetrical profiles, as seen from Fig. 22. Lozano Torrubia et al. (2015) studied the influence of the jet fluctuations on the machined surface quality. To do so, the authors proposed an enhanced model based on the one developed by Anwar et al. (2013b) by bringing some modifications. In order to be able to predict the deviations from a theoretical shape generated by the stochastic nature of the machining, the fixed values of the processing parameters characterising the jet were here replaced by probability distributions using Monte Carlo methods (results of several simulations for the same

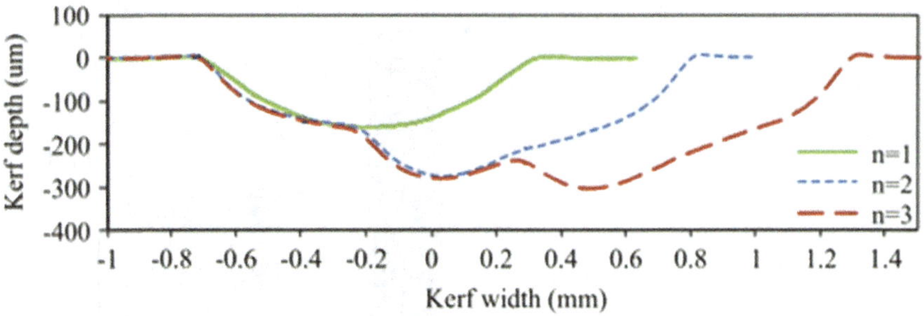

Fig. 22 Simulated effect of overlapped jet on the footprints; P = 345 MPa, V = 2 m/min, SS = 0.5 mm (Anwar et al. 2013b)

given set of conditions). Thus, for each particle, the following probability distributions were implemented into the model: the probability distributions of the particle's velocity and relative position within the jet were considered as Gaussian, its mass was taken pseudo-randomly following a probability distribution found experimentally and its orientation was considered as completely random. The model gave a very good prediction of the overlapped footprint's shape (error < 4%) performed by normal milling but showed that the standard deviation increases with the increasing number of passes (by a factor four between one and three passes). Hence, using a small scan step, though generating a smoother average profile, leads to higher shape fluctuations along the jet path. Although the simulation approach gives satisfying results, its main problem is the long computational time which can be incompatible with the industrial requirements (e.g. around 22 h on a 2 × 8-core 2.6 GHz processor for one simulation (Lozano Torrubia et al. 2015)).

Cénac et al. (2013) and Zitoune et al. (2013) proposed empirical models in order to determine the optimal abrasive mass flow rate (leading to the highest depth of cut for a given set of machining parameters) and the depth of cut for AWJ open pocket milling in aluminium (Cénac et al. 2013; Zitoune et al. 2013) and carbon/epoxy composite laminate (Zitoune et al. 2013). In both studies, the authors considered the jet pressure P, the orifice diameter d_O, the abrasive mass flow rate D_a and the targeted pocket depth H as variable parameters. The grit size, the stand-off distance and the scan step were set and the traverse speed was selected to obtain the targeted depth of cut by using the optimal abrasive mass flow rate (OAMFR) $D_{a,opt}$. The originality of these models lies in the choice of making the depth of cut an input parameter, contrarily to the usual prediction methods (Alberdi et al. 2010; Gupta et al. 2015; Lozano Torrubia et al. 2015), which considered this variable as an output. This decision is supported by the fact that the machined depth influences the predominant erosion mechanisms which cannot be represented by the same model (Fowler et al. 2009). Besides, this is more in accordance with the needs of manufacturers who have a target depth of cut and want the corresponding machining parameters to achieve it. The developed methodology to determine the empirical models was the same for both the studies. Several pockets were milled with various sets of pressure, orifice diameter, targeted depth and abrasive mass flow rate. The traverse speed was adjusted so the depth of cut given by the OAMFR corresponds to the targeted depth (±5%). These tests permitted to propose a prediction model for the OAFMR following Eq. (1):

$$D_{a,opt} = A_1 \cdot P^{A_2} \cdot (r \cdot d_O^2)^{A_3} \tag{1}$$

where:–A_1, A_2, A_3 represent interpolation constants depending on the material and the set of machining conditions (abrasive size, SOD, step-over distance).

–r corresponds to the nozzle efficiency defined by the measured water flow D_w over the theoretical one from the Bernoulli equation $r = \frac{D_w}{\sqrt{2P/\rho}}$ (ρ: water density).

The small differences in the interpolation constants between the models of aluminium developed in (Cénac et al. 2013) and (Zitoune et al. 2013) (30% for the proportionality constant, less than 11% for the other constants) tend to prove the limited influence of the SOD (60 mm in (Cénac et al. 2013) and 100 mm in (Zitoune et al. 2013)) on the OAMFR. Because the interpolation constants are similar for the two tested materials, the authors decided to merge the two prediction models into a third one following Eq. (2):

$$D_{a,opt} = 2.116 \cdot P^{0.815} \cdot (r \cdot d_O^2)^{0.49} \tag{2}$$

Though this model is a bit less accurate than the ones for a single material, the relative errors between the measured and predicted values of $D_{a,opt}$ are below 10%.

Similarly, the prediction models of the milling depth for the two materials are merged into a third equation by introducing the machinability U of the material:

$$f = U \cdot P^{2.136} \cdot (r \cdot d_O^2)^{0.905} \cdot \Omega^{0.415} \cdot H^{-1.027} \tag{3}$$

where:–$U = 1.616 \times 10^{-3}$ for aluminium and $U = 1.871 \times 10^{-2}$ for composite laminate.

–Ω represents the OAMFR ratio defined by $\Omega = \frac{D_a}{D_{a,opt}}$

It can be seen from Eq. (3) that the depth of cut H is inversely proportionnal to the traverse speed f, which is consistent with other works (Chillman et al. 2010b; Bui et al. 2017). This equation gave accurate predictions ($R^2 = 0.997$) of the needed traverse speed to reach a targeted pocket depth H. However, some limitations have to be noticed concerning the results presented below. Firstly, Eq. (3) has been validated for two materials and it could be interesting to check if this equation can be used for machining of other materials. Secondly, the equation gives truthfull results for single-pass milling but does not consider the possibility of multi-pass machining.

Though the empirical models presented above give satisfying results, they consider material removal at a macro scale. Sourd et al. (2020) proposed another empirical prediction model for the prediction of depth of cut when milling 3D woven CFRP composite at a smaller scale. This model considers the milled pocket as the algebraic sum of elementary trenches separated from each other by a distance called scan step. It also takes into account the regime of erosion depending on the depth of the elementary passes. This prediction model presents a good correlation with the experimental measurements (mean error of 5%) and can be easily calibrated with the milling of elementary trenches, contrarily to the other model presented before which needed complete pockets to be adjusted.

2.4 Conclusion of the Part

At first sight, complex combinations of process parameters influence the geometrical characteristics—depth of cut and surface quality—of PWJ/AWJ machined specimens. However, when having a closer look, it seems that some of these parameters are particularly important. Let us introduce the degree of influence of a machining parameter, defined as the evolution ratio of an output over the evolution ratio of this machining parameter, the other milling parameters being kept constant. For instance, as seen from Table 1, the average degree of influence of the waterjet pressure is 1.50 for the depth of cut in both cases of titanium or CFRP milling: it means that the ratio of amplitude (increase or decrease) of the depth of cut is 1.50 times more important than the ratio of the pressures. Thus, the machining parameters influencing the most the depth of cut seem to be the jet pressure, the traverse speed and the scan step. Other milling parameters, such as the abrasive size or the stand-off distance, appear to have a more limited effect on the depth of cut. One interesting thing to highlight is the relative independency of the milled material on the degree of influence. This is consistent with the work of Zitoune et al. (2013), in which the interpolation constants of the milling depth prediction models are similar for the two tested materials (aluminium and CFRP). The obtained classification is coherent with the parameters chosen by several studies (Anwar et al. 2013b; Billingham et al. 2013; Cénac et al. 2013; Zitoune et al. 2013; Bui et al. 2017) for their prediction models of the depth of cut. In addition, the choice of the process parameters leads to the appearance of various surface morphology features, such as craters and grooves which respectively change the surface roughness and waviness. In general, the recommendations for a high MRR and a good surface quality (e.g. low Ra) are contradictory. For example, machining with high pressure or large abrasive particles leads to an important MRR but a poor surface quality. Moreover, the tool path used during an AWJ milling operation must be carefully designed in order to respect the dimension tolerances of the machined part. This can be achieved mainly by minimizing as much as possible the fluctuations within the jet speed, hence limiting under-erosion or over-erosion. All these considerations lead to various surface qualities and textures, which are probably linked to different kind of defects and degrees of damage. The next parts will focus on the link between the machining parameters and the induced damages and their influence on the mechanical properties.

Table 1 Degree of influence of the main AWJ milling parameters on the depth of cut

Machining parameter	Material	Evolution ratio of the machining parameter	Evolution ratio of depth of cut	Degree of influence	Average degree of influence	Reference
Pressure	Ti6Al4V	2.25	3.75	1.67	**1.50**	(Bui et al. 2017)
		1.25	1.67	1.34		(Kanthababu et al. 2016)
	CFRP	1.67	2.17	1.30	**1.50**	(Hocheng et al. 1997)
		1.75	3	1.71		(Hejjaji et al. 2017)
Traverse speed	Ti6Al4V	3.5	3.25	0.93	**1.02**	(Bui et al. 2017)
		1.67	1.83	1.1		(Kanthababu et al. 2016)
	CFRP	2.22	2.74	1.23	**1.11**	(Hocheng et al. 1997)
		3	3	1		(Hejjaji et al. 2017)
Overlap (scan step)	Ti6Al4V	1.5	1.5	1	**1.12**	(Billingham et al. 2013)
		2	2.48	1.24		(Kanthababu et al. 2016)
	CFRP	3	2.8	0.93	**0.93**	(Hejjaji et al. 2017)
Abrasive size	Ti6Al4V	1.5	1.05	0.70	**0.63**	(Pal and Choudhury 2014)
		2.5	1.4	0.56		(Fowler et al. 2005b)
	CFRP	–	–	–	–	–
SOD	Ti6Al4V	2	1.03	0.51	**0.51**	(Pal and Choudhury 2014)
	CFRP	2	1.33	0.67	**0.55**	(Hocheng et al. 1997)
		3	1.33	0.44		(Hejjaji et al. 2017)

3 Damage Analysis Due to Water Jet Process

It is important to notice that each process of machining (conventional or non-conventional) induces different kind of defects. In addition, the defects are strongly influenced by the nature of the target material and the material removal mechanisms. The following sections present the different types of defects and damage consecutive to PWJ and AWJ machining of metallic and composite materials. Then, particular issue of grit embedment is studied with respect to the process parameters and some solutions to reduce this contamination are listed.

3.1 Machine-Induced Defects and Damage

3.1.1 Metals

Azhari et al. (2012) and Kong et al. (2010) performed PWJ milling of gamma titanium aluminide (γ-TiAl) and Ti6Al4V respectively. They observed that material removal occurs at two levels that Kong et al. called initial and evolved damage. The initial damage is provoked by the impact of water droplets with high velocity, which plastically deforms the material locally and initiates networks of cracks (cf. Fig. 23a). Moreover, Huang et al. (2012) found an additional proof of plastic deformation when PWJ milling Ti6Al4V. Indeed, during the first step of damage, observed for a single jet pass at low pressure (138 MPa), the effect of the starting plastic deformation is visible through some isolated damage occurring along the boundaries between α grains and β matrix, as shown from Fig. 23b by the contrast of the grain boundaries. It was explained that the small damage within the grains (microvoids and micropits) seen on the Scanning Electron Microscopy (SEM) image are probably due to stress concentration areas. A further investigation of the microstructure with Atomic Force Microscopy (AFM) shows that the changes in the surface topography are the consequences of grain tilting. Though the angles are small (maximum disorientation of 1.6°), the random orientation of the grains leads to a slightly roughened surface. Based on the authors analysis, it was mentioned that grain tilting is the consequence of a momentum induced by the water drops hitting the grain.

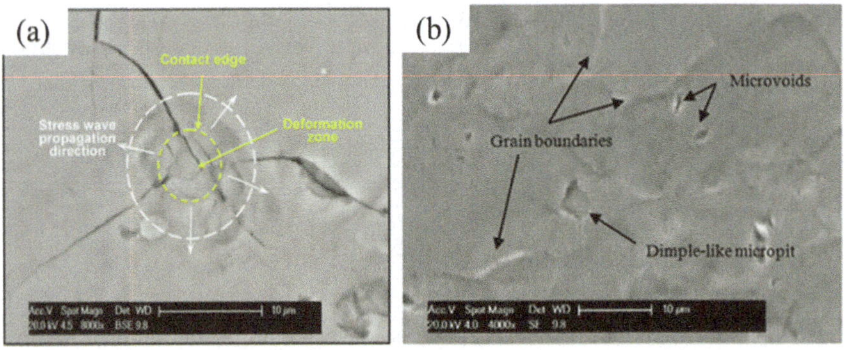

Fig. 23 Initial damage with **a** a deformation zone and networks of cracks (Kong et al. 2010) and **b** apparition of grain tilting (Huang et al. 2012)

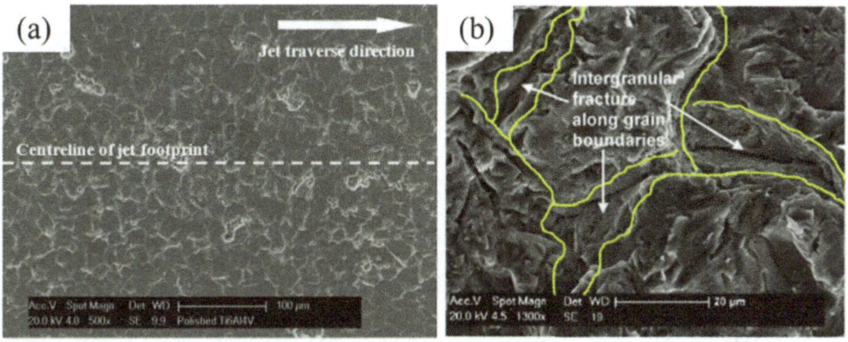

Fig. 24 Micrographs showing **a** the damage spreading along grain boundaries into a network (Huang et al. 2012) leading to **b** inter-granular fracture (Kong et al. 2010)

The damages propagation, appearing after several jet passes and/or at high jet pressure, are the consequence of stress waves, generated by the many droplets impacts of on the workpiece, which propagate. The cracks grow along the grain boundaries and join to form a network (cf. Fig. 24a), leading to inter-granular fracture (cf. Fig. 24b). Some cracks also appear inside the grains and propagates because of the penetration of water and the lateral outflow jetting. Their propagation generates fractures across and along the lamellae structure of the grains respectively called translamellar (cf. Fig. 25a) and interlamellar (cf. Fig. 25b) fractures. According to Huang et al. (2012), the rough and irregular topography of the material is the sign of ductile fracture. The removed material creates different kind of features like cavities or valleys.

As discussed in Sect. 2.1.2.1., the introduction of abrasive particles within the jet induces the development of different morphologies of the new generated surfaces which are strongly influenced by the process parameters (cf. Fig. 26). In addition to these morphologies, another type of defect called grit embedment is observed on the milled surfaces. This particular problem will be discussed in a further section.

3.1.2 Composites

One of the most critical damage when machining composite laminates with AWJ is delamination phenomenon. The conditions of apparition of this kind of defect have been studied by Cénac et al. (2008, 2009). The authors carried out AWJ normal pocket milling of two kinds of composite laminates. The first one is made of CFRP plies (referenced under HexPLY M21T700 GC) with 20 plies and quasi isotropic stacking. The second ones are made of GFRP woven ply serge 2 × 2 (referenced under HexFIT) with a stacking sequence of $[0°/90°]_4$. They found that the risks of delamination do not depend on the traverse speed and the focusing tube diameter but are influenced by the jet pressure, the orifice diameter and the abrasive flow rate. This led to the conclusion that delamination is linked to the repartition of the jet

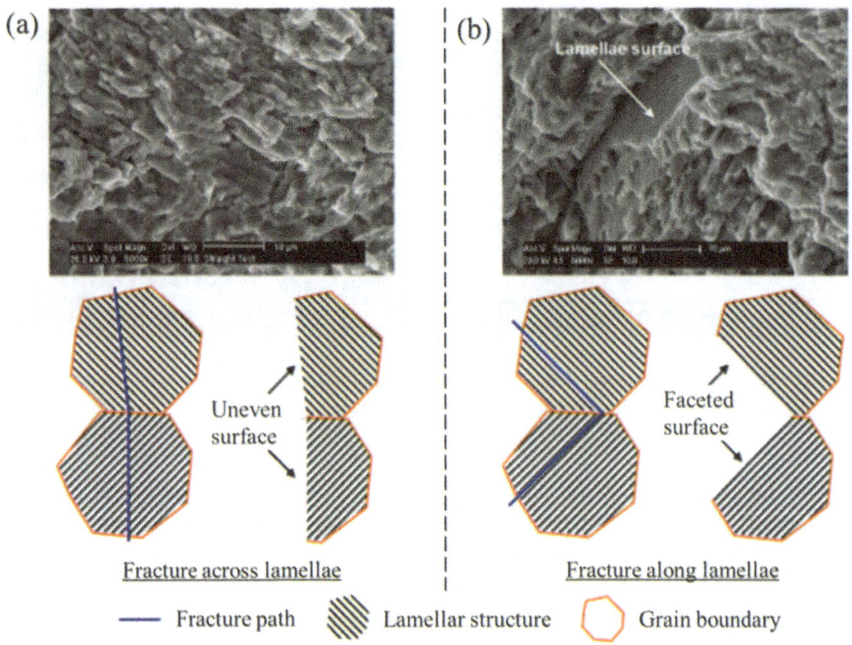

Fig. 25 Two types of fractures occurring within the grains in the final stage of material removal: **a** Translamellar and **b** interlamellar fractures (Kong et al. 2010)

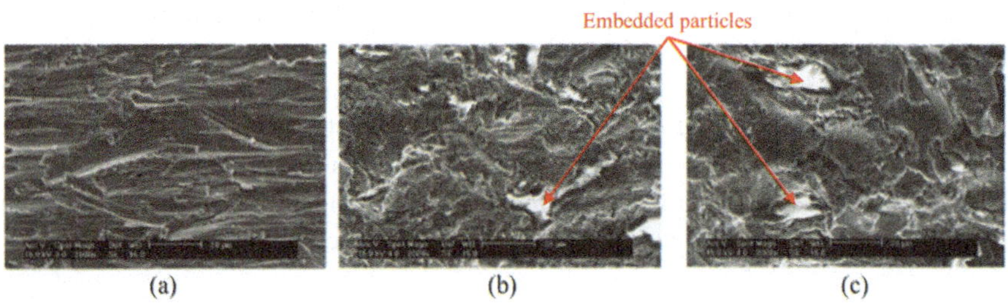

Fig. 26 **a** Grooved, **b** mixed grooved and cratered and **c** cratered morphologies developed during AWJ milling of Ti6Al4V (Fowler et al. 2005a)

energy between the water and the abrasive particles. Indeed, during machining, the water goes inside the micro-cracks and the porosities located on the top surface of the composite plate and makes them propagate until removing a piece of the material: it is called the wedge effect (cf. Fig. 27). Between plies, the cracks and porosities are more numerous due to a locally lower fiber volume fraction (especially in HexPLY M21T700) and need less stress to propagate than inside the plies themselves, making this wear mode an easy way to generate delamination. On the other hand, the abrasive particles only act locally and do not take part on the delamination process. Hence, the

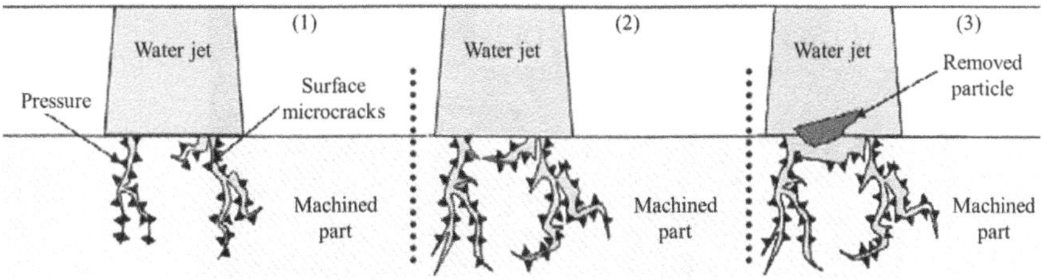

Fig. 27 The three steps of erosion by wedge effect (Cénac et al. 2008)

abrasive flow rate has to be optimised in order to remove enough water energy to avoid delamination. This is consistent with the work of Hashish (1995) who proscribed PWJ for composite milling applications. It has to be noticed that the threshold of wear energy avoiding delamination is material dependent. Indeed, though the influence of the machining parameters on the generated defects is quasi-similar between the CFRP unidirectional laminate and the GFRP 2D woven laminate, the size of the damaged area is different between these two materials. Moreover, to our knowledge, no study was led on AWJ milling of composite materials with other architectures like 3D weaving or else.

Shanmugam et al. (2008) also studied the delamination consecutive to CFRP machining with the AWJ process. The authors, as Cénac et al. (2008, 2009), observed that the delamination is initiated by water-wedging. However, they also found that the cracks induced by the effect of water are extended by the wedging action of the abrasive particles. Moreover, some grit embedment was detected in the cracks (cf. Fig. 28).

In other work, Hejjaji et al. (2017) performed AWJ normal pocket milling of unidirectional CFRP laminates (Hexply T700-M21.A). The analysis of the milled surfaces by SEM showed several types of damage depending on the process parameters and more especially on the kinetic energy of the jet (cf. Fig. 29). The presence of micro-craters, around few tens of microns in diameter, and broken fibers (cf. Fig. 29a, b and c2) on every milled specimen show that these two kinds of damage are the firsts to be generated by the machining process due to the impact of the abrasive particles. When increasing the pressure, craters of greater size (macro-craters), around the millimetre

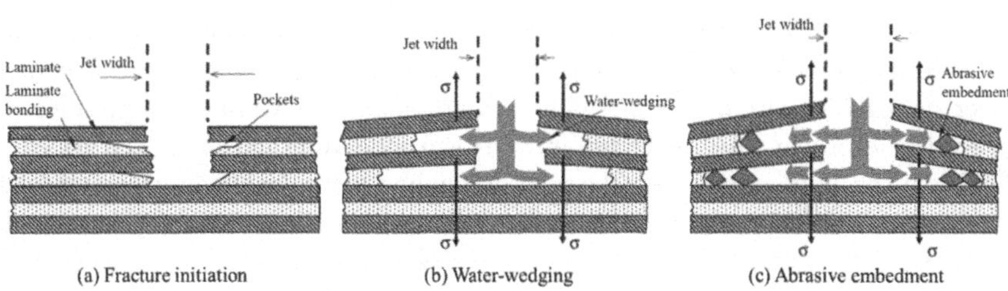

Fig. 28 The three steps leading to delamination of a laminate machined with AWJ (Shanmugam et al. 2008)

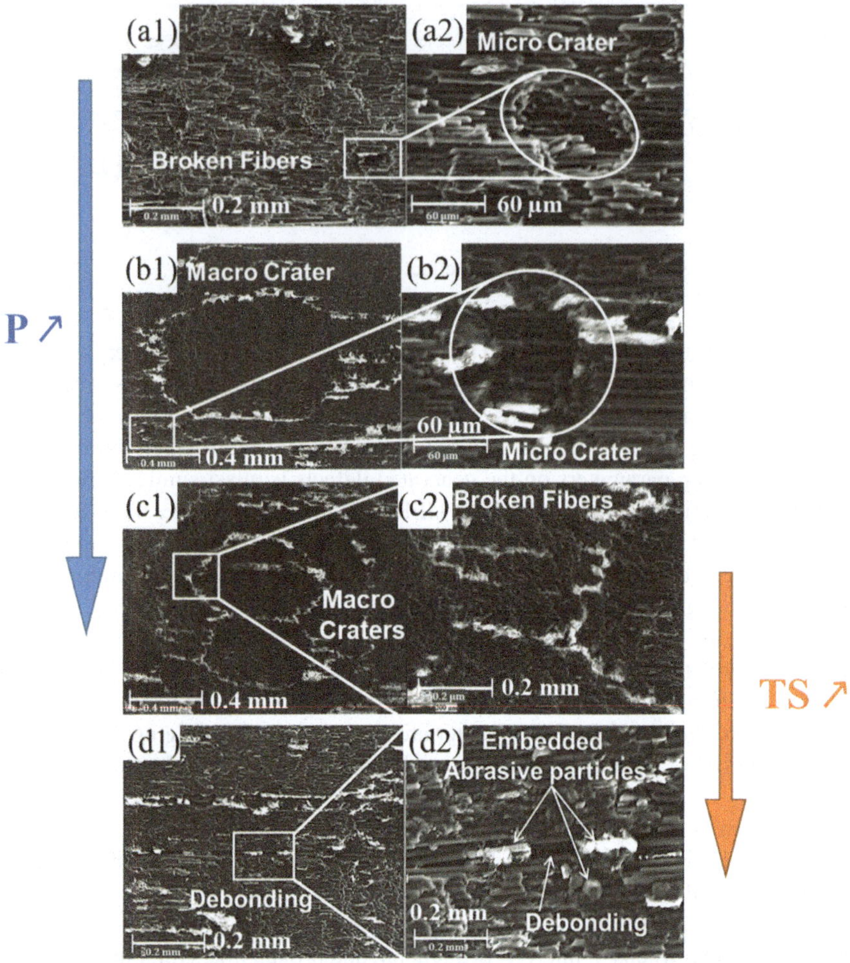

Fig. 29 SEM images showing the evolution of the damages generated by AWJ milling at SOD = 50 mm, scan step = 0.5 mm, **a** TS = 8 m/min, P = 80 MPa, **b** P = 100 MPa, **c** P = 140 MPa and **d** TS = 12 m/min, P = 140 MPa (Hejjaji et al. 2017)

in diameter (cf. Fig. 29b1 and c1), start to appear. They become more numerous with a further increase in jet pressure (cf. Fig. 29c1) and bigger when using a greater traverse speed. In case of milling with high pressure (140 MPa) and high traverse speed (12 m/min), the specimens suffer from local debonding and embedded grits (cf. Fig. 29d). The authors stated that the surface roughness is linked to micro-craters and broken fibers and the surface waviness to macro-craters. This statement is supported by the X-ray tomography images from Fig. 30. Moreover, in addition to craters and broken fibers, ridges can be observed when milling with a high value of scan step (1.5 mm i.e. no superposition of two consecutive jet passes), which contributes to the machined surface waviness (cf. Fig. 31).

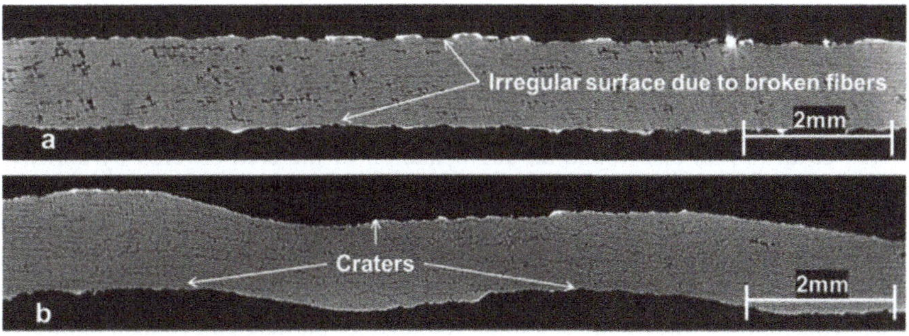

Fig. 30 Tomography images of specimen's cross section showing the surface quality induced by **a** broken fibers and **b** macro-craters (Hejjaji et al. 2017)

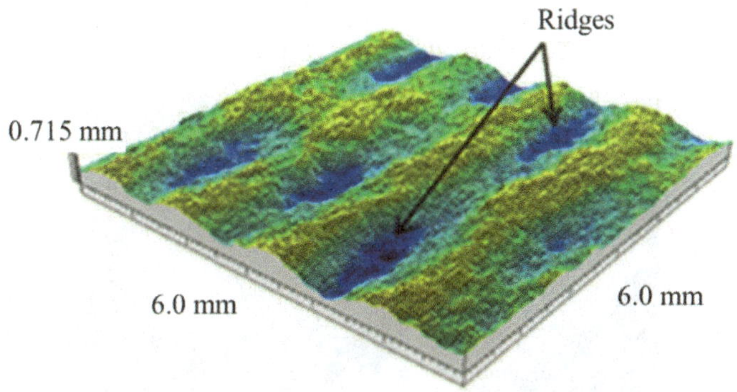

Fig. 31 Topography showing the presence of ridges after milling with a high scan step (1.5 mm) (Hejjaji et al. 2017)

3.2 Grit Embedment

During the AWJ machining process, abrasive particles with high velocity impinge the target material, leading to grit embedment. These particles, which are embedded into the material at various depths, could become stress concentration areas and may reduce the mechanical characteristics of the material. Moreover, the contamination of the machined surface by foreign bodies can alter their adhesive performances. The following section describes the influence of the process parameters on the degree of material contamination by abrasive particles. Then, several techniques developed to reduce the level of grit embedment are presented.

3.2.1 Influence of the Milling Conditions

Arola et al. (2000, 2002) noticed embedded abrasive particles consecutive to AWJ peening of Ti6Al4V (cf. Fig. 32), its frequency increasing with higher jet pressure. Rivero et al. (2018) also found an augmentation of the area covered by abrasive particles with a higher jet pressure in case of AWJ milling of Inconel alloy 718. They stated that this is due to the higher amount of energy available within the grit.

Fowler et al. (2005a) conducted a series of tests on Ti6Al4V in order to study the influence of the process parameters on the percentage of the milled surface subjected to grit embedment. They concluded that the level of particle embedment is piloted neither by the grit size nor by the number of passes. During milling, the abrasive particles are fractured into smaller grits of a few tens of microns (cf. Fig. 33). The embedded particles found on the milled surface are these fragments, explaining

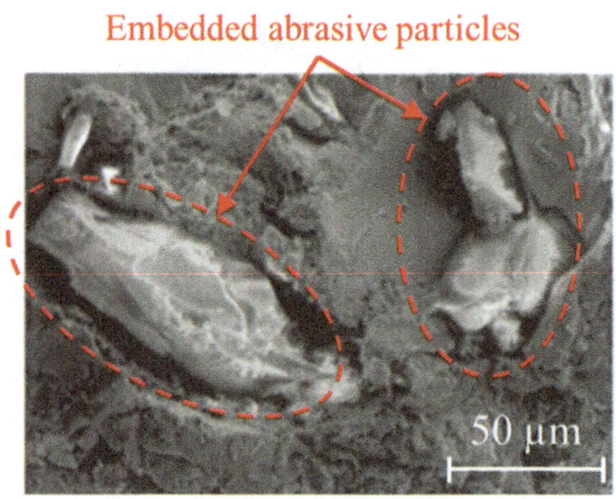

Fig. 32 Evidence of grit embedment resulting from AWJ peening of Ti6Al4V (from (Arola and McCain 2000))

Fig. 33 SEM images of 80# garnet grits: **a** in 'as-received' condition and **b** following AWJ milling (Fowler et al. 2005a)

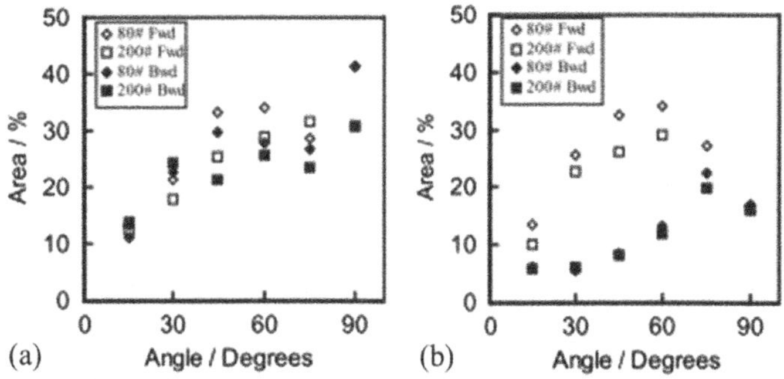

Fig. 34 Grit embedment development during AWJ milling of Ti6Al4V as a function of jet impingement angle with two garnet grit sizes (80# and 200#) and in both forward (Fwd) and backward (Bwd) milling mode. **a** Single pass of the jet with traverse speed of 0.166 $m.s^{-1}$ and **b** 0.003 $m.s^{-1}$ (Fowler et al. 2005a)

the insignificant influence of grit size on the percentage of area contaminated by embedded particles. The same conclusions were drawn by Shipway et al. (2005) on the same material. Concerning the effect of the number of passes on the level of grit embedment, the authors suggest that steady-state conditions are reached from the first passes. Indeed, the abrasive particles fragments embedded during a milling pass seem to be milled and replaced by the next one.

As for surface morphology, a mix of traverse speed, impingement angle and milling direction conditions influences the level of grit embedment. At high speed (0.166 m/s, cf. Fig. 34a), the machined surface presents the same increase of the level of embedment as the impingement angle increases on both forward and backward milling (10 to 35%). This trend is coherent with the work of Amada et al. (1999) on grit blasting of ceramics, who found a maximum abrasive presence on the treated surface at normal machining. The differences in level of grit embedment due to milling direction appear at low speed (0.003 m/s, cf. Fig. 34b). When milling forward, this level increases with higher impingement angles (from 10% at 15° to 35% at 60°) then decreases. In case of backward milling, though the evolution of level of grit embedment is similar—increase with the impingement angle (up to 75°) then decrease—the levels are much lower (only 5% at 15° of impingement up to 20% at 75°). Figure 35 shows the worst (a) and best (b) milling cases concerning grit embedment. Indeed, in case of forward milling with a high impingement angle (cf. Fig. 35a), the particles with high impulse are trapped against the pristine material, which promotes their embedment. On the contrary, in case of backward milling with a low jet angle (cf. Fig. 35b), the particles with lower impulse are free to escape after removing the material and flushed by the water jet, leading to less embedded grit.

According to Shipway et al. (2005), the influence of the traverse speed on the area covered with embedded particles is due to the secondary milling (redirection of the jet along the kerf walls when the jet loses its energy) which clears out the grits at low traverse speed.

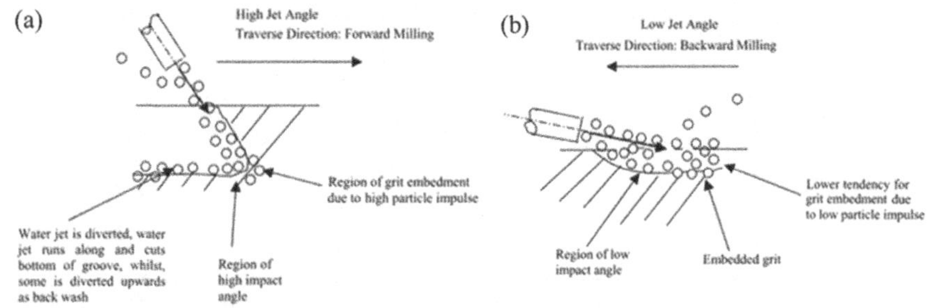

Fig. 35 Schematic diagram of flow patterns in AWJ milling with a low jet traverse speed: **a** high jet impingement angle, forward milling; **b** low jet impingement angle, backward milling (Fowler et al. 2005a)

Ramulu et al. (1993) found that the grit embedment depends on the ratio material hardness over particle hardness. The bigger this ratio, the more resistant to grit embedment the material.

3.2.2 Solutions to Reduce Grit Embedment

The first logical solution to come in mind in order to avoid the problem of grit embedment is to perform PWJ machining. However, as shown in the previous sections, AWJ offers some advantages for industrial applications (e.g. higher MRR) and is highly recommended in case of machining composite laminates. It has already been discussed about the solution of Fowler et al. (2005a) who proposed to perform forward milling with high speed and low impingement or backward milling at low speed and up to 45° impingement, inducing a reduction of the affected area from 40% to about 5%. Nevertheless, most of the solutions developed over the years to reduce grit embedment are post-machining ones.

Ultrasound cleaning seems to be the most common post-machining solution (Arola and McCain 2000; Kong et al. 2011; Liu et al. 2012). The machined specimen is immersed in a water tank and ultrasound waves remove the loose contamination. However, this technique is suitable for lightly embedded grits and is not able to take off the deeply submerged particles covered by material.

Hashish et al. (1998) submitted to clean the workpiece with PWJ after the AWJ milling process. This idea was put into practice by Huang et al. (2013) who performed a hybrid waterjet cleaning (HWJC) process composed of an AWJ milling operation to the near-desired depth followed by a PWJ cleaning process. The authors used this method for alpha case layer removal on titanium alloy Ti6Al4V. The goals of this study were to prove the effectiveness of the process and to analyse the influence of the PWJ machining parameters (SOD, traverse speed, impingement angle, number of passes and jet path) on surface roughness and grit embedment. As shown from Fig. 36, though all the PWJ cleaning parameters influenced the efficiency of the process, this cleaning method greatly reduces grit embedment (up to -95% with three passes using

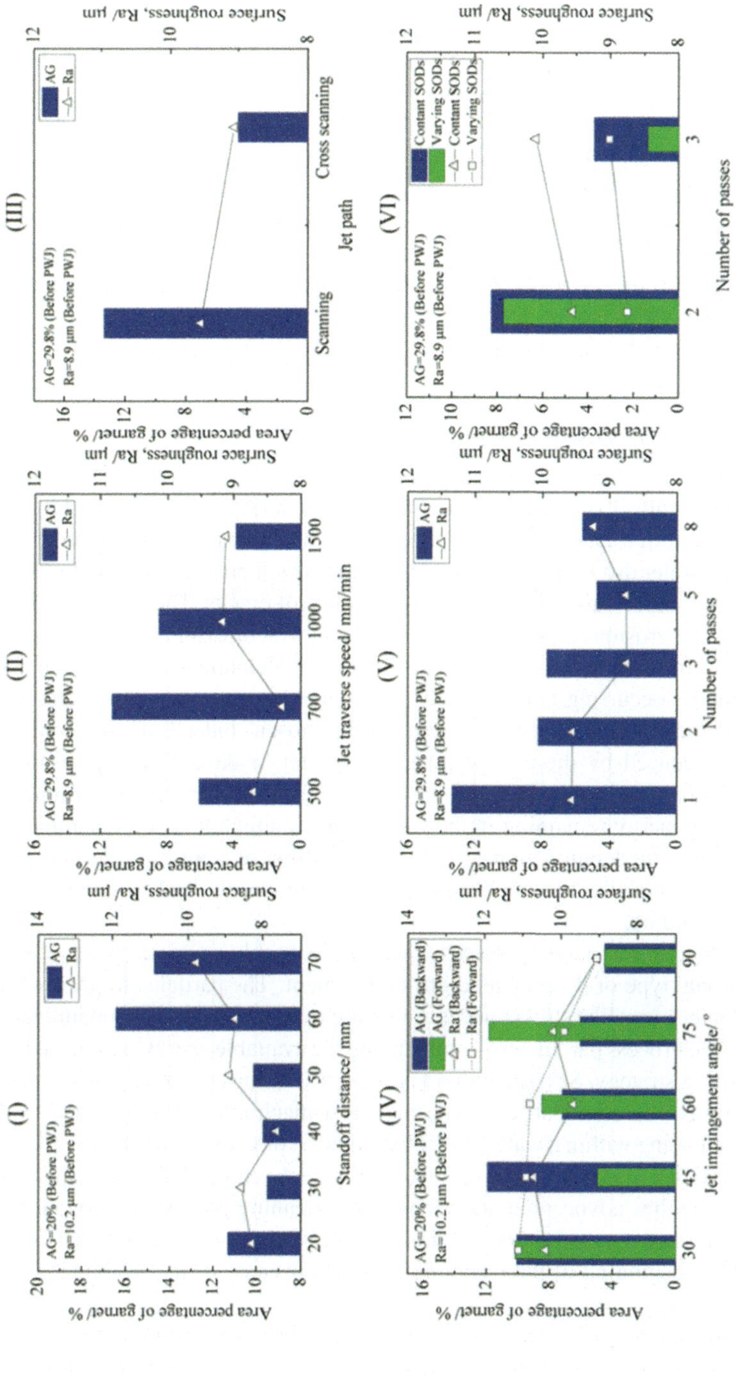

Fig. 36 Influence of PWJ parameters on grit removal performance (Huang et al. 2013)

the same jet path as AWJ but with different SODs for each pass). It also has various effects on the finished surface roughness (from −20% to + 15% depending on the cleaning parameters). The authors did not find any connection between the finished surface roughness and the percentage of area subjected to grit embedment at the end of the cleaning process. Of course, this cleaning process increased the final depth of material removal (from 50 to 110 μm depending on the PWJ cleaning operation parameters). Hence, the cleaning conditions must be selected before performing the AWJ machining in order to reach the wanted final depth of cut. The efficiency of the cleaning process is limited by the number of submerged particles covered by the material unless the waterjet power enables to remove it. Of course, as seen in Sect. 3.1.2, the wedge effect of water makes the use of this PWJ cleaning method maladjusted for composites.

3.3 Conclusion of the Part

The material removal of metallic materials with waterjet generates two levels of damage called initial and evolved damage. The initial damage is generated by the impact of the water droplets on the target material, which provokes grain tilting. The induced plastic deformation leads to the appearance of cracks. The evolved damage occurs at higher erosion rates around (inter-granular fracture) and inside (trans- and inter-lamellar fractures) the grains. On the other hand, delamination can be the main type of damage occurring consecutively to machining composites laminates with waterjets, if the process parameters are not well chosen. Indeed, the delamination is greatly influenced by the energy carried by water, making PWJ an unsuitable method for composite milling. Hence, the abrasive flow rate has to be determined so the fraction of energy remaining in the water is low enough to avoid delamination. However, even if no delamination occurs during milling of composite, several kinds of damage appear as the erosion becomes more pronounced: broken fibers, craters and local debonding.

The introduction of abrasive particles within the highly energetic waterjet leads to an additional type of defect called grit embedment. The particles trapped within the target material act like stress concentration areas and the degree of contamination depends on the process parameters—influencing the available energy within the jet—and the milling strategy. Several authors proposed solutions to reduce the degree of contamination by abrasive particles. The main post-machining technique seems to be ultrasound cleaning within a water tank. Nonetheless, though this method is efficient for loose particles, it is not able to remove more deeply embedded grit. In this case, a number of studies advocate an additional PWJ cleaning process (reduction up to 95% of the surface contamination). If this solution is chosen, it has to be taken into account at the beginning of the machining process because of the resulting greater depth of removal.

Machining metals with water jet generates compressive residual stress on the target. The stress amplitude reaches its maximum close to the material's surface and then exponentially decreases through a depth of few hundreds of microns. In case of AWJ machining, the amplitude of the residual stress decreases as the jet energy increases (e.g. higher pressure and/or lower traverse speed) due to material removal then stress relief. The works of Azhari et al. (2012, 2016) on Ti6Al4V showed that the evolution of the residual stress is linked to the subsurface microstructure and the material's hardness.

Both defects and residual stress may influence the mechanical behaviour of the machined specimens in static and fatigue. The following part will then focus on the link between the material integrity features and the post-machining mechanical behaviour.

4 Post-Machining Mechanical Behaviour

This section presents the results of works dealing with the consequences of machine-induced defects and damage on the mechanical behaviour of PWJ/AWJ machined specimens. Given that structures are subjected to dynamic loadings during service, the mechanical characterization must be conducted with taking into account these dynamic phenomena. More specifically, this section, which is divided into two parts, will focus on fatigue behaviour. The first part addresses the modifications in the microstructure induced by PWJ and AWJ machining of metallic materials going with the introduction of residual stresses. The influence of these features on the fatigue behaviour of the machined specimens is then discussed. The second part deals with the post-machining mechanical behaviour of composite materials.

4.1 Metallic Materials

4.1.1 Duality Microstructure/Residual Stress

Several methods have been proposed in order to measure the residual strains, hence, to estimate the residual stresses, within polymer matrix composite laminates. Indeed, destructive (compliance method (Montay et al. 2005) or incremental hole-drilling (Nobre et al. 2012)) and non-destructive techniques (X-ray diffraction thanks to the incorporation of aluminium particles within the composite laminate lay-up (Balasingh and Singh 1997)) have been considered as well as fiber Bragg grating embedded between plies (Mulle et al. 2007). However, these studies concern the residual stresses consecutive to the manufacturing of the composite laminates and, to our knowledge, there is no work focused on residual stress consecutive to PWJ/AWJ machining in composites. Thus, this section treats of studies led on metallic materials.

Arola et al. (2002), as previously found by Arola and Ramulu (1997) in AWJ cutting, determined that both PWJ and AWJ peening generate a biaxial in-plane compressive residual stress state on the machined surface. They recorded amplitudes

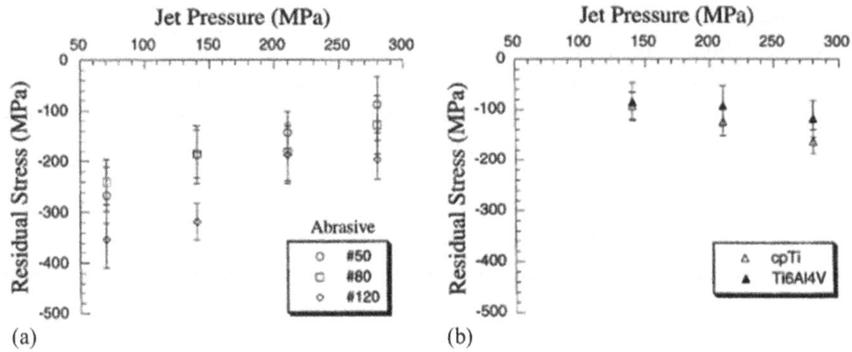

Fig. 37 Influence of the jet pressure on the residual stress. **a** AWJ peening of Ti6Al4V and **b** WJ peening of cpTi and Ti6Al4V (V = 3.81 m/min—SOD = 150 mm) (Arola et al. 2002)

of respectively between 60 and 200 MPa for commercially pure titanium (cpTi) and from 30 to over 400 MPa for Ti6Al4V. These stresses represent an average value calculated over the depth of X-Ray penetration, i.e. around 10 μm. It was observed that, in case of AWJ peening, the amplitude of the residual stress increases as the water jet pressure and the grit size decrease. As seen from Fig. 37a, the amplitude as the residual stress goes from 260 to 100 MPa when the pressure increases from 70 to 280 MPa (abrasive #50) and from 200 to 100 MPa when the abrasive size increases from #120 to #50 MPa (with a jet pressure of 240 MPa). Indeed, large grit and high pressure generate high energy for erosion, leading to stress relief in the material. The same trends were observed by Arola and McCain (2000). However, in the case of PWJ peening, the amplitude of compressive residual stresses slowly increases with the water jet pressure due to lower material removal, thus lower stress relief. As seen from Fig. 37b, the amplitude as the residual stress varies from 100to 160 MPa when the pressure increases from 140 to 280 MPa. Nonetheless, at the same jet pressure (e.g. 140 MPa) the amplitude of residual stresses induced by PWJ peening are lower than those induced by AWJ peening (respectively around 260 and 100 MPa). The link between MRR and the amplitude of residual stress seems to be supported by the results of Huang et al. (2015). In fact, they found that, such as MRR, the amplitude of the residual stress decreases as both the SOD (in the range of 5–50 mm) and the jet exposure time over the workpiece increase (i.e. the traverse speed decreases). As shown from Fig. 38, the residual stress loses 27% of its amplitude when decreasing the SOD from 50 to 5 mm (with an exposure time of 0.8 s) and 30% on average when the exposure time is shortened from 4.8 to 0.8 s. In another work, same evolution of the residual stress with the SOD was observed by Ramulu et al. (2000) in case of PWJ peening of several metals.

Huang et al. (2015) and Lieblich et al. (2016) determined that the amplitude of residual stresses consecutive to PWJ peening in Ti6Al4V decreases exponentially through the depth from the surface. The benefit of peening, i.e. the generation of bi-axial compressive stresses, is visible up to a small distance from the surface (around

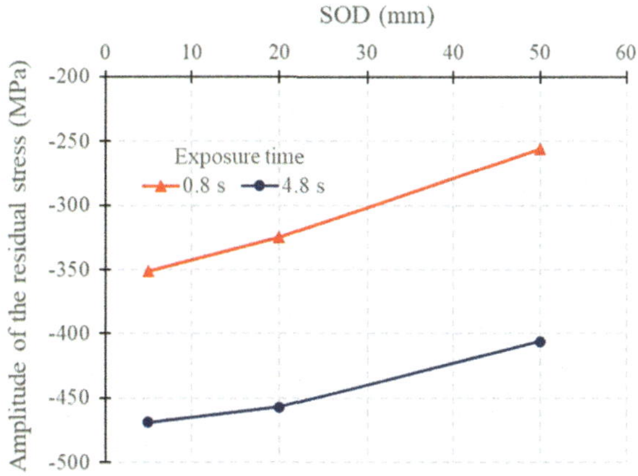

Fig. 38 Influence of the SOD and the exposure time on the amplitude of the residual stress (from results of (Huang et al. 2015))

40 μm from Fig. 39). Beyond this depth, the residual stresses are almost null. The same exponential decrease has been described by Tönshoff et al. (1997) in their study of PWJ peening on case hardened steel.

Azhari et al. (2012, 2016) linked the evolution of the compressive residual stress with the modifications in austenitic stainless steel subsurface microstructure and hardness in case of PWJ peening operation. They highlighted the plastic deformation of the grains through multiple systems of slip bands (cf. Fig. 40b), leading to a refinement of the microstructure (also observed by Lieblich et al. (2016)) then a hardening of the material. The subsurface hardness and the thickness of the hardening layer increase as both the water jet pressure and the number of passes increase but decrease with an increase in traverse speed. Indeed, two jet passes at a traverse speed of 2 m/min and a pressure of 200 MPa leads to a 12% increase of maximum hardness compared to the base material's one and a few tens of microns thick hardening layer

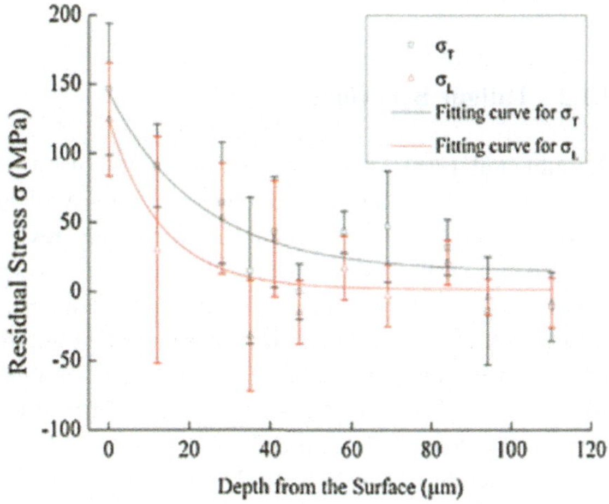

Fig. 39 Evolution of the residual stress amplitude (absolute values) through the depth of the machined workpiece (Huang et al. 2015)

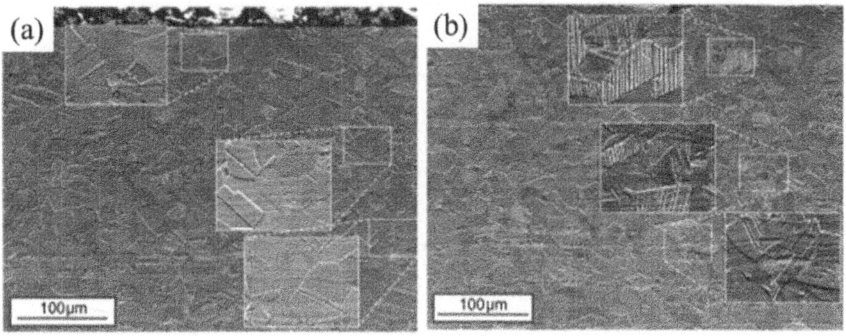

Fig. 40 Subsurface microstructures of austenitic stainless steel for respectively **a** two jet passes at a traverse speed of 2 m/min and a pressure of 200 MPa and **b** four jet passes at a traverse speed of 1 m/min and a pressure of 300 MPa. The insets show the magnified pictures at just below the eroded surface (top), at a depth of approximately 150 μm (middle), and at a depth of approximately 250 μm (bottom) (Azhari et al. 2012)

(cf. Fig. 40a). Similarly, four jet passes at a traverse speed of 1 m/min and a pressure of 300 MPa induce a 31% increase of maximum hardness compared to the base material's one and a 400 μm thick hardening layer (cf. Fig. 40b).

Rivero et al. (2018) showed that the AWJ milling process (Inconel 718) also induced a hardening of the machined surface by a factor two (40–45 HRC against 20–25 HRC for the untreated specimens). The hardening decreased exponentially to a depth of 300 μm below the machined surface, which is consistent with the results of Azhari et al. (2012) in the case of PWJ peening. However, the authors did not observe a significant influence of the machining parameters on the amplitude of the hardening, contrarily to Azhari et al. (2012). According to Boud et al. (2014), who carried out multiple waterjet milling passes on aluminium alloy (7475), this increase in hardness and residual stresses leads to a decrease of MMR per pass as more passes are performed.

4.1.2 Fatigue Behaviour

Ramulu et al. (2002) performed PWJ peening of 7075-T6 aluminium alloy in order to prove that this kind of machining can improve High Cycle Fatigue (HCF) life. To do so, the authors conducted a completely reversed rotating bending test ($R = -1$) on hourglass specimens with circular cross-section and concluded that the modifications in fatigue life were highly dependent on the peening conditions. Indeed, PWJ peening generally enhanced the fatigue strength up to 25% compared to the pristine specimens. The same range of fatigue strength improvement was found by Tönshoff et al. (1997) and Arola et al. (2006) with PWJ peening case-hardened steel (+30%) and AWJ peening Ti6Al4V (+25%, cf. Fig. 41) respectively. Moreover, these two studies (Tönshoff et al. 1997; Arola et al. 2006) showed that, contrary

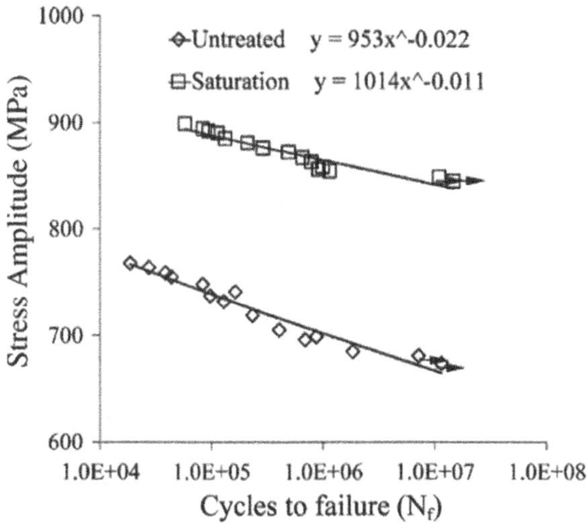

Fig. 41 Stress life diagrams for the control (untreated) and AWJ peened (saturation i.e. peening parameters giving the maximum compressive residual stresses) Ti6Al4V. Arrows indicate unbroken specimens (Arola et al. 2006)

to the untreated specimens where fatigue cracks are visible on the surface, AWJ peening treatment makes the fatigue cracks appear below the surface (150–200 μm) thanks to the residual stresses which prevent the appearance of cracks on the surface despite larger roughness. They concluded that fatigue life is theoretically maximized when the residual stresses are maximized, and the roughness minimized. However, the authors showed that the two parameters are linked as seen from Fig. 42. Thus, the objective is to find the operating parameters giving the optimum couple residual stress-roughness. However, a random distribution of the residual stress is observed when the roughness varies from 6 μm till 10 μm.

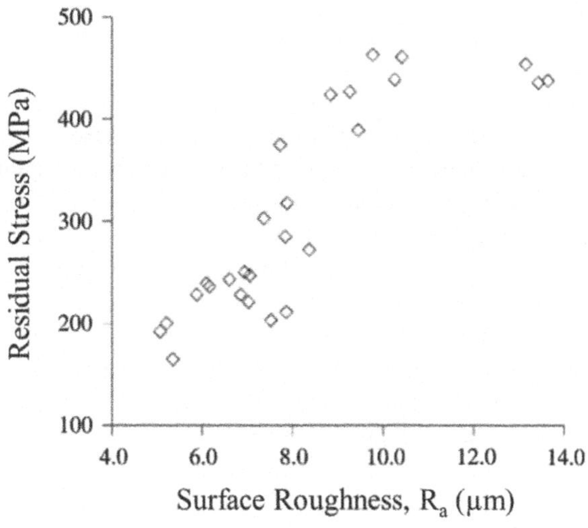

Fig. 42 Relationship between the average roughness and residual stress resulting from peening of the AISI 304 (Arola et al. 2006)

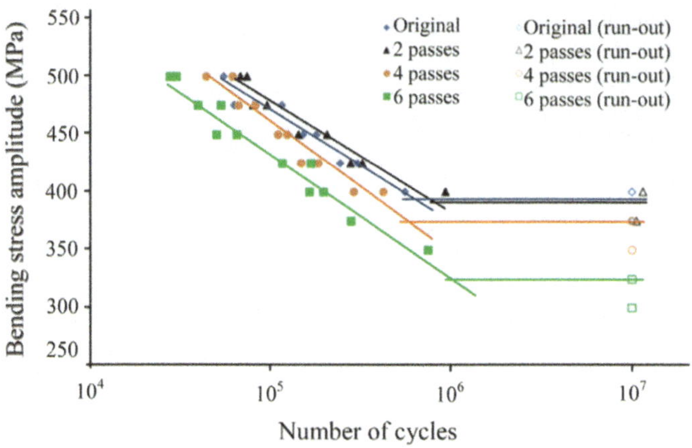

Fig. 43 S-N curves for the different treated specimens (Azhari et al. 2016)

Azhari et al. (2016) ran alternative bending fatigue tests of PWJ peened AISI 304 stainless steel specimens up to failure or 10^7 cycles. Authors have mentioned that the fatigue life of the specimens decreases when the number of peening passes increases, the fatigue strength at 10^5 cycles being 475 MPa after two peening passes and 425 MPa after six passes (cf. Fig. 43). Furthermore, the gap between two S–N curves increases as the number of peening passes increases, meaning that the fatigue life is even more reduced as the number of passes becomes important. The fact that the fatigue life decreases though the residual stresses are greater is in opposition with the observations of the previously cited works (Tönshoff et al. 1997; Ramulu et al. 2002; Arola et al. 2006). Nonetheless, these studies were conducted on only single pass peening and the differences in trends could be explained by the effect of multiple peening passes. By analysing the fractured surfaces of the specimens, Azhari et al. (2016) have noticed that the more passes are made on the specimens, the more cracks are observed. On the untreated material, the only zones of crack initiation are the corners because of their high stress concentration. After two passes, an additional crack initiation site appears at the surface where a notch is generated by the erosion. The surface defects extend as the number of passes, thus the erosion, increases. The differences between the conclusions of (Tönshoff et al. 1997; Ramulu et al. 2002; Arola et al. 2006) and of (Azhari et al. 2016) could be explained by the fact that the negative effects of surface defects created by peening surpass the beneficial effects of the compressive stresses as the number of passes increases. Boud et al. (2014) also found opposite influences of roughness and residual stresses on fatigue life as the number of passes increases. In conclusion, the machining conditions have to be selected to find a compromise between the beneficial residual stresses and the noxious crack initiation induced by the process.

Lieblich et al. (2016) have studied the effect of polishing, grit blasting with Al_2O_3particles and normal PWJ peening (stand-off distance of 2 mm, traverse speed of 0.05 m/s and water jet pressures of 240 and 360 MPa) on the fatigue behaviour

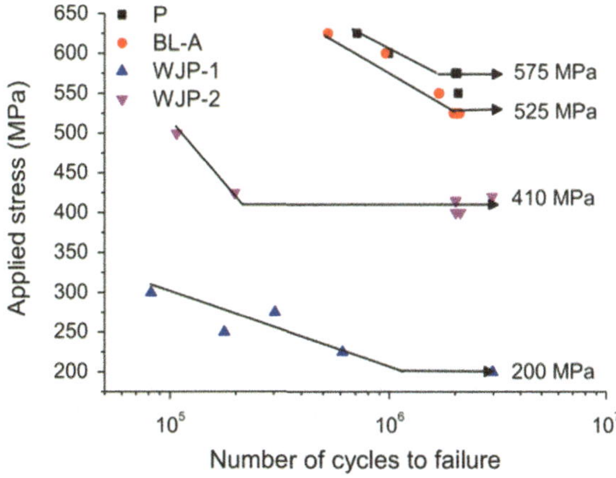

Fig. 44 S-N curves in the unblasted condition (P), after blasting with alumina particles (BL-A) and after waterjet peening at two pressures (respectively 360 MPa and 240 MPa for WJP-1 and WJP-2) (Lieblich et al. 2016)

of titanium alloy (Ti6Al4V). A tensile fatigue test ($R = 0.1$) has been performed on specimens until failure or 2×10^6 cycles. The maximum stress corresponds to the yield stress of the polished specimen (814 MPa) obtained by tensile test. The S–N curves of the different specimens are gathered in Fig. 44. In comparison with the polished specimen (P), the grit blasted specimen and the water jet peened specimen (pressure of 360 MPa, WJP-1) show a decrease of 9% and 65% in fatigue strength respectively. Nonetheless, a quite remarkable improvement of the fatigue strength can be made with the proper processing parameters (only 29% of reduction when PWJ penning with a pressure of 240 MPa—WJP-2). Arola et al. (2002) concluded that machining with a higher waterjet pressure occasioned higher damages to the specimens, which seems to explain the loss of fatigue strength between WJP-1 and WJP-2. By comparing the S–N curves and the amplitude of residual stress for the different specimens, this last parameter does not tend to be the only one influencing the value of fatigue strength. Lieblich et al. (2016) drew the curves of fatigue resistance as a function of roughness and waviness (cf. Fig. 45) and noted that fatigue resistance decreases linearly with the increase of waviness, which is not the case for roughness. However, it has to be mentioned that the conclusions have been drawn with only four values. Additional tests could be recommended in order to confirm or invalidate these trends. Moreover, the conclusions from Fig. 45 seem to be in contradiction with the results of Arola et al. (cf. Fig. 42). Indeed, these latter have shown that an increase in Ra leads to an augmentation of the amplitude of the residual's stresses. The authors also proved that the fatigue life of the specimens is improved when the compressive residual stresses are high. These opposite conclusions with the work of Lieblich et al. (2016) tend to show that the parameter Ra is not recommended for the characterization and correlation between the surface quality and mechanical behaviour.

In another study, Ramulu et al. (2002) have investigated the effect of PWJ peening of 7075-T6 aluminium alloy specimens on the fatigue crack growth. To do so, the authors conducted a uniaxial tensile test ($R = -0.1$) on single-edge-notch (SEN) specimens. The analysis of crack growth (cf. Fig. 46) showed that PWJ peening

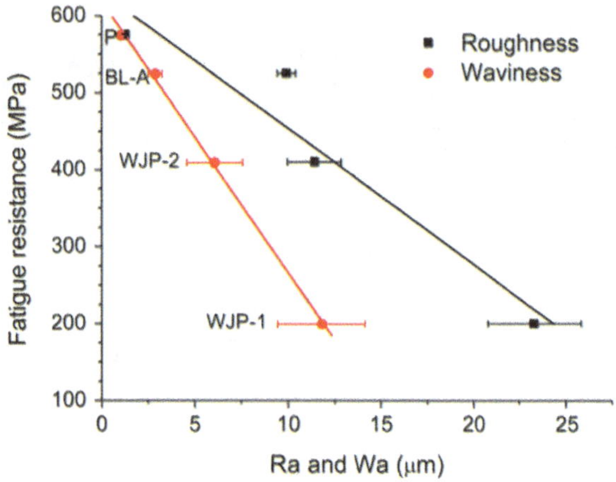

Fig. 45 Fatigue resistance as a function of roughness and waviness (Lieblich et al. 2016)

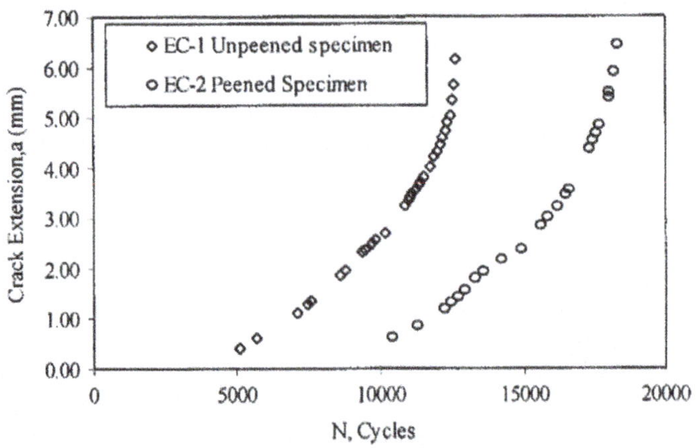

Fig. 46 Fatigue crack extension with respect to number of cycles for unpeened and PWJ peened SEN tensile test specimens (Ramulu et al. 2002)

noticeably delayed the crack initiation (from 5,000 cycles for the unpeened specimens to 10,000 cycles for the peened ones) and enhanced the fatigue crack growth life by 44% (from 12,500 to 18,000 cycles). In order to quantify the benefits of PWJ peening on the crack growth behaviour, curve fitting with Paris power low equation $\frac{da}{dN} = C.(\Delta K)^m$ (which links the crack extension per cycle $\frac{da}{dN}$ and the stress intensity range ΔK) was performed for both unpeened and peened specimens. An 18% decrease of the 'm' coefficient for the peened specimens proved that waterjet peening has slowed down the crack propagation thanks to the inducted plastic deformation. This improvement occurred at relatively low jet pressure (69 MPa) and might be even greater when increasing it.

4.2 *Composites*

4.2.1 Static Tests

Hejjaji et al. (2017) performed tensile tests, following the ASTM D3039M specifications, in order to study the influence of the damage induced by AWJ milling of unidirectional CFRP laminates (Hexply T700-M21.A) on their tensile behaviour. The authors compared three specimens with different surface quality (called "good", "medium" and "poor") based on the total volume of the craters (i.e. voids under the mean plan of the milled surface) generated by the milling process (cf. Table 2). The obtained results have clearly shown that both the tensile strength and modulus decrease when the total volume of craters increases (i.e. the machining quality worsens) (cf. Fig. 47). Even with a good machining quality, the tensile strength is still lower than the one of pristine specimens (respectively 1575 and 2000 MPa). This difference has mainly been attributed to the discontinuities in the fibers, induced by the micro-craters and the broken fibers, which reduce their load carrying ability. This assumption tends to be proven by the fact that when macro-craters appear on medium machining quality specimen, resulting in a higher total volume of craters, the tensile strength continues to decrease (-25% between good and medium qualities).

The further mitigation of the tensile behaviour of the poor machining quality specimens (-32% on tensile strength between good and poor qualities) is explained by the presence of embedded grits—which cannot be quantified by Cv—acting like stress concentration areas, and debonding in addition to the craters and broken fibers. The same conclusions can be drawn from the evolution of the tensile modulus, confirming that the machining quality greatly influences the mechanical behaviour of the milled coupons. Moreover, the total volume of craters seems to be a more suitable indicator to predict the mechanical behaviour of the milled material than surface roughness (cf. Table 2). Indeed, though the values of surface roughness of the medium and poor machining quality specimens are almost similar, their tensile behaviour is different (discard of around 10% for tensile strength and modulus).

The same conclusions have been drawn by Nguyen-Dinh et al. (2019a) who considered the influence of conventional edge trimming quality of CFRP composite laminates on their compressive strength. Indeed, the authors plotted the surface roughness and the crater volume with respect to the compressive strength of the specimens. They concluded that there is a linear correlation between the crater volume

Table 2 Quality parameters of the specimens used for tensile tests and resulting tensile behaviour (Hejjaji et al. 2017)

Quality	Crater volume (mm^3)	Ra (μm)	Tensile strength (MPa)	Tensile modulus (GPa)
Good	0.432	6.74	1574.47	126.80
Medium	1.152	10.3	1182.06	112.19
Poor	2.088	9.8	1072.14	102.27

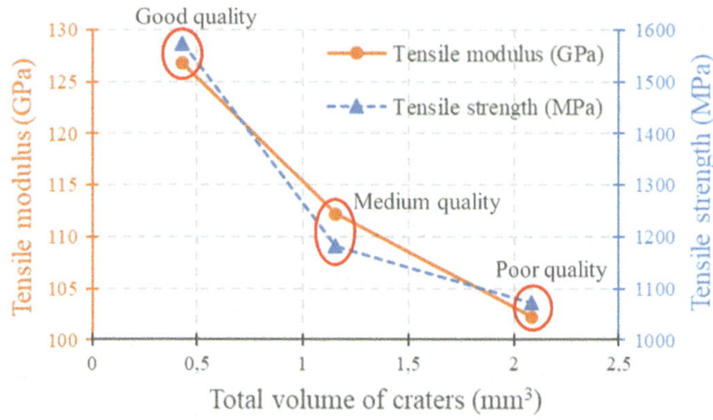

Fig. 47 Influence of the milling quality on the tensile behaviour of unidirectional CFRP specimens (from results of (Hejjaji et al. 2017))

Cv and the compressive strength ($R^2 = 0.9$), which is not the case when considering the surface roughness ($R^2 = 0.4$).

Though crater volume Cv can be an efficient criterion to predict the post machining mechanical behaviour of composite specimens, one can wonder if the crater volume is also a suitable parameter to estimate the fatigue behaviour of the machined specimens.

4.2.2 Fatigue Behaviour

In another work, Hejjaji et al. (2019) performed fatigue tests consecutive to AWJ milling of composites in order to link the post-machining surface quality and the mechanical behaviour of the milled specimens. The laminates were made of unidirectional prepregs of Hexply T700-M21 with 22—group A—and 24—group B—plies with stacking sequences [90°/90°/90°/−45°/0°/45°/90°/−45°/90°/45°/90°]s and [90°/90°/90°/90°/−45°/0°/45°/90°/−45°/90°/45°/90°]s respectively. They were then milled on both sides so the final specimens contain 20 plies. The machining strategy was a raster scan pattern and the milling direction was set parallel to the 90° fiber orientation. The variable machining parameters (jet pressure, traverse speed, scan step and SOD) were selected in order to obtain different surface qualities and damage levels. All the specimens were submitted to tension-tension fatigue tests ($R = 0.1$, $f = 10$ Hz), following the load protocol shown on Fig. 48, with F_{uts} represents the ultimate failure stress during static tensile tests. The obtained results clearly shown that the worst is the post-machining surface quality—quantified thanks to the crater volume Cv—the greater is the loss of endurance limit of the milled specimens. Indeed, as seen from Fig. 49, an increase in Cv by a factor of 3.2 (from 1.2 to 3.8 mm^3/cm^2) leads to a 18% decrease of the endurance limit of the corresponding specimens (from 450 to 380 MPa) in case of two plies removal on each side of the specimens—group B. Another interesting point that can be discussed from Fig. 49

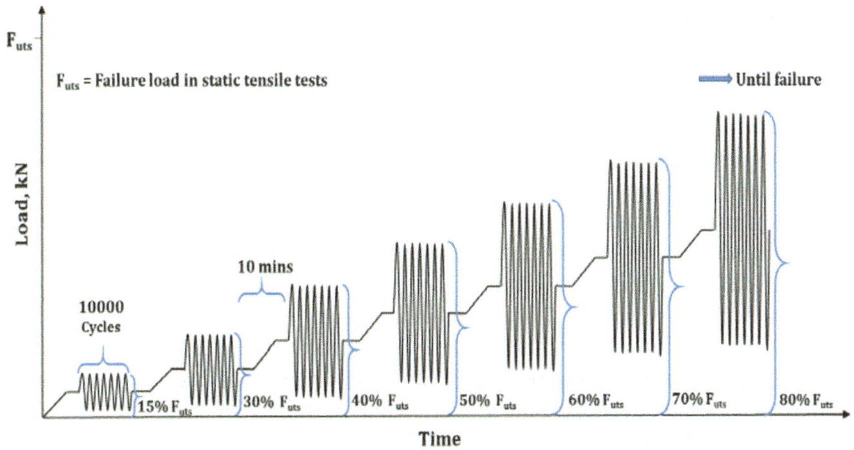

Fig. 48 Load protocol followed by (Hejjaji 2018) for their fatigue tests

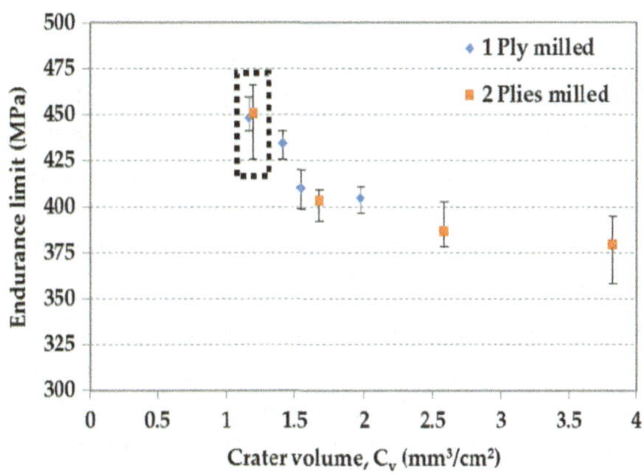

Fig. 49 Influence of the surface quality (Cv) on the endurance limit of AWJ milled specimens (Hejjaji 2018)

is located in the dotted rectangle. Though the crater volume is the same for specimens of group A (one ply milled on both sides) and group B (two plies milled on both sides) the deviation of the endurance limit is greater for group B specimens (40 MPa against 25 MPa for specimens of group A). This might be due to the greater presence of embedded particles when machining with high pressures as shown from studies on metallic materials (Arola and McCain 2000; Arola et al. 2002; Rivero et al. 2018). As this issue cannot be detected by Cv, two specimens with the same amount of crater volume can have different number of embedded particles, then various endurance limits. Thus, the endurance limit is greatly influenced by the types and dimensions of the machine-induced defects and damage. Thereby, the influence of the machining parameters of the AWJ process on the rate of embedded particles in CFRP and their impact on the mechanical behaviour as well as the crater volume (Cv) will be investigated in the future.

4.3 Conclusion of the Part

Static and fatigue tensile tests performed on CFRP after AWJ milling highlighted a link between the machining quality and the mechanical behaviour of the machined specimens. Indeed, the different features representative of the mechanical behaviour of the milled specimens (tensile strength, endurance limit) decrease as the volume of craters generated during the machining process increases. Nonetheless, it has to be noticed that the mechanical properties of the milled specimens worsen even in case of a good surface equality (e.g. drop of 25% in tensile strength).

A similar trend is observed on the post-machining fatigue behaviour of metals. However, the noxious effect of the machine-induced damage and defects is partially counteracted by the relatively beneficial effect of the residual stress, which delays the initiation of fatigue cracks and slows down their propagation.

5 General Conclusion and Perspectives

The main objective of this review is to perform a state of the art of PWJ and AWJ controlled-depth machining of isotropic (e.g. aluminium and titanium alloys) and anisotropic (CFRP composites) materials used in the aerospace field. The focus has been made on three main topics viz. the influence of the process parameters on the material removal, the defects and damage induced by waterjet technique and their impact on the post-machining mechanical behaviour. In this context, the following conclusions can be drawn:

- Complex combinations of numerous machining parameters influence the material removal, mainly quantified by the machined depth of cut or, from the industrial point of view, the Material Removal Rate (MRR). However, this review has shown that the list of parameters to consider when aiming a specific depth or MRR can be reduced to fewer features viz. jet pressure, traverse speed and scan step mainly. This reduction led to the development of prediction models of cutting depth for metallic materials and composite laminates made of unidirectional plies or 2D fabrics. These models gave satisfying results even for removal of small layers of material (few tens of microns). It might then be interesting to consider the waterjet process as a texturing technique for surface preparation in the context of adhesive bonding.
- Another issue when performing controlled-depth machining is the traverse speed fluctuation over the machining duration, leading to variable depths of cut. This is particularly the case when the jet changes of direction: the decelerations or accelerations of the nozzle modify the exposure time, which impact the energy transferred from the jet to the target workpiece, hence the machined depth. Indeed, the development of different jet path strategies shows that it is possible to master

the desired depth of cut. Then, this issue has to be considered at the beginning of the process.

- By using PWJ or AWJ technique to machine metals such as titanium or Inconel alloys, some issues such as thermal damage faced when considering conventional methods can be overridden. Likewise, the emission of harmful particles of composite material can be considerably reduced when performing waterjet machining. However, machining with PWJ or AWJ induces specific kinds of defects and damage depending on the process parameters as well as the properties of the target material. Indeed, when milling with waterjet, the material removal occurs both by water erosion and micro-cutting consecutive to the impact of the abrasive particles. This generates a texturing (craters, streaks) of the target surface depending on the machining parameters. In case of milling metallic materials, the observed damage is mainly ductile fracture consecutive to the growth of cracks initiated by the impact of water droplets and abrasive grits, forming craters throughout the surface of the machined specimens. However, in case of milling Fiber Reinforced Polymer (FRP) composite laminates, damage appear in the form of craters, broken fibers, fiber/matrix debonding as well as delamination. Several studies (Hashish 1995; Cénac et al. 2008, 2009) have proven that the use of PWJ machining of laminates is prohibited because of the important rate of delamination recorded. To our knowledge, no study was led on the observation and quantification of damage induced by PWJ or AWJ milling of 3D woven composites, which are specifically designed to avoid delamination. Another kind of defect generated by AWJ and common to both types of materials is grit embedment. The level of contamination of the machined surfaces as well as the depth of embedment depend on the process parameter. This defect can be an issue if considering AWJ machining as a surface preparation technique prior to adhesive bonding, the substrates needing to be free from any foreign bodies. The embedded particles can also act as stress concentration zones leading to a drop of the post-machining mechanical behaviour of the specimens. Several studies (Arola and McCain 2000; Fowler et al. 2005a; Kong et al. 2011; Liu et al. 2012) proposed solutions to reduce the level of grit contamination. However, these techniques are mainly effective for lightly embedded particles. The use of PWJ machining as a cleaning operation after AWJ milling of metallic materials, though suggested several years ago by Hashish (1998), has scarcely been studied (Huang et al. 2013). It might be interesting, in the context of surface preparation, to analyse the evolution of the surface texturing prior and after this cleaning operation.
- Machining with PWJ or AWJ modifies both the microstructure (metallic materials) and the mechanical properties (both metallic materials and OMC laminates) of the specimens. In case of milling metallic materials, the plastic deformation of the surface, due to the impact of droplets and abrasive particles, leads to a refinement of the microstructure, then a hardening of the subsurface. This can be linked to the generation of compressive residual stresses. Moreover, studies

(Tönshoff et al. 1997; Ramulu et al. 2002; Arola et al. 2006) have proven the beneficial effect of these stresses which mitigate the drop of fatigue life consecutive to damage generated by waterjet machining. Indeed, this compressive stress state delays the cracks initiation and slows their propagation. However, these conclusions were drawn considering PWJ or AWJ peening. This operation being more a surface treatment than a machining operation, it might be interesting to analyse the evolution of the material's microstructure and residual stress state when a harsher removal (i.e. milling) is performed. In case of milling FRP laminates, the alteration of the mechanical properties of the machined specimens can be linked to the level of damage generated by the process. Indeed, the works of Hejjaji et al. (2017, 2018) seem to be the firsts to quantify the level of damage on the milled surface of CFRP composite laminates made of unidirectional plies with a parameter called the crater volume. The authors also find a good correlation between the total volume of craters and the loss of mechanical properties (both with static and fatigue behaviours) of the specimens. To our knowledge, no such study has been performed on 3D woven composites.

- Finally, this review can be put into perspective by considering multi-material (such as metal/composite) machining using PWJ or AWJ process. Indeed, there exists a limited amount of works dealing with this problematic and mainly focusing on AWJ cutting of stacks of titanium alloys and carbon fibers composites (Escobar-Palafox et al. 2012; Pahuja et al. 2014, 2016; Alberdi et al. 2016). The main issue faced when performing multi-material machining seems to be the difference in material machinability leading to dissimilar levels of erosion from one material to the other. The two stacks being bolted in the previous studies, it might be interesting to consider adhesively bonded stacks in order to analyse the modifications consecutive to machining within the adhesive film. Moreover, it seems that waterjet milling of multi-material stacks is not yet investigated and could help to find alternative solutions to machined parts in the context of repairs application.

References

Alberdi A, Artaza T, Suárez A et al (2016) An experimental study on abrasive waterjet cutting of CFRP/Ti6Al4V stacks for drilling operations. Int J Adv Manuf Technol 86:691–704. https://doi.org/10.1007/s00170-015-8192-x

Alberdi A, Rivero A, Lopez De Lacalle LN (2011) Experimental study of the slot overlapping and tool path variation effect in abrasive waterjet milling. J Manuf Sci Eng 133:034502–1–034502–4. https://doi.org/10.1115/1.4004320

Alberdi A, Rivero A, Lopez De Lacalle LN et al (2010) Effect of process parameter on the kerf geometry in abrasive water jet milling. Int J Adv Manuf Technol 51:467–480. https://doi.org/10.1007/s00170-010-2662-y

Amada S, Hirose T, Senda T (1999) Quantitative evaluation of residual grits under angled blasting. Surf Coatings Technol 111:1–9. https://doi.org/10.1016/S0257-8972(98)00482-4

Anwar S, Axinte DA, Becker AA (2013a) Finite element modelling of abrasive waterjet milled footprints. J Mater Process Technol 213:180–193. https://doi.org/10.1016/j.jmatprotec.2012.09.006

Anwar S, Axinte DA, Becker AA (2013b) Finite element modelling of overlapping abrasive waterjet milled footprints. Wear 303:426–436. https://doi.org/10.1016/j.wear.2013.03.018

Arola D, Alade AE, Weber W (2006) Improving fatigue strength of metals using abrasive waterjet peening. Mach Sci Technol 10:197–218. https://doi.org/10.1080/10910340600710105

Arola D, McCain ML (2000) Abrasive waterjet peening: a new method of surface preparation for metal orthopedic implants. J Biomed Mater Res 53:536–546

Arola D, McCain ML, Kunaporn S, Ramulu M (2002) Waterjet and abrasive waterjet surface treatment of titanium: a comparison of surface texture and residual stress. Wear 249:943–950. https://doi.org/10.1016/S0043-1648(01)00826-2

Arola D, Ramulu M (1994) Machining-induced surface texture effects on the flexural properties of a graphite/epoxy laminate. Composites 25:822–834

Arola D, Ramulu M (1997) Material removal in abrasive waterjet machining of metals—a residual stress analysis. Wear 211:302–310. https://doi.org/10.1016/S0043-1648(97)00131-2

Axinte DA, Srinivasu DS, Billingham J, Cooper M (2010) Geometrical modelling of abrasive waterjet footprints: a study for 90° jet impact angle. CIRP Ann—Manuf Technol 59:341–346. https://doi.org/10.1016/j.cirp.2010.03.112

Azhari A, Schindler C, Godard C et al (2016) Effect of multiple passes treatment in waterjet peening on fatigue performance. Appl Surf Sci 388:468–474. https://doi.org/10.1016/j.apsusc.2015.11.195

Azhari A, Schindler C, Kerscher E, Grad P (2012) Improving surface hardness of austenitic stainless steel using waterjet peening process. Int J Adv Manuf Technol 63:1035–1046. https://doi.org/10.1007/s00170-012-3962-1

Balasingh C, Singh V (1997) Measurement of residual stresses in CFRP laminates by X-ray diffraction method. Bull Mater Sci 20:325–332. https://doi.org/10.1007/BF02745098

Billingham J, Miron CB, Axinte DA, Kong MC (2013) Mathematical modelling of abrasive waterjet footprints for arbitrarily moving jets: part II—overlapped single and multiple straight paths. Int J Mach Tools Manuf 68:30–39. https://doi.org/10.1016/j.ijmachtools.2013.01.003

Bitter JGA (1963a) A study of erosion phenomena. Part I. Wear 6:5–21

Bitter JGA (1963b) A study of erosion phenomena. Part II. Wear 6:169–190. https://doi.org/10.1016/0043-1648(63)90073-5

Boud F, Loo LF, Kinnell P (2014) The impact of plain waterjet machining on the surface integrity of aluminium 7475. Procedia CIRP 13:382–386. https://doi.org/10.1016/j.procir.2014.04.065

Bui VH, Gilles P, Sultan T et al (2017) A new cutting depth model with rapid calibration in abrasive water jet machining of titanium alloy. Int J Adv Manuf Technol 93:1499–1512. https://doi.org/10.1007/s00170-017-0581-x

Cénac F, Collombet F, Zitoune R, Déléris M (2008) Abrasive-water-jet blind-machining of polymer matrix composite materials. In: ECCM

Cénac F, Collombet F, Zitoune R, Déléris M (2009) Respect de l'intégrité dans l'usinage non-débouchant par jet d'eau abrasif de matériaux composites = Material integrity maintain while milling composite using abrasive water jet. In: JNC16. Toulouse, France

Cénac F, Zitoune R, Collombet F, Deleris M (2013) Abrasive water-jet milling of aeronautic aluminum 2024–T3. Proc Inst Mech Eng Part L J Mater Des Appl. https://doi.org/10.1177/1464420713499288

Chillman A, Ramulu M, Hashish M (2007) Waterjet peening and surface preparation at 600 MPa: a preliminary experimental study. J Fluids Eng 129:485–490. https://doi.org/10.1115/1.2436580

Chillman A, Ramulu M, Hashish M (2010) Waterjet and water-air jet surface processing of a titanium alloy: a parametric evaluation. J Manuf Sci Eng 132(011012):1–10. https://doi.org/10.1115/1.4000837

Chillman A, Ramulu M, Hashish M, Cantrell A (2010) High pressure waterjets—an innovative means of alpha case removal for superplastically formed titanium alloys. Key Eng Mater 433:103–111. https://doi.org/10.4028/www.scientific.net/KEM.433.103

Crouzeix L, Davila Y, Collombet F et al (2012) Study of double step lap composite repairs on CFRP evaluators. ECCM 15:24–28

Escobar-Palafox GA, Wika KK, Gault RS, Ridgway K (2012) Characterisation of abrasive water-jet process for drilling titanium and carbon fibre reinforced polymer stacks. Surfaces

Finnie I (1958) The mechanism of erosion of ductile metals. In: 3rd US national congress of applied mechanics

Fowler G (2003) Abrasive water-jet—controlled depth milling titanium alloys

Fowler G, Pashby IR, Shipway PH (2009) The effect of particle hardness and shape when abrasive water jet milling titanium alloy Ti6Al4V. Wear 266:613–620. https://doi.org/10.1016/j.wear.2008.06.013

Fowler G, Shipway PH, Pashby IR (2005a) A technical note on grit embedment following abrasive water-jet milling of a titanium alloy. J Mater Process Technol 159:356–368. https://doi.org/10.1016/j.jmatprotec.2004.05.024

Fowler G, Shipway PH, Pashby IR (2005b) Abrasive water-jet controlled depth milling of Ti6Al4V alloy—an investigation of the role of jet-workpiece traverse speed and abrasive grit size on the characteristics of the milled material. J Mater Process Technol 161:407–414. https://doi.org/10.1016/j.jmatprotec.2004.07.069

Gupta TVK, Ramkumar J, Tandon P, Vyas NS (2013) Role of process parameters on pocket milling with abrasive water jet machining technique 7:348–353

Gupta TVK, Ramkumar J, Tandon P, Vyas NS (2015) Application of artificial neural networks in abrasive water jet milling. Procedia CIRP 37:225–229. https://doi.org/10.1016/j.procir.2015.08.076

Hashish M (1989) An investigation of milling with abrasive-waterjets. J Eng Ind 111:158. https://doi.org/10.1115/1.3188745

Hashish M (1984) A modeling study of metal cutting with abrasive waterjets. J Eng Mater Technol 106:88–100. https://doi.org/10.1115/1.3225682

Hashish M (1998) Method and apparatus for abrasive water jet milling

Hashish M (1995) Waterjet machining of advanced composites. Mater Manuf Process 10:1129–1152. https://doi.org/10.1080/10426919508935098

Hejjaji A (2018) Abrasive waterjet milling of CFRP composites and its influence on the mechanical behavior and patch adhesion intended for repair application. Université Paul Sabatier (Toulouse—France)

Hejjaji A, Zitoune R, Crouzeix L et al (2017) Surface and machining induced damage characterization of abrasive water jet milled carbon/epoxy composite specimens and their impact on tensile behavior. Wear 376–377:1356–1364. https://doi.org/10.1016/j.wear.2017.02.024

Hejjaji A, Zitoune R, Toubal L et al (2019) Influence of controlled depth abrasive water jet milling on the fatigue behavior of carbon/epoxy composites. Compos Part A Appl Sci Manuf 121:397–410. https://doi.org/10.1016/j.compositesa.2019.03.045

Hloch S, Hlaváček J, Vasilko K et al (2014) Abrasive waterjet (AWJ) titanium tangential turning evaluation. Metalurgija 53:537–540

Hocheng H, Tsai HY, Shiue JJ, Wang B (1997) Feasibility study of abrasive-waterjet milling of fiber-reinforced plastics. J Manuf Sci Eng 119:133–142. https://doi.org/10.1115/1.2831088

Huang L, Folkes J, Kinnell P, Shipway PH (2012) Mechanisms of damage initiation in a titanium alloy subjected to water droplet impact during ultra-high pressure plain waterjet erosion. J Mater Process Technol 212:1906–1915. https://doi.org/10.1016/j.jmatprotec.2012.04.013

Huang L, Kinnell P, Shipway PH (2013) Parametric effects on grit embedment and surface morphology in an innovative hybrid waterjet cleaning process for alpha case removal from titanium alloys. Procedia CIRP 6:594–599. https://doi.org/10.1016/j.procir.2013.03.077

Huang L, Kinnell P, Shipway PH (2015) Removal of heat-formed coating from a titanium alloy using high pressure waterjet: Influence of machining parameters on surface texture and residual stress. J Mater Process Technol 223:129–138. https://doi.org/10.1016/j.jmatprotec.2015.03.053
Kanthababu M, Rajes RM, Emannuel PN et al (2016) Experimental investigations on pocket milling of titanium alloy using abrasive water jet machining. FME Trans 44:133–138. https://doi.org/10.5937/fmet1602133K
Kong MC, Anwar S, Billingham J, Axinte DA (2012) Mathematical modelling of abrasive waterjet footprints for arbitrarily moving jets: part I—single straight paths. Int J Mach Tools Manuf 53:58–68. https://doi.org/10.1016/j.ijmachtools.2011.09.010
Kong MC, Axinte DA, Voice W (2011) An innovative method to perform maskless plain waterjet milling for pocket generation: a case study in Ti-based superalloys. Int J Mach Tools Manuf 51:642–648. https://doi.org/10.1016/j.ijmachtools.2011.04.006
Kong MC, Axinte DA, Voice W (2010) Aspects of material removal mechanism in plain waterjet milling on gamma titanium aluminide. J Mater Process Technol 210:573–584. https://doi.org/10.1016/j.jmatprotec.2009.11.009
Lieblich M, Barriuso S, Ibáñez J et al (2016) On the fatigue behavior of medical Ti6Al4V roughened by grit blasting and abrasiveless waterjet peening. J Mech Behav Biomed Mater 63:390–398. https://doi.org/10.1016/j.jmbbm.2016.07.011
Liu H-T, Hovanski Y, Dahl ME (2012) Machining of aircraft titanium with abrasive-waterjets for fatigue critical applications. J Press Vessel Technol 134:011405. https://doi.org/10.1115/1.4004834
Lozano Torrubia P, Axinte DA, Billingham J (2015) Stochastic modelling of abrasive waterjet footprints using finite element analysis. Int J Mach Tools Manuf 95:39–51. https://doi.org/10.1016/j.ijmachtools.2015.05.001
Montay G, Sicot O, Gong XL et al (2005) Determination of the residual stresses in composite laminate using the compliance method. Mater Sci Forum 490–491:533–538. https://doi.org/10.4028/www.scientific.net/msf.490-491.533
Mulle M, Zitoune R, Collombet F et al (2007) Thermal expansion of carbon-epoxy laminates measured with embedded FBGS—comparison with other experimental techniques and numerical simulation. Compos Part A Appl Sci Manuf 38:1414–1424. https://doi.org/10.1016/j.compositesa.2006.08.008
Nguyen-Dinh N, Bouvet C, Zitoune R (2019) Influence of machining damage generated during trimming of CFRP composite on the compressive strength. J Compos Mater. https://doi.org/10.1177/0021998319883335
Nguyen-Dinh N, Zitoune R, Bouvet C, Le Roux S (2019) Surface integrity while trimming of composite structures: X-ray tomography analysis. Compos Struct 210:735–746. https://doi.org/10.1016/j.compstruct.2018.12.006
Nobre JP, Batista AC, Nau A, et al (2012) Using the Incremental hole-drilling technique for measuring residual stresses in fibre-reinforced polymer composites. In: ECCM 2012—Compos Venice, Proceedings of the 15th European conference on composite materials 24–28
Pahuja R, Ramulu M, Hashish M (2014) Abrasive water jet machining (AWJ) of hybrid titanium/graphite composite laminate: preliminary results. In: BHR Group—22nd international conference on water jetting. pp 83–95
Pahuja R, Ramulu M, Hashish M (2016) Abrasive waterjet profile cutting of thick titanium/graphite fiber metal laminate. In: ASME 2016 international mechanical engineering congress and exposition. pp 1–11
Pal VK, Choudhury SK (2014) Surface characterization and machining of blind pockets on Ti6Al4V by abrasive water jet machining. Procedia Mater Sci 5:1584–1592. https://doi.org/10.1016/j.mspro.2014.07.346
Phapale K, Singh R, Patil S, Singh RKP (2016) Delamination characterization and comparative assessment of delamination control techniques in abrasive water jet drilling of CFRP. Procedia Manuf 5:521–535. https://doi.org/10.1016/j.promfg.2016.08.043

Rabani A, Madariaga J, Bouvier C, Axinte DA (2016) An approach for using iterative learning for controlling the jet penetration depth in abrasive waterjet milling. J Manuf Process 22:99–107. https://doi.org/10.1016/j.jmapro.2016.01.014

Ramulu M, Arola D (1994) The influence of abrasive waterjet cutting conditions on the surface quality of graphite/epoxy laminates. Int J Mach Tools Manuf 34:295–313. https://doi.org/10.1016/0890-6955(94)90001-9

Ramulu M, Kunaporn S, Arola D et al (2000) Waterjet machining and peening of metals. Analyzer 122:90–95. https://doi.org/10.1115/1.556155

Ramulu M, Kunaporn S, Jenkins M et al (2002) Fatigue performance of high-pressure waterjet-peened aluminum alloy. J Press Vessel Technol 124:118. https://doi.org/10.1115/1.1398553

Ramulu M, Raju SP, Inoue H, Zeng J (1993) Hydro-abrasive erosion characteristics of 30vol.%SiCp/6061-T6 Al composite at shallow impact angles. Wear 166:55–63. https://doi.org/10.1016/0043-1648(93)90279-U

Rivero A, Alberdi A, Artaza T et al (2018) Surface properties and fatigue failure analysis of alloy 718 surfaces milled by abrasive and plain waterjet. Int J Adv Manuf Technol 94:2929–2938. https://doi.org/10.1007/s00170-017-0979-5

Saleem M, Toubal L, Zitoune R, Bougherara H (2013) Investigating the effect of machining processes on the mechanical behavior of composite plates with circular holes. Compos Part A Appl Sci Manuf 55:169–177. https://doi.org/10.1016/j.compositesa.2013.09.002

Shanmugam DK, Nguyen T, Wang J (2008) A study of delamination on graphite/epoxy composites in abrasive waterjet machining. Compos Part A Appl Sci Manuf 39:923–929. https://doi.org/10.1016/j.compositesa.2008.04.001

Shipway PH, Fowler G, Pashby IR (2005) Characteristics of the surface of a titanium alloy following milling with abrasive waterjets. Wear 258:123–132. https://doi.org/10.1016/j.wear.2004.04.005

Sourd X, Zitoune R, Crouzeix L et al (2020) New model for the prediction of the machining depth during milling of 3D woven composite using abrasive waterjet process. Compos Struct 234. https://doi.org/10.1016/j.compstruct.2019.111760

Srinivasu DS, Axinte DA (2014a) Surface integrity analysis of plain waterjet milled advanced engineering composite materials. Procedia CIRP 13:371–376. https://doi.org/10.1016/j.procir.2014.04.063

Srinivasu DS, Axinte DA (2014b) Mask-less pocket milling of composites by abrasive waterjets: an experimental investigation. J Manuf Sci Eng 136:041005–1–41013. https://doi.org/10.1115/1.4027181

Tönshoff HK, Kroos F, Marzenell C (1997) High-pressure water peening—a new mechanical surface-strengthening process. Ann CIRP 46:113–117

van Luttervelt ICA (1989) On the selection of manufacturing methods illustrated by an overview of separation techniques for sheet materials. CIRP Ann—Manuf Technol 38:587–607. https://doi.org/10.1016/S0007-8506(07)61127-5

Vasanth S, Muthuramalingam T, Vinothkumar P et al (2016) Performance analysis of process parameters on machining titanium (Ti-6Al-4V) alloy using abrasive water jet machining process. Procedia CIRP 46:139–142. https://doi.org/10.1016/j.procir.2016.04.072

Zain AM, Haron H, Sharif S (2011) Genetic algorithm and simulated annealing to estimate optimal process parameters of the abrasive waterjet machining. Eng Comput 27:251–259. https://doi.org/10.1007/s00366-010-0195-5

Zitoune R, Collombet F, Cénac F (2013) Méthodologie spécifique à l'usinage des composites par jet d'eau abrasif. Tech l'ingénieur

Damage to Carbon Fiber Reinforced Polymer Composites (CFRP) by Laser Machining: An Overview

Sharizal Ahmad Sobri, Robert Heinemann, David Whitehead, Mohd Hazim Mohamad Amini, and Mazlan Mohamed

Abstract Carbon fiber reinforced polymer or CFRP composite is the most sort-after material for aircraft manufacturing today and commercial aircraft manufacturers are critically attempting to solve the delamination problem as well as other damages. On the other hand, implementation of laser technology in cutting and drilling composites is now becoming more important as an alternative solution for machining of composites. The major obstacles in laser machining of composites for industrial applications are typically related to inferior kerf width (due to tapered cut), heat-affected zone (HAZ), charring, matrix recession, protruding fibers and potential delamination. Therefore, it is desirable to optimize the machining parameters in order to control the level of heat-induced in machining of CFRP composites. Currently, no comparative study has been conducted between the effects of ring shape and spiral trepanning, which this study might be the potential to apply on drilling thick CFRP composites. High powered lasers are also important in drilling thick CFRP composites. This article reviews the experimental progress in laser machining of CFRP composites, which is intended to help readers to obtain the latest views in order to develop the appropriate machining parameters.

Keywords Carbon fiber reinforced polymer composites (CFRP) · Lasers · Machining parameters · Heat-affected zone (HAZ)

S. A. Sobri (✉) · M. H. M. Amini · M. Mohamed
Faculty of Bioengineering and Technology, Universiti Malaysia Kelantan, Jeli Campus, 17600 Jeli, Kelantan, Malaysia
e-mail: sharizal.s@umk.edu.my

R. Heinemann · D. Whitehead
Department of Mechanical, Aerospace and Civil Engineering, The University of Manchester, Pariser Building, Sackville Street, Manchester M13 9PL, UK

M. T. Hameed Sultan et al. (eds.), *Machining and Machinability of Fiber Reinforced Polymer Composites*, Composites Science and Technology,
https://doi.org/10.1007/978-981-33-4153-1_10

1 Introduction

Manufacturing companies around the globe have no choice but need to improve the efficiency of their manufacturing processes. Due to competitive market demands, commercial aircraft manufacturers like Airbus and Boeing are now attempting to change the structures of aircrafts to make them lighter and more efficient (Garrick 2007). According to Garrick (2007) and Mangalgiri (1999), fiber-reinforced plastics including carbon fiber reinforced plastic/polymer (CFRP) are the most prominent and sought-after materials for aircraft manufacturing today. The latest generation of large passenger/commercial aircrafts contains a large proportion of composite materials; for example the Airbus A350 and Boeing B787 have a composite contents of over 30% and 50%, respectively (Reza 2010). For the components made of CFRP, certain difficulties such as fiber pull-out, delamination, and the decomposition of the matrix material are the most frequent problems that occur during the machining processes, all of which affect the quality of the machined surfaces and the material properties (Pecat et al. 2012; Santhanakrishnan et al. 1988; Komanduri 1997; Hintze et al. 2011). Thus, the interest in composite machining has been growing, particularly in reducing the extent of damage incurred by various machining processes (Garrick 2007; Pecat et al. 2012; Santhanakrishnan et al. 1988; Komanduri 1997; Hintze et al. 2011; Lau et al. 1995).

Apart from the research work carried out so far on conventional machining, the amount of research conducted on unconventional machining of composites is increasing steadily. Examples of unconventional machining technologies include laser and water jet cutting, electro-discharge, electrochemical and ultrasonic machining. Machining composites by laser appears to be a viable approach (Abrate and Walton 1992) since one of the composite phases is often a type of polymer, and polymers in general exhibit a very high absorption coefficient for infrared radiation as well as a low thermal conductivity, which causes the thermal energy to remain highly localized (Abrate and Walton 1992; Singh and Maurya 2017; Kumar 2014; Steen and Mazumder 2010). An ideal cut for laser machining would be very narrow, completely straight and vertical, perfectly clean and the cut faces would be smooth (Singh and Maurya 2017; Kumar 2014; Sheikh-Ahmad 2009; Yilbas 2013). However, a typical laser cut is characterised by striations (regular straight lines caused by the laser beam axis on the cutting surface), spatter on the top surface, kerf width, edge roundness, microstructure defect (Heat-Affected Zone or HAZ), dross at the bottom surface and unwanted taper (Reza 2010; Abrate and Walton 1992; Singh and Maurya 2017; Kumar 2014; Steen and Mazumder 2010; Li et al. 2002,2007; Shanmugam et al. 2002; Yilbas 2013).

The objective of this review article is to highlight the laser machining strategies and damage to CFRP composites by laser machining in order to assist in the process of developing on machining parameters. Section 2 describes some laser machining strategies that commonly applied by the manufacturing industries. The essential aspects on damage to CFRP composites by laser beam machining are described in Sect. 3. Some concluding remarks are contained in Sect. 4.

2 Laser Machining Strategies

Composite materials such as carbon fiber reinforced polymer (CFRP) require high temperature to heat up the fibers, which can lead to undesirable surface quality, and may produce taper if the machining involves deep cuts. Adverse effects on the properties of the machined materials may lead to inferior surface quality; it is important to investigate in order to identify any deterioration of material performance. For instance, machining CFRP composites usually creates a HAZ, due to high localization of temperatures during vaporization, and this may affect the structure of the lay-up strength (Sheikh-Ahmad 2009; Li et al. 2008). In order to find the optimum parameters in laser machining, strategies of how to achieve fast, retaining acceptable hole quality with good repeatability have led to many variations in laser machining strategies by industry players (Steen and Mazumder 2010). Currently, there are three common types of laser machining strategies used in industry (Steen and Mazumder 2010; Sheikh-Ahmad 2009), as shown in Fig. 1.

The first type of laser machining strategy is called single/double-shot drilling; the common factor is the use of a focused beam spot without any movement of the laser beam spot and workpiece during the drilling process. The most common drilling process is single-shot drilling, which can produce a cut through hole with a single pulse. For a double-shot drilling process, the energy is basically divided between two pulses, and it will interact with the plasma expeditiously. In addition, these pulses will follow each other in sequence during the interaction process. Next is called percussion drilling usually uses single or multiple shots by utilizing a focused laser spot to heat, melt and vaporize the desired hole with no movement of the laser beam or workpiece. Moreover, with the intense laser burst during drilling, the hole size produced is based on the size of the beam, which is varied by a focusing mechanism.

Percussion drilling is the preferred method in many aerospace applications for cooling holes, owing to its speed, sufficient accuracy and repeatability. The method requires pulses of between 10^5 and 10^7 W/cm^2 (Steen and Mazumder 2010; Low and Li 2002); according to Steen and Mazumder (Steen and Mazumder 2010), it is critical to consider the energy and peak power of a pulse to determine how much of the energy will be utilized for evaporation as opposed to melt. Furthermore, more

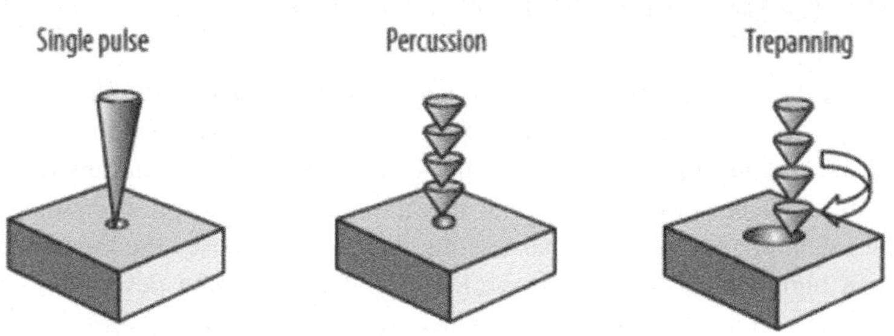

Fig. 1 Schematic illustration of different styles of laser beam machining (Adapted From Steen and Mazumder (2010), Sheikh-Ahmad (2009))

melt will occur with longer pulses of low energy applied, as compared with short sharp pulses. In addition, pulses which are longer than 0.3 ms with a very fine 50 μm spot size will create significant melt. In laser-material interaction, shorter pulses cause heating of a smaller thickness of the treated zone, and require sufficient energy to be supplied to evaporate the affected volume, in order to have a larger spot. Thus, as opposed to percussion drilling—including single/double-shot drilling—these are still not as precise as trepanning (Steen and Mazumder 2010; Jacobs 2008; Naeem 2017; Choudhury et al. 2012; Stock et al. 2012). It is unfavorable to be utilised for machining CFRP, due to high localized temperature in a single spot, even though these strategies are considered as the quickest. The most obvious problem, particularly for single-shot drilling, is that of kerf taper. Double-shot and percussion drilling may also contribute towards this problem, due to the possibility of variations in the power and focus between shots (Steen and Mazumder 2010; Choudhury et al. 2012; Stock et al. 2012).

The third type of laser machining strategy is called Trepanning–see Fig. 1, which a method used to remove a cylindrical core, or circular disc, from a substrate. This strategy is significantly different than percussion and single/double-shot laser drilling, in terms of the position of the beam or substrate. The beam initially pierces the workpieces and then it moves around the perimeter of the hole in order to cut out any desired shape such as round, square or even a star shape (Steen and Mazumder 2010; Jacobs 2008). The trepanning path can be performed several times, and if it is done twice, the first orbit will usually trepan and the next orbit will just eliminate any residues in the hole. There is a set pattern for trepanning, known in the industry as 'overlap', where the laser spot is moved by rotating the beam around the perimeter of the hole to achieve the desired edge quality and production throughput (Steen and Mazumder 2010; Jacobs 2008; Ashkenasi et al. 2011). In laser trepanning, a small overlap is used to increase the material removal rate level, but it may create irregular indentations on the edge, whereas a larger overlap creates a finer hole resolution and edge quality.

3 Damage to Carbon Fiber Reinforced Polymer Composites (CFRP) by Laser Machining

Abrate et al. (1992) pointed out that laser cutting also has a problem when the thickness of the part increases; parts of the material to be removed interfere with the laser beam and limit its ability to cut. Nevertheless, the laser cutting of plastics is very efficient because this material has a very high absorption coefficient for infrared radiation, and low thermal conductivity, which causes the thermal energy to remain highly localised (Vanderwert 1983). The following descriptions are of the experimental work that was performed so far on the laser cutting of carbon fiber reinforced polymer composites. Most of these works focused on the utilisation of CO_2 and Nd: YAG lasers, which are common in the machining of composite materials. Fiber lasers, on the contrary, are lacking in research, and offer a number of advantages compared with these two lasers. In addition, other researchers who utilised CO_2, Nd: YAG and other types of lasers are considered the priority concerns relating to the effect of process parameters (i.e. laser power, cutting speed, type of gas and its pressure and material thickness—on the quality of the workpiece). In this section, both machining processes (i.e. cutting and drilling) are covered in order to understand the relationships between machining parameters related to laser cutting/drilling and cut/hole quality.

Research by Reza and Li (2013) investigated the laser cutting of carbon fiber-reinforced polymer composites using an IPG YLR-1000-SM single-mode 1 kW fiber laser. Experimental studies were conducted to investigate the thermal damage due to different assist gas types and pressures, as well as focal plane positions or abbreviated as FPP. Reza and Li (2013) adopted a method called Design of Experiments or DoE in their analysis, which narrowed down the process parameter windows like laser power, scanning speed, gas pressure and FPP so that only three responses were considered, such as the matrix recession, kerf width and the cut depth. A power level of 340 W and a cutting speed of 20 mm/s were chosen. They concluded that beam power and scanning speed are the most influential factors in single-pass mode. Based on the DoE analysis, when FPP is below the surface of the workpiece with a laser power of 340 W and at scanning speed of 20 mm/s, it was shown that these optimal conditions successfully cut CFRP in a single-pass operation. The fiber laser system used for the cutting experiments is at the beam focussed below the material, which according to the authors proved to be effective in reducing thermal damage. The authors observed that the workpiece material could not be cut through using a power level below 230 W, even when at low scanning speeds of 1 mm/s. Lau et al. (1995) had already reached the same verdict like Reza and Li (2013) regarding it being essential to understand the interaction between a laser beam and workpiece, in which laser cutting quality is strongly influenced by laser power and cutting speed/feed rate, as highlighted by other researchers as well (Abrate and Walton 1992; Li et al. 2008; Naeem 2017; Klotzbach et al. 2011; Mathew et al. 1999). Furthermore, each individual combination of laser power and speed can produce significantly different cutting quality; for instance, as high power and low speed is applied, it may effectively remove the material in large amounts, but at the same time, it creates a significant HAZ effect due to high energy of the laser and the interaction time between the laser beam and workpiece. Based on the research attempts, the typical range for laser power is between 10 W and 3 kW and for scanning speed the range is between 1 mm/s and 1000 mm/s.

The effect of assist gas observed by Reza and Li (2013) indicates that increasing the gas pressure leads to a decrease of the matrix recession as well as the kerf width, i.e. at the entrance and exit sides. This result occurred when using inert gasses, such Argon and Nitrogen. However, for the case of Oxygen, by increasing the gas pressure, the matrix recession can be reduced while the kerf widths at both sides increased when the gas pressure increased. The authors claimed that this was due to an oxidation process, which caused accelerated decomposition or vaporisation of the material. Based on the experimental results by Klotzbach et al. (2011) and

Naeem (2017), the surface of the sample was covered by a thin layer of thermally-decomposed polymer matrix by using Nitrogen gas after the drilling operation. In the work of Mathew et al. (1999), they used Argon gas in their research attempts and found out that this particular type of gas has more effective outcomes during laser machining, giving a better quality cut, which was also experienced by Reza and Li (2013).

The next observation of Reza and Li (2013) is the effect of FPP, in which the authors set the range of −2.38 mm to +2.38 mm at constant parameters of 340 W laser power, 20 mm/s scanning speed and 8 bar gas pressure. The effects can be seen when the FPP was increased gradually from –2.38 mm. At the beam entrance, heat damage was seen, whereas, at the end of the cut depth, the thermal damage was diminished. Moreover, this upward movement caused incomplete cuts when it reached 2.38 mm. This effect was also observed when the FPP was at zero, positioned on the top surface, with no through cuts. Based on the findings that were highlighted above, changing the FPP can influence the power density at the surface, together with the change of beam spot size. The effect of FPP influences the thermal damage on the material. However, this research was attempted on 2 mm-thick CFRP, and it is believed that the variation of FPP will be different due to thickness concerns, as it relates to the change of power density when changing FPP. A smaller beam spot produces effective and highly focused energy density towards the material. Consistency of focusing straightness and stability of energy distribution allow the laser beam spot to prolong the surface quality during machining. This experiment also indicated the superiority of a fiber laser's beam focusing position compared to the experiments of CFRP composites machining by using CO_2 and Nd:YAG lasers (Lau et al. 1995; Mathew et al. 1999; Wahab et al. 2012). At the moment, this is the only research attempt that deals with variation of FPP specifically for machining of thick CFRP composites by using a fiber laser.

Reza and Li (2013) suggested an alternative strategy by applying multiple-pass cutting; they furthered their research on this strategy that thermal damage and depth of cut were found to be substantially influenced by variation of interaction time—i.e. scanning speed—based on experimentation through the energy per unit length analysis. They considered that cutting with the beam spot focused below the material (FPP < 0) at low power levels and high scanning speeds are the most suitable combination to minimise delamination formation. It is interesting to note that—for a complete cut—it is necessary to decrease the power and/or increase the scanning speed, and consequently to increase the number of passes in order to reduce the energy per unit length. In addition, the material thickness and power density are the major factors that affect the complete cutting process. Li et al. (2008) also introduced a similar strategy but the difference is the laser beam penetrates in a multi-ring pattern (see Fig. 2) compared to Reza and Li's (2013) approach, which only penetrates at the circumference of the hole. Furthermore, using multiple passes Li et al. (2008) observed that with an increase in spacing between two subsequent passes, the total number of passes required to cut through the material could be reduced. Using a spacing of 100 μm, the 1 mm-thick workpiece could be penetrated with only two passes, whereas when using a spacing of 75 μm, three passes were required; this was

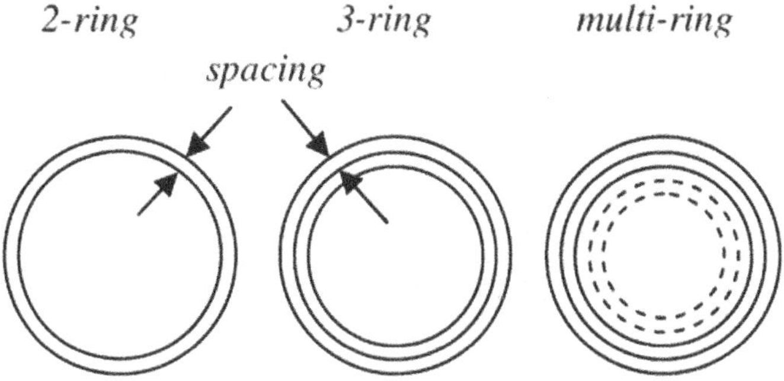

Fig. 2 Sketches to illustrate laser beam scanning using multiple traces (Adapted From Li et al. (2008))

assumed to be brought about by the wider opening, which is beneficial for material removal in a given thickness laminate. The wider opening provides more room for the expanding plume and plasma generated during the laser ablation process. As a consequence, more incident laser energy reached the material and led to a higher material removal rate. It was also noticed that the beam spacing in these experiments, in most cases, it was larger than the effective beam size (35 μm). While the surface quality at the outer rings was strongly affected by the scanning speed, the scanning speed had only a marginal effect on the internal rings. Using this spacing strategy apparently decreases the effect of plasma and plume and also allows fibers to be chopped into small pieces and ejected layer by layer. The findings from both research attempts might be far more interesting if the authors had addressed the question of which carbon fiber orientation is the most affected by delamination because this is unavoidable in continuous wave (CW) laser cutting due to the anisotropic high thermal expansion coefficient of CFRPs. Moreover, for drilling thick CFRP composites, the drilling process theoretically takes a longer time in a wider opening area, which is why more laser traces are required.

There are numerous applications of laser systems in the industry today. An experimental study was conducted by Li et al. (2010) to minimise the severity of the heat affected zone (HAZ) in the machining of CFRP, by using a diode pumped solid state (DPSS) UV laser. Using a short pulsed UV laser and optimising the drilling process, the authors managed to reduce the HAZ to around 50 μm. They observed that the heat accumulation was the promising element for avoiding damage, and exploited this as a potential way to reduce the HAZ. CFRP with a thickness of 0.3 mm was cut in directions of 0°/90° and 45°/45° relative to fiber orientations, and the resultant edges were inspected. The scanning speed of the focused laser beam was 100 mm/s, which was separated in parallel traces. The research revealed that, in the 0°/90° cut direction, inferior heat conductivity of the polymer matrix within the 'island' material, i.e. the un-irradiated area between the laser traces, resulted in leftover fibers at the edges with lengths of around 1.3 mm—see Fig. 3a. Conversely, with a cutting

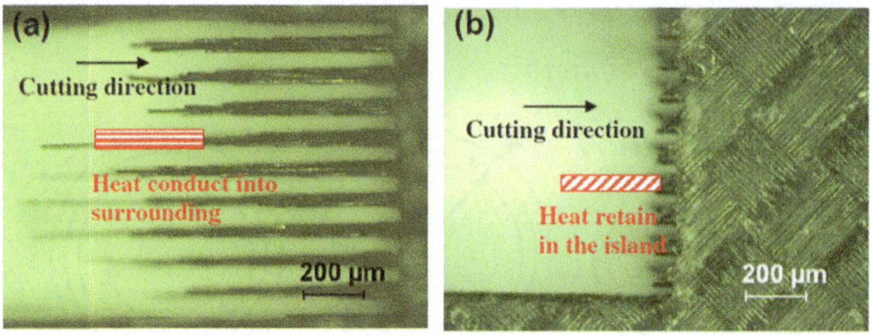

Fig. 3 Cutting edges of laser cutting direction: **a** 0°/90° and **b** 45°/45° relative to fiber orientation (Li et al. 2010)

direction of 45° relative to the fiber orientation, shorter leftover fibers were noticed—see Fig. 3b—which was believed to be due to the heat being conducted to the 'island' material, helping to disintegrate the polymer matrix. With the laser cutting through the edge of a CFRP sample in a 60°/30° direction to the fibers—see Fig. 4—the fiber ejection that is visible at the corner of the top portion of the cut line suggests that the polymer matrix disintegrated due to heat accumulation. The work of Li et al. (2010) also applied the laser machining strategy as developed by Li et al. (2008). This strategy is suitable to be used in cutting a large hole in a thick laminate in order to produce a wide cut kerf for the focusing beam to go deeper into the laminate. Figure 5 shows an illustration of the results of heat accumulation in CFRP during laser machining. The authors from both study attempts have listed the advantages of this strategy, which include (a) block heat conduction to surrounding material and hence a limitation to the heat damage to the component only to the first outline cutting, not to the subsequent inward cutting; and (b) heat is accumulated within the 'island' (to be removed), which enhances the cutting efficiency.

On the contrary to the strategy applied by the authors (Li et al. 2008, 2010; Naeem 2017) claimed that if the composites are cut at a thickness greater than 1 mm,

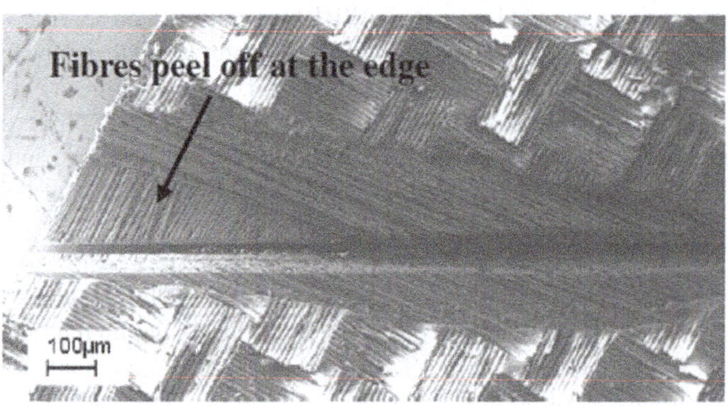

Fig. 4 Cutting qualities at edge of a sample in direction of 60°/30° (Li et al. 2010)

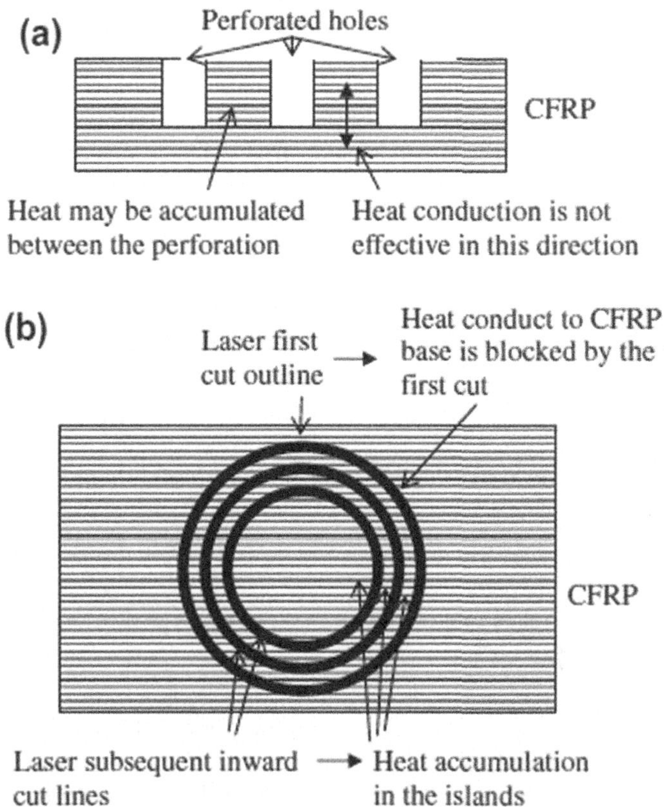

Fig. 5 Illustration of results of heat accumulation in CFRP during laser machining: **a** Cross-sectional view of perforation–potential thermal damage between holes caused by heat accumulation; **b** top view if laser scan traces in drilling a hole–enhancement of cutting efficiency due to heat accumulation (Adapted From Li et al. (2010))

this approach leads to undesirable cut quality. In order to achieve a better cut, the author suggested a new strategy by implementing laser spiral drilling; during the laser drilling process, the material needs to escape efficiently without the matrix cohering onto the freshly-cut surface, and this can only be realised with the creation of a larger kerf. This is a reason why the laser spiral drilling is a very useful strategy, and the results showed that the edge quality on the surface had less thermal damage compared to percussion drilling results by the Nd: YAG laser, as it can be seen that only a few tens of microns of top surface suffered burn-back. In addition, the fiber laser is able to reduce the damage to the matrix material through the assistance of laser processing with a scanning head combined with the open architecture of the hole geometry. The author concluded that thermal management is the key to process and control the HAZ on CFRP. For a fiber laser, the small spot size carefully targets the heat input to the material, which the author believed creates a significant power density to the material. Apart from adjusting the machine operational parameters, the author also suggested laser spiral drilling as a potential strategy in future laser materials processing.

Klotzbach et al. (2011) conducted laser machining of CFRP experiments by using three high brilliant laser beam sources–a single-mode fiber laser, a multi-mode fiber laser and CO_2 slab laser. The research mainly deals with elementary analysis of the remote laser cutting process and gas assisted laser cutting (conventional) of CFRP, as shown in Fig. 6 and Table 1 for machining parameters. This research mainly highlights the benefits of using conventional gas assisted laser cutting and of remote laser cutting. A common term for a remote laser cutting tool in the laser materials processing industry is a galvanometer scanning head. Both laser cutting heads give significantly different laser machined surface cut qualities. At the microscopic

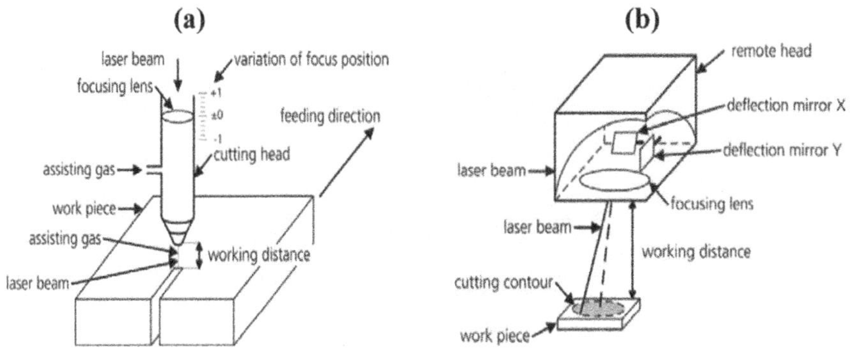

Fig. 6 Schematic diagram of **a** gas assisted laser cutting and **b** remote laser cutting (Klotzbach et al. 2011)

Table 1 Range of parameters for gas assisted (G) and remote laser cutting (R) (Klotzbach et al. 2011)

Parameter	Unit	Process	Gas-assisted laser cutting (g)	Remote laser cutting (r)
Approximately spot diameter	μm	g, r	240	240/50
Laser power	kW	g, r	1…2.85	1…2.6
Focus position	mm	g, r	±1.0/ ±2.0	±1.0/±2.0
Nozzle distance	mm	g	1.0…3.0	–
Feeding velocity	m/min	g	1…100	–
Assisting gas	–	g	N_2, H_e, air	–
Assisting gas pressure	bar	g	1…20	–
Laser spot velocity	m/s	r	–	0.25…10
Maximum cycles	–	r	–	40
Time between cycles	S	r	–	>1

range view of 100–150 μm, by using a raster electron beam microscope, the matrix displacement can be seen clearly at this range for the results of the remote cutting head, whereas for the gas assisted cutting head, it can be seen at a range of more than 500 μm. This shows that by using the remote cutting head, it is possible to create only a small amount of matrix reduction due to the short interaction between a laser beam and workpiece. On the other hand, the gas assisted cutting head produces a significant amount of matrix displacement. This occurs in the work of Li et al. (2010), in which they also applied a galvanometer scanning head. The authors from both research attempts identified that the number of cycles plays a significant role in terms of HAZ reduction when increasing the laser spot velocity during the remote-processing. The higher the velocity applied, the higher the cyclic rate is required until the desirable kerf width is achieved. Cutting material with smaller focus diameters using comparable optical configurations requires the use of high-brilliance laser sources such as fiber lasers. Although the authors demonstrated this by increasing the intensity of the laser spot with the maximum processing speed, they admitted that the benefit is not as high as expected because it results in a thinner cutting kerf, which results in an increase in the aspect ratio between material thickness and kerf width. In summary, these results indicate that it is essential to realise a good cut quality by optimising the laser spot velocity/cutting speed and laser power; it is believed that these parameters are the keys to thermal management in laser machining without involving the 'hardware' (changing the scanning head or enhancing laser optical system) of the laser machine. However, there are no further explanations for the effect of these parameters elsewhere in the paper from Klotzbach et al. (2011), and their findings would have been more interesting if the authors had assessed these parameters. For instance, when the authors (Klotzbach et al. 2011) commented on the visibility of the matrix displacement, no arguments were made about the influence of gas type and pressure, in which it is significant to understand the mechanism of a gas-assisted cutting head, and that it has different outcomes compared to the remote cutting head. Furthermore, focus position and nozzle distance are other parameters that the authors have not covered in much detail. It is also crucial to identify the limitations of material removal capability because each distance contributes a significantly different interaction between a laser beam and workpiece.

Lau et al. (1995) conducted the experiments on the performance of the Nd: YAG laser and the Excimer laser on the machining of CFRP. The results indicated that intermittent cutting produces an inferior surface finish and scanning speed is the major contributor for material removal, as claimed by the authors. Moreover, one of the critical problems in pulsed Nd: YAG lasers was the maximum depth of cut in laser drilling, which is characterised by its limited peak pulse power. They agreed that the Excimer laser performs better than Nd: YAG laser in terms of surface quality, and it is only used in the machining of GFRP and MMC. However, there is no explanation of the material removal mechanism in the Excimer laser and hence, no conclusion is drawn at this stage. As concluded by the authors, there is a strong possibility that the optimal cutting rate and the extent of material damage are important factors in achieving the best surface quality. On the contrary, without consideration of variation of laser power, it is impossible to understand the exact nature of the HAZ and the authors lack this in their investigation. Mathew et al. (1999) conducted parametric studies on pulsed Nd: YAG lasers with a 300 W capacity on machining 2 mm thick CFRP, by studying the HAZ and the kerf taper as their outcomes. There were five process parameters considered: cutting speed, pulse energy, pulse duration, pulse repetition rate and gas pressure; these are applied in a predictive model created by the authors. They found that all parameters significantly influence the existence of HAZ and the distortion of kerf width and taper. In addition, there are two properties of the fibers that control the cutting performance, principally the thermal properties of the constituent material, and the volume fraction. In conclusion, they claimed that the vicissitude of achieving good quality cuts through CW mode laser cutting is due to the vast dissimilarity in the thermal properties of the carbon fiber and the matrix material in the workpiece. On the contrary, they believed that pulsed Nd: YAG lasers could provide a smaller thermal load during cutting that is characterised by high beam intensity and better focusing behaviour that assists in laser cutting of CFRP composites with the least amount of defects. The major obstacle in pulse settings is the divergence of the laser beam; this usually occurs along the keyhole with a decreasing trend of power density beyond the focal plane. Furthermore, it happens during the incident radiation in which some power by absorption and reflection from the plasma recedes in the keyhole. This implies that the energy absorbed by the material decreases along with the depth of the keyhole. The key problem of pulse setting is when the energy supplied during the machining process is not sufficient for a complete cut due to limited peak pulse power. This finding is in agreement with Lau et al. (1995) who showed that inferior cut quality by applying pulse-mode laser machining is the key issue. However, the research does not take into account the focal plane position, and it might be interesting to understand the efficiency of the laser beam spot when it cuts a deeper hole.

Wahab et al. (2012) investigated the laser cutting operation of CFRP using a Nd: YAG laser. As concluded by Wahab et al., they claimed that a pulsed Nd: YAG laser satisfactorily cuts the workpiece at a cutting speed of 30 mm/min and pulse duration 0.5 ms. Furthermore, the significant effects on kerf width, HAZ and taper angle are contributed to by the pulse energy and the pulse repetition rate, where a high pulse energy leads to a high HAZ, and a high pulse repetition rate produces a narrow kerf width. As suggested by the authors (Wahab et al. 2012), pulse energy and pulse repetition rate play an important role during laser cutting. However, the researchers managed to cut a 1.5 mm thickness of CFRP, which is still challenging, despite the authors being able to achieve the optimum parameters. This result shows that pulse-mode operation requires considering additional parameters, such as pulse duration, pulse energy and pulse repetition rate to optimise the best laser beam quality and power density, and full penetration cuts are almost impossible due to the available energy supplies during the laser machining process (Lau et al. 1995; Singh and Maurya 2017; Kumar 2014; Sheikh-Ahmad 2009; Mathew et al. 1999; Wahab et al. 2012; Yung et al. 2002). Future research needs to find the optimal parameters, such as shorter interaction time and higher effective laser power. These are the most fundamental parameters to achieve the best cut quality. Other parameters that are not involved with the manipulation of pulses may also enhance the surface quality

Fig. 7 Initial laser drilling result (laser power: 900 W; scanning speed at 10 mm/s; single pass) (Sobri et al. 2018)

and save more time in penetrating the hole area. Thus, these research attempts (Lau et al. 1995; Singh and Maurya 2017; Kumar 2014; Sheikh-Ahmad 2009; Mathew et al. 1999; Wahab et al. 2012; Yung et al. 2002) indicated the drawbacks of utilising pulse-mode laser machining, which helps to promote the superiority of fiber laser in machining CFRP. The common range of pulse off-time is between 1 and 90 ms while pulse repetition rate is ranged between 0.5 Hz and 20 kHz.

Sobri et al. (2018) performed a laser drilling experiment with thick CFRP, i.e. 25 mm thick. As shown in Fig. 7, the drilling procedure failed to reach the workpiece even though the tests were carried out with differences in laser power and scanning rates. Significant amount of HAZ existed in the outer area of the hole diameter. Certain noticeable disruption, i.e. matrix contraction induced spatter, also occurred around the perimeter of the opening. Due to a failure attempt to drill the whole sample, Fig. 8a, b display a half-hole shape (180°) in order to study the fundamental injury. In comparison to the spatter and the HAZ, which can be easily seen the presence of charring along the cut path together with a few disoriented fiber layers. This indicates that the depth at the beginning of the cut, i.e. at the left side, is greater than at the end of the cut, i.e. at the right side. Although the depth tends to be significantly deeper, as the vertical cuts below the half-hole outline indicate, these are simply just surface cuts that do not extend into a substrate of significant depth. It was thought that when the laser pulse had stopped, the flow had ceased and the melt had dropped back as a splatter around the edge of the cutting path or had even sunk into the kerf, as the assist gas could not effectively blow away the dust and vapor produced during the cutting process. Figure 9 demonstrates the laser drilling of CFRP by a spiral trepanning technique in a single-pass cutting process regardless of the quality of the cavity. The picture showed that the laser beam will penetrate roughly at a depth of 9.01 mm and that observable damage, i.e. matrix dissolution, disoriented fiber layers and charring,

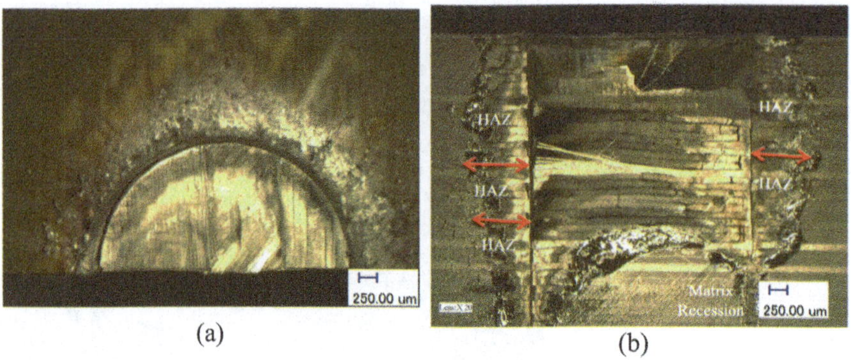

Fig. 8 180° hole shape (laser power: 800 W; scanning speed at 5 mm/s; single pass): **a** top view and **b** kerf view (Sobri et al. 2018)

is significantly reduced relative to Fig. 8. Nevertheless, HAZ and protruding fibers appeared dramatically at the entrance to the perimeter of the tube. It can be clarified that by increasing the pressure of the assist gas, it can potentially reduce the defect as well as eliminate dust and heat dissipation from the cutting zone effectively. In comparison, a spiral trepanning technique created a deep cut relative to single ring trepanning because the initial laser-material contact began at the center point of the hole compared to the standard trepanning, beginning at the edge of the hole. An interesting fact that laser drilling by spiral trepanning strategy can penetrate more than 5 mm by a fiber laser. This is because at the moment, Goeke et al. (2010) (Goeke and Emmelmann 2010) highlighted that a fiber laser (λ = 1070 nm; P = >1 kW) at scanning speed of 83 mm/s can possibly cut CFRP up to 5 mm.

4 Conclusion and Future Perspectives

This review work was devoted to identify the potential of machining parameters to be implemented in laser machining of CFRP composites, and provided some alternatives in order to avoid/reduce potential damages after machining. Previous research works had identified that laser power and cutting speed are the most significant parameters affecting surface roughness. In addition, for kerf width, the laser power, cutting speed, and gas pressure are the most significant parameters, while taper angle has the same factors as kerf width but with additional concern of pulse frequency. For the HAZ, the most significant parameters are laser power, cutting speed, gas pressure, pulse repetition rate, and pulse duration. Other parameters such as multi-pass cutting and focal plane position (FPP) are still lacking in references, and these could be some potential of new research areas in order to enhance the quality of drilled holes by laser machining. The work of Sobri et al. (2018) showed that laser penetration struggled to reach the maximum thickness of CFRP composites even though the fiber laser

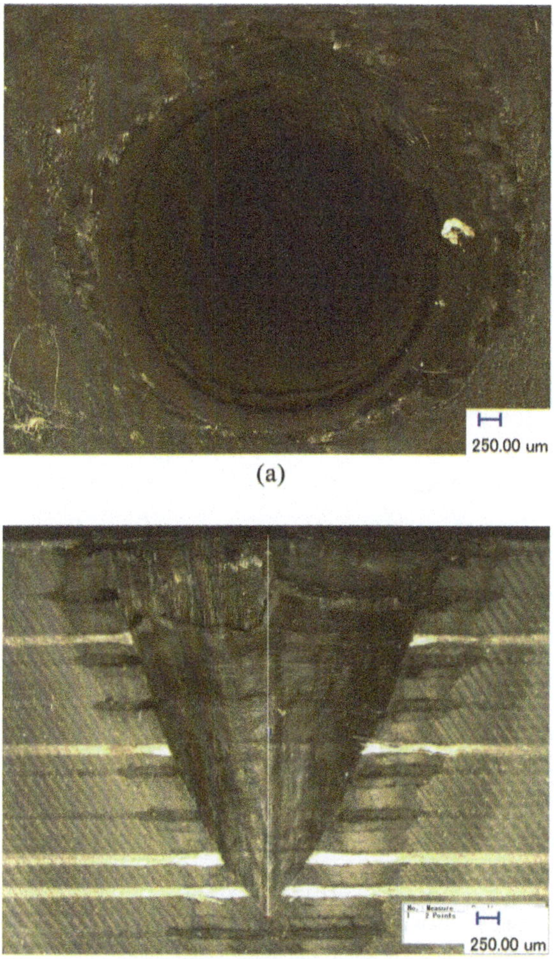

Fig. 9 Laser drilling by using spiral trepanning strategy (laser power: 900 W; scanning speed at 10 mm/s; gas type: N_2; gas pressure: 7 bar; single-pass): **a** top view and **b** kerf view (Sobri et al. 2018)

was mounted at the highest power and lowest scanning speed. There was, indeed, a possible way to overcome this issue by improvising a laser drilling technique, i.e. a spiral drilling strategy. The laser beam was centered only on the inner region, with a spiral-cutting path; this meant that the laser energy could be optimized and more material could be effectively removed by spiraling the laser beam.

Moreover, the discovery of feasibility in drilling process strategy is not being explored extensively and this is also considered as a potential for laser drilling experiments. At the moment, no comparative study has been conducted between the effects of ring shape (Li et al. 2008) and spiral trepanning (Sobri et al. 2018), thus, this study might be the potential to apply on drilling thick CFRP composites. It is interesting to note that no attempts were successful in penetrating a complete hole in CFRP thicker than 5 mm. It is believed that, with the advantages of a fiber laser machine, it will be feasible to drill beyond 5 mm deep, due to successfully drilling of 50 mm thick steel using a 700 W fiber laser machine (Steen and Mazumder 2010). It is important to understand that the process-material interaction effects require the machining operation parameters to be carefully selected in order to avoid any strong negative effect of adopting the hybrid or sequential machining process, due to the combination of two or more different machining technologies (i.e. Laser-Mechanical or Laser-Electric-Discharge or etc.).

Acknowledgements We would like to express my gratitude to the Malaysian government authorities, i.e. Ministry of Higher Education (MOHE) and Universiti Malaysia Kelantan (UMK), for funding this Ph.D. research project (SLAI-810925146285).

References

Abrate S, Walton DA (1992) Machining of composite materials. Part II: non-traditional methods. Compos Manuf 3(2):85–94. https://doi.org/10.1016/0956-7143(92)90120-J

Ashkenasi D, Kaszemeikat T, Mueller N, Dietrich R, Eichler HJ, lling G (2011) Laser trepanning for industrial applications. Phys Procedia 12(Part B):323–331, ISSN 1875–3892. https://doi.org/10.1016/j.phpro.2011.03.140.

Choudhury IA, Chong WC, Vahid G (2012) Hole qualities in laser trepanning of polymeric materials. Opt Lasers Eng. https://doi.org/10.1016/j.optlaseng.2012.02.017

Garrick R (2007) Drilling advanced aircraft structures with PCD (Poly-Crystalline Diamond) drills. SAE Technical Paper 2007–01–3893. https://doi.org/https://doi.org/10.4271/2007-01-3893.

Goeke A, Emmelmann C (2010) Influence of laser cutting parameters on CFRP part quality. Phys Procedia 253–258

Hintze W, Hartmann D, Schutte C (2011) Occurence and propagation of delamination during the machining of carbon fibre reinforced plastics (CFRPs)-an experimental study. Compos Sci Tecnol 71:1719–1726. https://doi.org/10.1016/j.compscitech.2011.08.002

Jacobs P (2008) Precision trepanning with a fiber laser. Smithfield, Republic of Ireland, LFI Incorporated's research and development paper

Klotzbach A, Hauser M, Beyer E (2011) Laser cutting of carbon fibre reinforced polymers using highly brilliant laser beam sources. Phys Procedia 12:572–577. https://doi.org/10.1016/j.phpro.2011.03.072

Komanduri R (1997) Machining fiber-reinforced composites. Mech Eng (ISSN 0025–6501) 115(4): 58–64

Kumar S (2014) Laser cutting process-a review. Int J Darshan Inst Eng Res Emerg Technol 3(1):44–48

Lau WS, Yue TM, Lee TC, Lee WB (1995) Un-conventional machining of composite materials. J Mater Process Technol 48:199–205. https://doi.org/10.1016/0924-0136(94)01650-P

Li L, Low DKY, Ghoreshi M., Crookall JR (2002) Hole taper characterisation and control in laser percussion drilling. CIRP Annu-Manuf Technol 51 (1):153–156. https://doi.org/10.1016/S0007-8506(07)61488-7.

Li L, Sobih M, Crouse P (2007) Striation-free laser cutting of mild steel sheet. CIRP Annu, Manuf Technol 56(1):193–196. https://doi.org/10.1016/j.cirp.2007.05.047.

Li ZL, Zheng HY, Lim GC, Chu PL, Li L, Marimuthu S, Negarestani R, Sheikh M, Mativenga P (2008) Process development of laser machining of carbon fibre reinforced plastic composites.

ICALEO. In: international congress on applications of lasers and elctro-optics, Temecula, CA, USA
Li ZL, Zheng HY, Lim GC, Chu PL, Li L (2010) Study on UV laser machining quality of carbon fibre reinforced composites. Compos Part A 41:1403–1408. https://doi.org/10.1016/j.compositesa.2010.05.017
Low DKY, Li L (2002) Hydrodynamic physical modelling of laser drilling. Trans ASME 124:852–862. https://doi.org/10.1115/1.1510518
Mangalgiri PD (1999) Composite materials for aerospace applications. Bull Mater Sci 22(3):657–664. https://doi.org/10.1007/BF02749982
Mathew J, Goswami GL, Ramakrishnan N, Naik NK (1999) Parametric studies on pulsed Nd:YAG laser cutting of carbon fibre reinforced plastic composites. J Mater Proc Technol 89–90:198–203. https://doi.org/10.1016/S0924-0136(99)00011-4
Naeem M (2017) Laser machining fiber-reinforced composites. .https://www.industrial-lasers.com/articles/print/volume-26/issue-5/features/laser-machining-fiber-reinforced-composites.html.
Pecat O, Rentsch R, Brinksmeier E (2012) Influence of milling process parameters on the surface integrity of CFRP. Procedia CIRP 1:466–470. https://doi.org/10.1016/j.procir.2012.04.083
Reza N (2010) Laser cutting of carbon fibre-reinforced polymer composite materials. PhD thesis, The University of Manchester
Reza N, Li L (2013) Fibre laser cutting of carbon fibre-reinforced polymeric composites. Proc Inst Mech Eng Part B: J Eng Manuf 0954405413490513.
Santhanakrishnan G, Krishnamurthy R, Malhotra SK (1988) Machinability characteristics of fibre reinforced plastics composites. J Mech Work Technol 17:195–204. https://doi.org/10.1016/0378-3804(88)90021-6
Shanmugam DK, Chen FL, Siores E, Brandt M (2002) Comparative study of jetting machining technologies over laser machining technology for cutting composite materials. Compos Struct 57(1–4):289–296. https://doi.org/10.1016/S0263-8223(02)00096-X
Sheikh-Ahmad JY (2009) Machining of polymer composites. Springer, New York. https://doi.org/10.1007/978-0-387-68619-6
Singh SK, Maurya AK (2017) Review on laser beam machining process parameter optimization. Int J for Innov Res Sci Technol 3(8):34–38
Sobri SA, Heinemann R, Whitehead D, Shuaib N (2018) 2018) Preliminary investigation of drilling thick carbon fibre reinforced polymer composite (CFRP). AIP Conf Proc 2030:020014. https://doi.org/10.1063/1.5066655
Steen WM, Mazumder J (2010) Laser material processing. Springer. https://doi.org/10.1007/978-1-84996-062-5
Stock J, Zaeh MF, Conrad M (2012) Remote laser cutting of CFRP: improvements in the cut surface. Phys Procedia 39:161–170. https://doi.org/10.1016/j.phpro.2012.10.026
Vanderwert TL (1983) Machining plastics with lasers. Manuf Eng 55–58
Wahab MS, Rahim EA, Rahman NA, Uyub MF (2012) Laser cutting characteristic on the laminated carbon fiber reinforced plastics (CFRP) composite of aerospace structure panel. Adv Mater Res 576:503–506. https://doi.org/10.4028/www.scientific.net/AMR.576.503
Yilbas BS (2013) Laser drilling-practical applications. Springer-Verlag, Berlin Heidelberg
Yung KC, Mei SM, Yue TM (2002) A study of the heat-affected zone in the UV YAG laser drilling of GFRP materials. J Mater Process Technol 122:278–285. https://doi.org/10.1016/S0924-0136(01)01177-3

Damage Response of Hybrid Fiber Reinforced Polymer Composite via SPH for Abrasive Water-Jet Cutting/Piercing

Irina Ming Ming Wong and Azwan Iskandar Azmi

Abstract A hole piercing operation is typically part of the early stage of the manufacturing of fiber reinforced composites and occurs prior to the cutting or trimming stages. Unfortunately, a shock wave impact is induced by this operation and the propagation of abrasive water-jets causes serious delamination damage both on the top and the bottom of the surface during piercing. In the present work, the mechanism of the piercing hole operation is described, and the phenomenon is modelled by combining Smoothed Particle Hydrodynamic (SPH) with Finite Element Analysis (FEA). Results of this numerical model indicate that stagnation of abrasive particles at impact point have generated delamination (crack initiation and propagation) and, subsequently, shear-out mechanisms. In addition, it was found that the numerical model agrees well with qualitative and quantitative experimental results at the setting of hydraulic pressure of 320 MPa; abrasive flow rate of 120 g/min; stand-off distance of 2 mm.

Keywords Piercing behaviour · Smoothed particles hydrodynamic · Finite element analysis · Composite laminates

1 Introduction

Abrasive water-jet machining (AWJM) technology offers several advantages over conventional machining methods for the cutting of fiber-reinforced (FRP) composites. Typically, the water-jet process is used for unique shape-cutting, trimming and drilling of these FRP composites. The process starts with hole piercing prior to the cutting or trimming stages. Instead of the usual mechanical drill for pre-drill a starting hole, it is now quite common for AWJM to be used to create holes as an initial part of the cutting sequence. However, the piercing operation often results in severe damage,

---

I. M. M. Wong · A. I. Azmi (✉)
Faculty of Mechanical Engineering Technology, Universiti Malaysia Perlis, Pauh Putra Campus, 02600 Pauh, Perlis, Malaysia
e-mail: azwaniskandar@unimap.edu.my; aazm007@aucklanduni.ac.nz

M. T. Hameed Sultan et al. (eds.), *Machining and Machinability of Fiber Reinforced Polymer Composites*, Composites Science and Technology,
https://doi.org/10.1007/978-981-33-4153-1_11

such as fiber plugging and delamination (Schwartzentruber et al. 2018). It is, therefore, essential to understand the mechanisms of damage initiation and propagation of composite laminates under a continuous high-velocity water-jet stream.

The AWJ machining process capability is controlled by a number of process parameters (Anwar 2013). The water-jet-material interaction depends on the nature of the material begin cut, being either brittle or ductile materials (Gudimetla and Yarlagadda 2007). A kerf is usually created from the abrasive grains entrained in a high-velocity water-jet and this aids the material removal process. The erosion process not only depends on the water-jet energy but also the kinematic parameters and properties of the target material (Srinivasu et al. 2009). The profile of a single jet footprint represents an actual cutting edge of the jet stream. This is important when employing abrasive water-jet trimming for the generation of a complex geometric surface. Moreover, the machining of FRP composite material requires a comprehensive understanding of cutting processes due to its high stiffness and highly specific strength (Azmi 2013). The brittle behaviour of FRP composite results from machining via an abrasive water-jet cutting and this often leads to the generation and propagation of micro-cracks due to the impact.

A number of simulation modelling approaches for abrasive water-jet impact have been developed using finite element analysis (FEA) (Ma et al. 2008; Anwar et al. 2013; Schwartzentruber et al. 2018). A finite element model for pure water-jet cutting by Maniadaki et al. (2007), for example, simulated the erosion of the target material caused by the high-velocity water-jet flow. Data input for the water-jet flow for the present study was separated into three distinct areas: the interior of the water-jet nozzle, the water-jet flow into the air and the water-jet impact on a specific material. The velocity profile, von-Misses stress and erosion stages presented in the target material at various time intervals were obtained from this model. Another study by Jianming et al. (2010) proposed the smooth-particle hydrodynamics (SPH) method to model the abrasive particles as spherical balls equivalent to the average diameter of the abrasive particle. For that study, one size was adopted for the abrasive particles, in spite of the fact that the size distribution of the particles is also influenced the depth of cut. This model was further improved by Anwar et al. (2013), who included the mass distribution of the abrasive particles within the jet stream. Another SPH model for a single impact study of angular particles on a ductile target through abrasive water-jet was developed by Takaffoli and Papini (2012). The findings showed that the multiple abrasive particles impacted on the target material at different impact angles, which led to the formation of erosion with non-overlapping and overlapping conditions. This model can be used to predict the accumulation of material removal rate and surface damage.

In the literature, various damage models were constructed to assess the material response of composite laminates through AWJM. Schwartzentruber et al. (2018) proposed a numerical modelling method to evaluate the delamination damage of carbon fiber composite through abrasive water-jet machining using computational fluid dynamics (CFD) coupled finite element model (FEM) analysis. The study successfully modelled the AWJ cutting action using a Fluid-structure interaction model and results were validated experimentally using novel moisture uptake testing.

Chiu et al. (2015) implemented a three-dimensional continuum damage model for predicting the crush response of composite damage structures using the user-defined material subroutine (VUMAT) in Abaqus/Explicit. In a recent study. Lui et al. (2018) presented a mesoscale damage model using VUMAT subroutine to predict the low-velocity impact damage response of hybrid laminates with two woven five-harness satin weave and unidirectional carbon fibers. This intra-laminar damage model was successfully validated by the experimental results and was shown to accurately determine the post-impact delamination.

However, previous numerical modelling responses for composite damage structures were unsuccessful to fully capture the detail of damage response for an initial AWJ piercing process. In fact, existing techniques have only focused on the morphological characteristics of impact or crush damage of composite laminate. The present study, in contrast, delineates 3D solid finite-element based damage for the AWJ piercing process. This approach enables a comprehensive damage mechanism simulation and shows detail of the interactions within the laminated fiber-matrix structure or interface. However, complete or comprehensive simulation of the abrasive water-jet trimming process is very challenging due to multiple complexities such as fluid-structure interaction and large deformation problems. The present study outlines a fundamental implementation of the continuum damage model for hybrid FRP composite machining under the AWJ process. This was established using an intra-laminar damage model, along with an abrasive water-jet simulation, which was established using smoothed particle hydrodynamics. It is noteworthy that this approach discretises abrasive water-jets into particles for simulation of the cutting process. The aim of this chapter was to bridge the gap between a numerical model and experimental observations, as well as to extend the understanding of the mechanism of hybrid composite laminates during the AWJ piercing process.

2 Modelling of Piercing Behaviour- Theory and Implementation

2.1 Model Implementation

One of the difficulties in the finite element modelling (FEM) of fiber reinforced composites is that the damage criteria, based on Hashish and Puck, is not available in the ABAQUS finite element program. Correct implementation of the proposed method and material model is perhaps the most difficult and challenging part of developing the finite element model. Fortunately, the ABAQUS package provides a series of interfaces that allow users to implement or customise required constitutive equations to suit user-specific problems that are not available in the ABAQUS standard material library. The user-defined material behaviour subroutine for ABAQUS/Explicit is known as VUMAT. In particular, VUMAT makes it possible

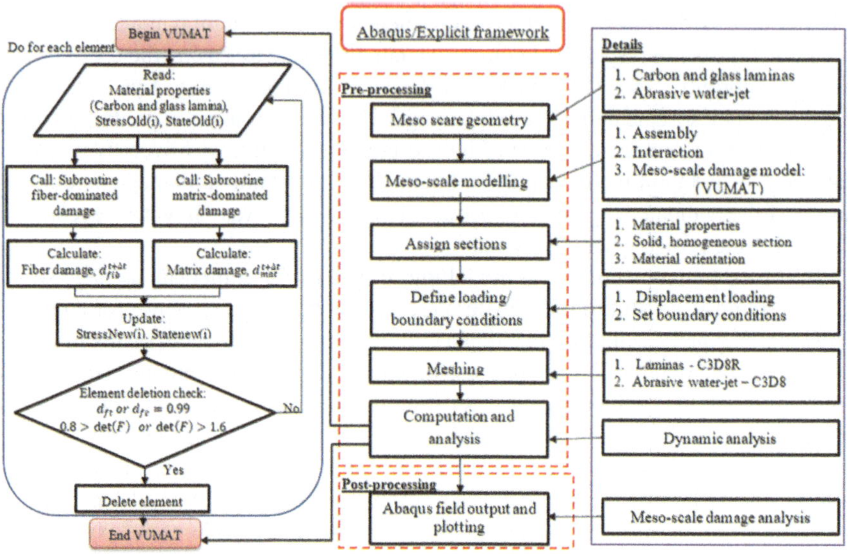

Fig. 1 Flow diagram of mesoscopic computation of hybrid FRP composite through AWJM

to define any constitutive model or criterion for finite element modelling. The overall mesoscopic computation flowchart is depicted in Fig. 1.

2.2 *Theoretical Background*

2.2.1 Abrasive Water-Jet Simulation Method via Smoothed Particle Hydrodynamics

The mesh-free particle method is a method by which the discretisation continuum particles are proposed as a set of nodal points without any mesh constraints. This method is mainly for alleviating the problems associated with the distorted finite element. Among different computational mechanics the Smoothed Particle Hydrodynamics (SPH) is considered the earliest mesh-free method. Notably, by introducing the SPH, Gingold and Monaghan (1997) and Lucy (1977) were the first to study hydrodynamic problems in astrophysics. Recently, this method has been widely implemented in the simulation of fluid flow. In order to consolidate the understanding of SPH, a review of the estimation of kernel density using the mesh-free method, as well as the properties of kernel functions is necessary. SPH is a mesh-free particle method in which each set of particles carries respective material properties that conform to the required governing conservation equation. By its nature, this method is highly adaptive as it allows for field variable approximation, for specific time intervals and relies on the current local set of arbitrarily distributed particles. Due

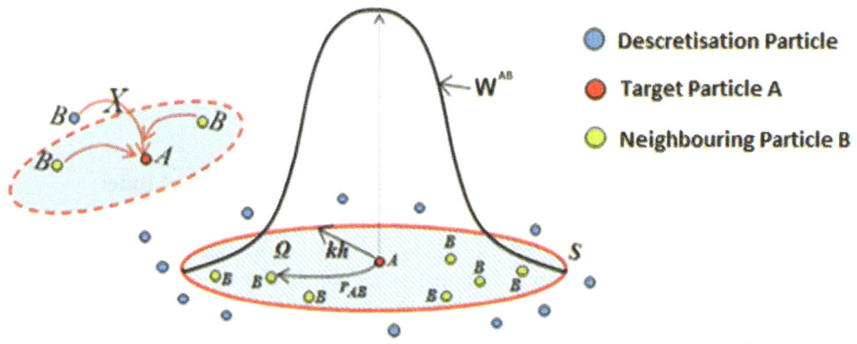

Fig. 2 Approximations for particles under the influence domain of the smoothing function W for particle A

to this advantage, SPH is independent of the arbitrary nature of particle distribution, which allows the study of model with extremely large deformations. In addition, SPH interpolates the principle and converts the partial differential form into an integral form through kernel estimation (Hongxiang et al. 2014).

Considering a problem in domain Ω, which is discretised by a group of particles as a collection of particles, and by assuming kernel weight or smoothing function, $W\left(x^A - x^B, h\right)$ has a compact supporting domain with a radius of h. Herein, the area of kernel smoothing function affected is defined by the position vector x and the smoothing length h. Figure 2 shows the particle approximation using particles within the influence domain of the smoothing function W for particle A; such that κ is the influence domain of the kernel estimation, which is related to the smoothing function for point at x. The approximation $f\left(x^A\right)$ is denoted by its neighbouring particles in a domain Ω, respectively expressed as:

$$f\left(x^A\right) = \int f\left(x^B\right) W\left(x^A - x^B, h\right) dx^B \tag{1}$$

The calculation domain in the SPH method is characterised by a finite number of particles that contains mass and field variables such as density and velocity. To discretise the summations over all the arbitrarily distributed particles in the influence domain area, the continuous integral equation becomes:

$$f\left(x^A\right) = \sum_{B=1}^{N} \frac{m^B}{\rho^B} f(x^B) W\left(x^A - x^B, h\right) \tag{2}$$

where, m^B and ρ^B represent the mass and density of particle B respectively.

This equation is highly effective for the process of particle approximation for different time steps. Variable smoothing length plays a major role in maintaining the neighbouring particles at a particular distance. With large deformations, the particles tend to move randomly and there will be no interactions between each other. To ensure the existence of interactions, the smoothing length needs to be updated as per averaged density or to integrate the smoothing length with respect to the time of smoothing function in terms of a continuity equation (Naib 2015).

2.2.2 Theoretical Background of the Intra-Laminar Damage Model of a Hybrid Composite Material

From the woven fiber reinforced composites shown in Fig. 3, it can be seen that the in-plane material response (along the 0° and 90° directions) is controlled by the fiber reinforcements. This stands in contrast to the through-thickness material response, which is dominated by the matrix. As a result, the model for the intra-laminar damage explains two types of damage: fiber dominated and matrix-dominated damage. While fiber dominated damage is characterised by fiber pull-out, fiber-matrix damage suffers debonding and fiber breakage. Matrix–dominated damage is, it should also be said, often characterised by matrix cracking.

Fiber-dominated failure modes

As shown in Fig. 4, fiber-dominated damage is induced by the damage due to the longitudinal behaviour in the fiber direction.

A bilinear stress-strain law, as depicted in Fig. 5, describes the material response in the direction of the fiber. Damage initiation can be determined by contrasting the strain to longitudinal damage initiation strain. The criteria for controlling damage initiation in the fiber direction is expressed in Eq. 3 (tensile load) and Eq. 4 (compressive load) respectively.

$$F_{fib}^{T} = \left(\frac{\varepsilon_{11}}{\varepsilon_{11}^{OT}}\right)^2 \quad (3)$$

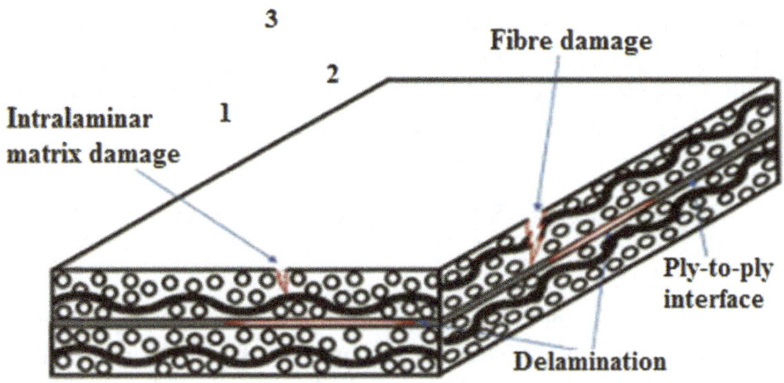

Fig. 3 Failure modes in woven fiber reinforced composite laminates (Chiu et al. 2015)

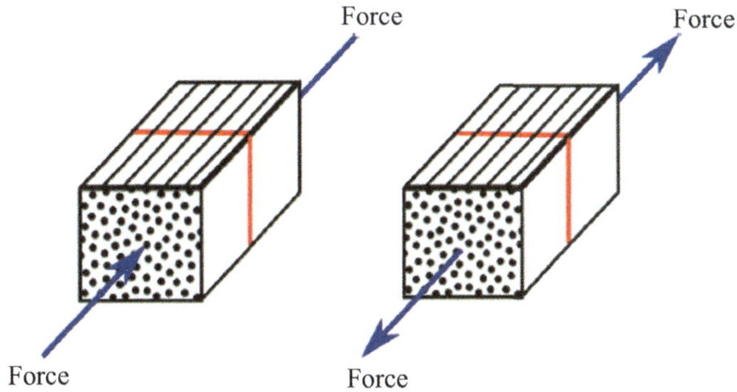

Fig. 4 Fiber-dominated damage with associated fracture plane (Chiu et al. 2015)

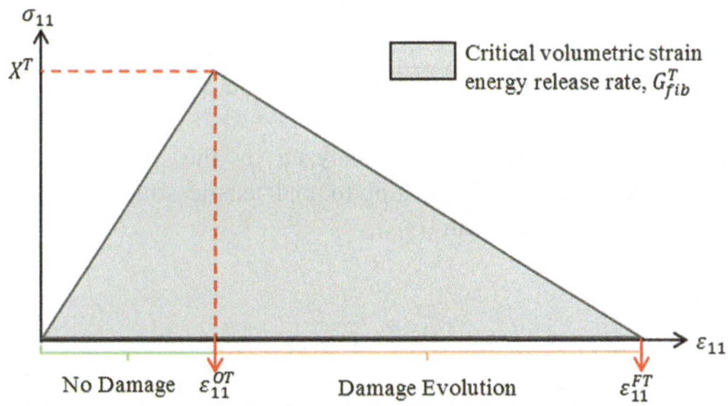

Fig. 5 Bilinear stress-strain law (Chiu et al. 2015)

$$F_{fib}^T = \left(\frac{\varepsilon_{11}}{\varepsilon_{11}^{OC}}\right)^2 \tag{4}$$

where F_{fib}^T and F_{fib}^C denote the failure indices of tensile and compressive loads fiber damage respectively, whereas the failure initiation (ε_{11}^{OT} for tension and ε_{11}^{OC} for compression) are analysed by the strengths (X^T and X^C) and longitudinal elastic modulus (E_{11}) in the corresponding directions.

$$\varepsilon_{11}^{OT} = \frac{X^T}{E_{11}} and \varepsilon_{11}^{OC} = \frac{X^C}{E_{11}} \tag{5}$$

For the present work, the strain of fiber tensile and compressive damage initiation for woven carbon composite were 1.65% and 1.58%, respectively. The strain of fiber tensile and compressive failure initiation strains for woven glass composite were 3.98% and 3.88%, respectively. The damage begins to propagate once it reaches the unity of any damage mode for initiation-function and meet the initiation criterions. The bilinear model is ideally used to determine the fiber dominated tension due to the brittle behaviour of fiber reinforcement, while it is just an estimation of fiber dominated compression. A positive linear stiffness indicates the relationship between stress-strain prior to damage initiation. The tangent modulus drops to negative once the damage has been initiated. It is attributed to the degradation of the elastic modulus by the damage parameter of fiber (d_{fib}). There are three monotonically increasing damage variables that control the damage, which are tensile damage in the direction of fiber d^T_{fib}, the compressive damage in the direction of fiber d^C_{fib} and the matrix cracking, d_{fib}, attributable to a combination of transverse tension/compression loading. Equation 5 describes the damage parameter due to the load in the longitudinal fiber dominated direction. The tensile fiber-dominated damage parameter (d^T_{11}) can be analysed by contrasting the current strain (ε_{11}) and the tensile failure strain (ε^{FT}_{11}):

$$d^T_{11}(\varepsilon_{11}) = \frac{\varepsilon^{FT}_{11}}{\varepsilon^{FT}_{11} - \varepsilon^{OT}_{11}}\left(1 - \frac{\varepsilon^{OT}_{11}}{\varepsilon_{11}}\right) \tag{5}$$

The tensile failure strain (ε^{FT}_{11}) is analyzed by the fiber-dominated tensile critical energy release rates (G^T_{fib}), longitudinal tensile strengths (X^T), and the corresponding characteristic length (l_{fib}).

$$\varepsilon^{FT}_{11} = \frac{2G^T_{fib}}{X^T l_{fib}} \tag{6}$$

The area under the bilinear traction-separation curve is the tensile critical energy release rate (G^T_{fib}). It defines the energy consumption in generating an area of crack under uniaxial tensile loading in a longitudinal direction. A similar method is applied for compressive loading, which leads to:

$$d^C_{11}(\varepsilon_{11}) = \frac{\varepsilon^{FC}_{11}}{\varepsilon^{FC}_{11} - \varepsilon^{OC}_{11}}\left(1 - \frac{\varepsilon^{OC}_{11}}{\varepsilon_{11}}\right) \tag{7}$$

$$\varepsilon^{FC}_{11} = \frac{2G^C_{fib}}{X^C l_{fib}} \tag{8}$$

Before initiation of damage, both the tensile and compressive loads experience elastically unloading and load reversal. An introduction of unloading and load reversal after damage initiation can induce damage due to the combination of tensile and compressive loads. To ease understanding, the growth of damage in the tensile mode is assumed to have minimal impact on the compression response when the load is reversed. The surrounding matrix is still assumed to be capable of sustaining the fiber when it experiences composite loading even in the presence of fiber breakage. The compressive modulus is, therefore, maintained. The result is presented in Fig. 6.

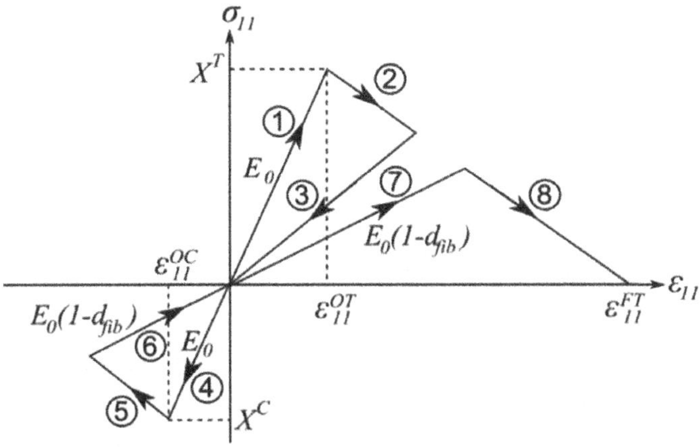

Fig. 6 Stress-strain response during fiber direction loading/unloading: path 1–2, tensile load; path 3, unload; path 4–5, load in compression; path 6–8, compressive unload and tensile reload until failure (Chiu et al. 2015)

Composite laminate experiences compression when it is unloaded along path 3 and the compressive load is reversed to lead the initial elastic stiffness response, which is represented in path 4. However, the reverse is not true. The fiber breakage will occur with a kink band formation in the damage due to compressive loading. The stiffness of the composite material is, consequently, reduced under the tensile load. Therefore, there is a linear relationship between the compressive damage parameter and the tensile damage parameter. Moreover, the growth of a compressive damage parameter can soften the stiffness of the composite (paths 6–7).

Equation 9 expresses the interaction shown in Fig. 6, where the modulus is reduced with respect to the longitudinal damage parameter:

$$d_{fib} = \begin{cases} d_{11}^{C}\varepsilon_{11} < 0 \\ d_{11}^{T}\varepsilon_{11} \geq 0 and d_{11}^{T} > d_{11}^{C} \\ d_{11}^{C}\varepsilon_{11} \geq 0 and d_{11}^{T} < d_{11}^{C} \end{cases} \quad (9)$$

A similar fiber dominated damage interaction mechanism has been employed by other authors (Puck and Schurmann 1998). Indeed, Hooke's law can be used to determine the effective stress vector $\{\bar{\sigma}\}$ before the application of the softening effect of damage.

$$\{\bar{\sigma}\} = [C]\{\varepsilon\} \quad (10)$$

where $[C]$ is defined as the stiffness matrix of the orthotropic laminate that is obtained from the elastic properties in the fiber (11), transverse (22) and thickness (33) directions. In matrix notation, Eq. 10 can be written as

$$\begin{bmatrix}\sigma_{11}\\ \sigma_{22}\\ \sigma_{33}\\ \sigma_{12}\\ \sigma_{23}\\ \sigma_{31}\end{bmatrix}=\begin{bmatrix}\frac{1-\upsilon_{23}\upsilon_{32}}{E_{22}E_{33}\Delta} & \frac{\upsilon_{21}-\upsilon_{23}\upsilon_{31}}{E_{22}E_{33}\Delta} & \frac{\upsilon_{31}-\upsilon_{21}\upsilon_{32}}{E_{22}E_{33}\Delta} & 0 & 0 & 0\\ \frac{\upsilon_{21}-\upsilon_{23}\upsilon_{31}}{E_{22}E_{33}\Delta} & \frac{1-\upsilon_{13}\upsilon_{31}}{E_{11}E_{33}\Delta} & \frac{\upsilon_{32}-\upsilon_{12}\upsilon_{31}}{E_{11}E_{33}\Delta} & 0 & 0 & 0\\ \frac{\upsilon_{31}-\upsilon_{21}\upsilon_{32}}{E_{22}E_{33}\Delta} & \frac{\upsilon_{32}-\upsilon_{12}\upsilon_{31}}{E_{11}E_{33}\Delta} & \frac{1-\upsilon_{12}\upsilon_{21}}{E_{11}E_{22}\Delta} & 0 & 0 & 0\\ 0 & 0 & 0 & 2G_{12} & 0 & 0\\ 0 & 0 & 0 & 0 & 2G_{23} & 0\\ 0 & 0 & 0 & 0 & 0 & 2G_{13}\end{bmatrix}\begin{bmatrix}\varepsilon_{11}\\ \varepsilon_{22}\\ \varepsilon_{33}\\ \varepsilon_{12}\\ \varepsilon_{23}\\ \varepsilon_{31}\end{bmatrix}$$

$$where\Delta=\frac{1-\upsilon_{12}\upsilon_{21}-\upsilon_{23}\upsilon_{32}-\upsilon_{13}\upsilon_{31}-2\upsilon_{21}\upsilon_{32}\upsilon_{13}}{E_{11}E_{22}E_{33}} \tag{11}$$

The damage stress in the longitudinal direction can be expressed based on Eq. 10 and Eq. 11:

$$\sigma_{11}=\left(1-d_{fib}\right)\bar{\sigma}_{11}$$

Matrix-Dominated Damage

Matrix dominated failure, as illustrated in Fig. 7, is a combination of transverse tensile and compressive and shear loading cases; a more complex type of failure than fiber-dominated failure. In uniaxial tension, the fracture plane forms perpendicular to the principal loading direction, as illustrated in Fig. 7a and b. However, for compressive and shear loads, see Fig. 7c and d, where the fracture occurs via shear cracking along a rotated fracture plane.

Puck and Schűrmann (1998) used a fracture plane at an angle θ_f, along an axis parallel to the fiber direction to introduce a set of damage criterion, shown in Fig. 8. In Eq. 12, the function of the standard transformation matrix $T(\theta)$ is to convert between the fracture plane coordinate system (FPCS) and the material coordinate system (MCS). Moreover, $T(\theta)$ is used to rotate the stress (Eq. 13) and strain (Eq. 14) tensor on the fracture plane.

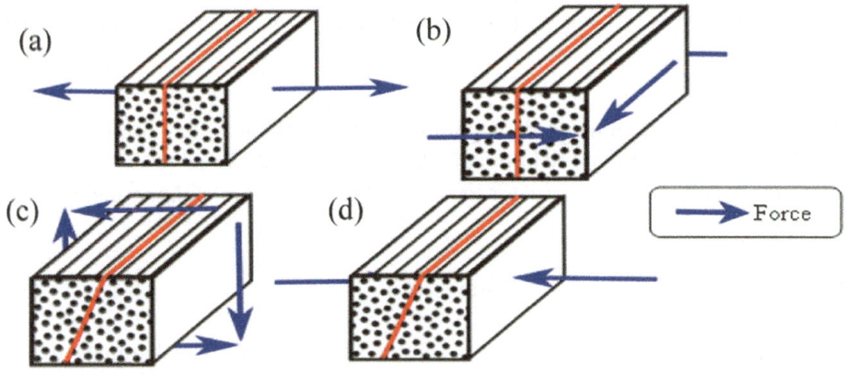

Fig. 7 Matrix-dominated fracture with associated fracture plane (Chiu et al. 2015)

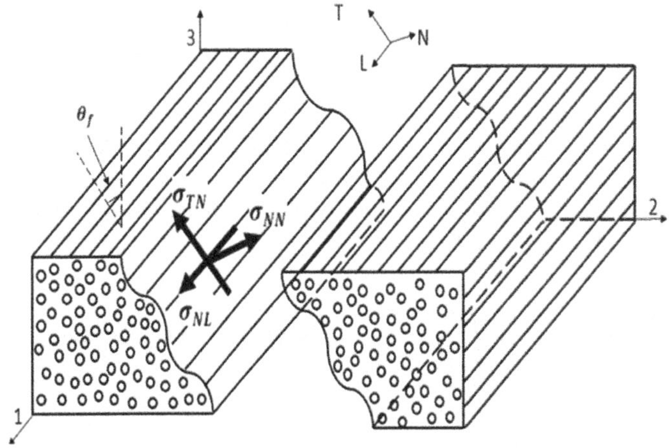

Fig. 8 Coordinate system attached to the fracture plane (1, N, T) relative to the material coordinate system (1, 2, 3) (Chiu et al. 2015)

$$T(\theta) = \begin{bmatrix} 1 & 0 & 0 \\ 0 & cos(\theta) & sin(\theta) \\ 0 & -sin(\theta) & cos(\theta) \end{bmatrix} \tag{12}$$

Stress tensor:

$$\{\sigma_{FPCS}\} = [T(\theta_f)]\{\sigma_{MCS}\}[T(\theta_f)]^T \tag{13}$$

Strain tensor:

$$\{\varepsilon_{FPCS}\} = [T(\theta_f)]\{\varepsilon_{MCS}\}[T(\theta_f)]^T \tag{14}$$

where θ_f is the angle of the potential fracture surface.

A close relationship exists between the criterion of the matrix damage initiation for unidirectional composites and the stress state on the fracture plane which contains linear normal stress, σ_{NN}, and non-linear stresses, σ_{LN} and σ_{TN}, (expressed Eq. 15 and Eq. 16). The failure indices, F^T_{mat} and F^C_{mat} are functions of the fracture plane normal stress (σ_{NN}) and in-plane shear stress (σ_{LN} and σ_{TN});

$$F^T_{mat} = \left(\frac{\sigma_{NN}}{Y^T}\right)^2 + \left(\frac{\sigma_{NT}}{S_{23}}\right)^2 + \left(\frac{\sigma_{LT}}{S_{12}}\right)^2 \text{for } \sigma_{NN} > 0 \tag{15}$$

$$F^C_{mat} = \left(\frac{\sigma_{NT}}{S_{23} - \mu_{NT}\sigma_{NN}}\right)^2 + \left(\frac{\sigma_{LT}}{S_{12} - \mu_{LT}\sigma_{NN}}\right)^2 \text{ for } \sigma_{NN} \le 0 \tag{16}$$

The function of damage initiation is the comparison between the load opposing resistance along the fracture plane, comprising of transverse tensile, compressive strengths (Y^T and Y^C), longitudinal and transverse shear strengths (S_{12} and S_{23}) and the transverse friction coefficients (μ_{NT} and μ_{LT}). In addition, the Hashin-Rotem criteria was used to govern the intra-laminar matrix damage initiation in through-thickness direction by considering the interaction between normal stress (σ_{33}) and shear stress (σ_{13} and σ_{23}) on the plane perpendicular to the through thickness direction. It can be expressed by:

$$\left(\frac{\sigma_{33}}{\sigma_{33}^{OT(C)}}\right)^2+\left(\frac{\sigma_{13}}{\sigma_{13}^{O}}\right)^2+\left(\frac{\sigma_{23}}{\sigma_{23}^{O}}\right)^2-1\geq 0 \text{for inter} - \text{laminar failure on plane 3} \tag{17}$$

where $\sigma_{ij}(i,j=1,2,3)$ are the stresses acting on the fracture surfaces and $\sigma_{ij}^{OT(C)}(i,j=1,2,3)$ represent the strengths for tension and compression respectively. σ_{13}^{O} and σ_{23}^{O} are the shear strength under corresponding shear loads.

It would be computationally extensive to determine the local ply thickness and local ply configuration, especially when the local condition is an element deletion that would have to be recalculated constantly. Moreover, the constraining effect of neighbouring plies can be reduced by damage. Restrained representation of the true local strength, therefore, relies on the unidirectional values. The θ_f, is not initially known for being a general loading condition. For initiation functions (Eq. 15 and Eq. 16), the stresses (σ_{NN}, σ_{LN} and σ_{TN}) are a function of the inclination angle of the fracture plane (θ) which must first be maximised with respect to θ. The initiation of damage starts if the maximised initiation function is greater than unity. The element for the remainder of the simulation will remain unchanged once the angle is determined as shear micro-cracking has occurred and any further fracture will preferentially occur on this plane. (Chiu et al. 2015)

The Mohr-Coulomb theory states that μ_{NT}, S_{23}, and μ_{LT} is the interpretation of the slope of the linear section of the damage initiation profile to a friction coefficient that supports resisting shear loading (Chiu 2015).

$$\mu_{NT}=-\frac{1}{tan(2\theta_f)} \tag{18}$$

The transverse shear strength (S_{23}) is defined in terms of the transverse compressive strength.

$$S_{23}=\frac{Y^C}{2tan(\theta_f)} \tag{19}$$

The same analysis is repeated for the thickness direction response via:

$$\mu_{LT} = \mu_{NT} \frac{S_{12}}{S_{23}} \tag{20}$$

Deletion of Elements with Physical Distortion

Element deletion was employed in this simulation due to the softening nature of the constitutive relationships. To prevent excessive distortion and premature termination of the analysis in simulations, the failed continuum shell elements were discarded from the model. As damage progresses in the element, the stiffness of the element decreases. Also, activation of the damage-based element deletion will occur at the presence of any damage variable along the fiber or the equivalent matrix strain as the shear deformation has reached a maximum specified value. For the deformation-based deletion criterion, it will only be considered when the tensile or compressive principal logarithmic strain has also reached its maximum or minimum specified value, respectively. Therefore, an element deletion method is used to remove the elements based on: (i) the damage parameter, or (ii) the determinant of deformation gradient, $\det(F)$ (Chiu et al. 2015; Liu et al. 2018).

$$deleteelementwheneither \begin{cases} (i)\ d_{ft} or d_{fc} = 0.99 \\ (ii)\ 0.8 > \det(F) or \det(F) > 1.6 \end{cases} \tag{21}$$

Condition (i) in Eq. 21 represents the fiber-dominated failure mode. It is known that composite behaviour basically depends on the reinforcing fiber strength and modulus since the fiber carries the majority load in the FRP composite. Therefore, total composite failure can be assumed to occur when the fibers have failed. In order to prevent the existence of elements with extremely low stiffness, the maximum allowable damage was set to 0.99. In addition, condition (ii) in Eq. 21 is a measurement of the volume changes in the element under deformation. Large volume changes are not permitted in order to prevent infinite and/or unphysical distortions to occur. This would be symptomatic of an element that has lost most of its stiffness and hence, it should be deleted.

Model Assumption and Limitations

The proposed numerical model for fiber reinforced composite is assumed to be homogenised composite laminate at the meso-mechanics level. A meso-mechanics model, as opposed to a micro-mechanics model, is more practical for structural modelling due to its advantages in analysing structures at high orders magnitude when compared to the individual fiber and matrix constituents. Furthermore, each composite laminate is modelled separately as a material of homogeneous nature and fiber direction is assumed to be orthotropic. Homogeneous laminate may not show any difference between unidirectional ply and woven fabric, so the model can be used to represent both types of composite plies. An assumption of irreversible damage parameters was made for any damage occurred and was constrained to be monotonically increased for the entire model. This would be physically manifested in the behaviour of the cracks. The element would be deleted when a crack is created. This model was developed with the use of 8-node brick linear hexahedral elements with reduced integration (C3D8R) in ABAQUS/Explicit software. These elements offer a good combination of adequate accuracy with reduced computational demands by using a single integration point (Hongxiang et al. 2014). On the other hand, it was assumed that the proposed abrasive water-jet model was evenly distributed over the volume of the element. An application of conventional finite element analysis is obviously more pronounced than exotic methods such as the extended finite element method due to the lack of discontinuities within the element. As abrasive particles play a main role in the entire impact event, the SPH particles were, therefore, modelled to be abrasive properties. The abrasive particles were assumed to be in spherical shape and applied the average diameter of the mesh 80 abrasive particle.

2.3 Finite Element Model

Validation of the predictive capability of the model for the piercing response of composite structure on the first ply laminate was taken based on the results obtained from the in-house cutting test on hybrid composite specimens. The simulation software ABAQUS 6.17/Explicit was used to simulate the piercing process and the FE model is shown in Fig. 9.

The abrasive water-jet was modelled as a cylindrical shape with a radius of 0.4 mm and a length of 70 mm. In this model, the FE model of abrasive water-jet was modelled by SPH elements, in which the abrasive water-jet was discretised into 9800 SPH abrasive water-jet particles with a reference lumped mass of 0.14 g in the type of C3D8. In this model, the abrasive particle was assumed to be traveling with the help of hydraulic pressure. Therefore, Bernoulli's law was applied to obtain the initial velocity, which would indicate the relationship between the water pressure

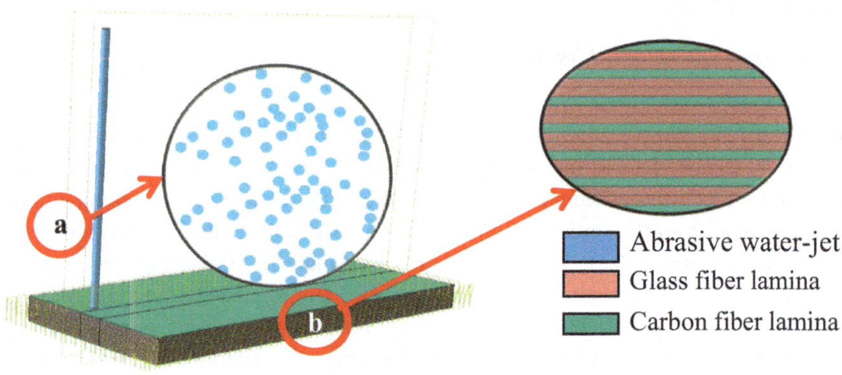

Fig. 9 FE model for a hybrid composite plate with abrasive water-jet tube, **a** full view, close-up of **b** abrasive water-jet and **c** hybrid composite plate

and abrasive water-jet velocity from the orifice:

$$V_A = \sqrt{\frac{2P}{\rho_W}} \tag{22}$$

where, V_A refers to abrasive velocity, P refers to hydraulic pressure and ρ_W refers to water density. These abrasive particles have the following properties (Feng et al. 2012; Tan et al. 2016): density, $\rho = 4000\ kg/m^3$; Possion's ratio = 0.27; Young's Modulus = 248 GPa. The rectangular $[CGGC]_6$ composite plate was modelled in the size of 20 mm × 40 mm with 19 plies fabric with 3.52 mm thickness, as shown in Fig. 9. This composite model was meshed together with a solid element layer; type C3D8R, with 30 × 120 being the number of elements in the vicinity of the cut area. At the same time, a coarser mesh of 10 × 120 elements was used in the area away from the zone of interest. The material properties used for the numerical computation were obtained from the experimental data and from the previous studies of Tan et al. (2016), Liu et al. (2018) and Phadnis et al. (2013), which are tabulated in Table 1. For a plain weave fabric, the mechanical properties in the weft and warp fiber direction are indistinct. Therefore, the input parameters associated with both the warp direction and the weft direction are the same. The Vumat subroutine was developed and implemented in the ABAQUS/Explicit.

Table 1 Material properties of the woven glass fiber laminates and woven carbon fiber laminates

Parameter	Symbol	Weave-glass fiber	Weave-carbon fiber
Longitudinal moduli	$E_{11} = E_{22}$	12.5 GPa	39 GPa
Transverse modulus	E_{33}	11.8 GPa	4 GPa
Poisson's ratio	$\upsilon_{12} = \upsilon_{13}$	0.26	0.33
	υ_{23}	0.35	0.04
Shear modulus	$G_{12} = G_{13}$	3.5 GPa	3.59 GPa
	G_{23}	3.5 GPa	3.27 GPa
Tensile failure	$X_{1t} = X_{2t}$	332 MPa	532 MPa
	X_{3t}	290 MPa	240 MPa
Compressive failure	$X_{1c} = X_{2c}$	288 MPa	497 MPa
	X_{3c}	200 MPa	220 MPa
Shear moduli	$G_{12} = G_{13}$	39 GPa	79 GPa
	G_{23}	30 GPa	61 GPa
Interlaminar critical strain energy release	G_{IC}	207 Jm^{-2}	176 Jm^{-2}

3 Results and Discussion

As aforementioned, the majority of the AWJM research on composite material has concentrated on trimming or drilling a hole, or on shape-cutting operations, with very little focus on the piercing. Piercing is the preliminary material removal process for AWJM to start the cutting process. In this study, a simulation of an abrasive water-jet of 0.8 mm diameter was impacted on the hybrid laminate with a high impact velocity up to 800 m/s (at hydraulic pressure of 320 MPa; abrasive flow rate of 120 g/min; stand-off distance of 2 mm). General-purpose FE software, ABAQUS/Explicit was used to simulate the effects of the high-velocity impact of multi abrasive particles on hybrid FRP laminates and the effects of the progressive damage during the piercing event. The VUMAT subroutine was employed by combining the Hashin-Puck criteria and the Hashin-Rotem criteria to predict the material response of composite laminates when subjected to a high-velocity impact. The general kerf width patterns obtained at the top and bottom surfaces of the laminates are shown in Fig. 10. It is apparent that the numerical results when compared with the experimental data of previous studies, show a good correlation.

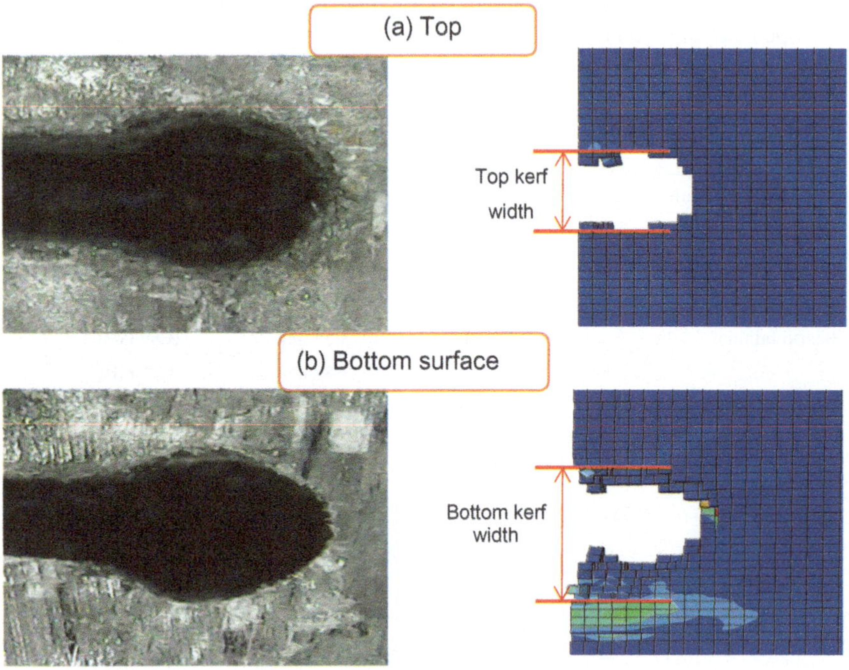

Fig. 10 Damage patterns on the top and bottom surfaces of composite panels after piercing by abrasive water-jet

3.1 Qualitative and Quantitative Results of AWJ Simulation

FEM is a powerful technique to solve many mechanical, manufacturing and design problems, which support many modern engineering practices and verify any design safety requirements. However, many of the parameters in FEM are uncertain, necessitating a verification of the model to ensure that the ideal premises, analyses and conclusions are valid. To verify the piercing damage model, the kerf width for the top ply of the composite laminates was compared with the experimental results. As assumed earlier, the abrasive water-jet model was evenly distributed over the volume of the element. Divergence of the water stream was, consequently, non-existent and no initial damage region was created on the entrance side of the composite laminate.

The kerf width of the top layer of composite laminate at the setting of hydraulic pressure of 320 MPa; abrasive flow rate of 120 g/min; and stand-off distance of 2 mm (Fig. 10a) was 0.9 mm in average. Under settings similar to previous experiments, but with a different traverse speed of abrasive water-jet of 1000 mm/min and 2500 mm/min, the top kerf widths were found to be 0.99 mm and 0.91 mm, respectively. The percentage errors for the settings of both experiments with different traverse speed, compared to that of numerical model results, were 10.2% and 1.2%, respectively. It can be claimed with confidence that the simulation results of the top kerf width are in agreement with previous experimental results for this study. In the case of Fig. 10b, however, the bottom kerf width of the composite laminate obtained from the SPH simulation was 1.4 mm, which showed a large deviation with previous experiments for this study, despite being under a different traverse speed of abrasive water-jet. The percentage of errors for both experiment settings were 47.9 and 53% respectively. This might be due to difficulties in properly determining the delamination, fiber pull-out, matrix-fiber debonding and other damages from the simulations or experiments. As aforementioned, the present numerical model performed a deformation-based deletion criterion, which means that when either the fiber or matrix elements reached the threshold fracture energy value, the element would be ready for element deletion. However, it is worth highlighting that the present results are somewhat in good agreement with the observation of Schwartzentruber et al. (2018) and Sheikh-Ahmad (2009). It can, then, be concluded that severe damage occurred on the bottom side when the piercing process was carried out. The mechanisms that lead to this phenomenon will be further explained in the following section.

Figure 11 compares the magnified image of the interior surface of a hole after the piercing process of the AWJ. Qualitatively, there is a positive agreement between the simulation and the experimental results when the piercing hole shape was compared. Moreover, the obtained results have less delamination damage in comparison with that of unidirectional carbon FRP composite as reported by several researchers (Schwartzentruber et al. 2018; Phadnis et al. 2013; Shanmugam et al. 2008). This could be attributed to the presence of braids between the fill and wrap tows of woven architecture laminate which can have the effect of reducing the crack propagation during the piercing event. Nevertheless, it is believed that the positive hybridisation

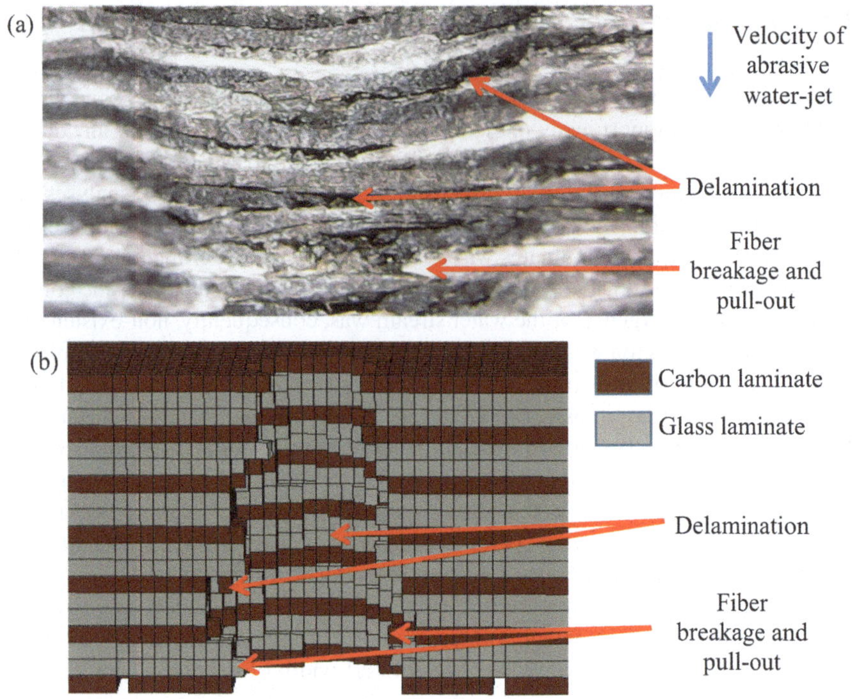

Fig. 11 Comparison of **a** experimental and **b** simulated piecing process of the hybrid composite laminate

effect has also provided a reasonable trade-off between the failure strain and the in-plain strength, resulting in better impact resistance properties compared to that of plain carbon FRP composites (Phadnis and Silberschmidt 2018). This is likely to be due to the strong bonding and braided effect in the optimal spacing between the carbon and glass fibers. Since glass fibers have a considerably higher strain-to-fracture rate in nature than that of the carbon fibers, glass fibers are able to hinder the propagation of crack as well as sustain higher deformations before erosion takes place (Tan et al. 2016).

3.2 Meso-Mechanism of AWJM for Hybrid FRP Composites Through Simulation Modelling

Figure 12 illustrates the numerical progress for the cutting sequences of the abrasive water-jet through the cross-section of the hybrid composite laminates at a meso-scopic level. It is apparent that the damage occurred during the initial piercing operations when the abrasive water-jet particles accumulated at the interface between

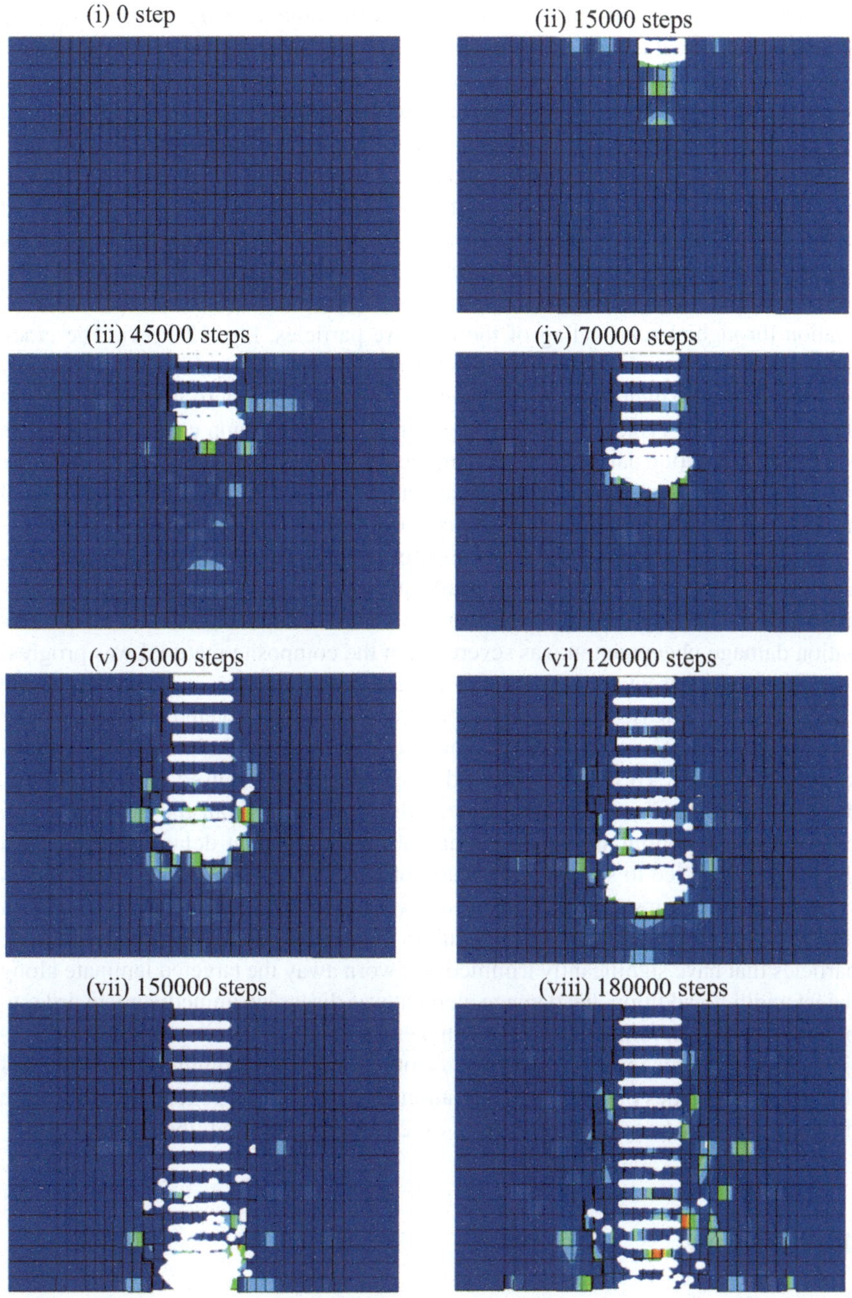

Fig. 12 The progress of the cutting sequence of the abrasive water-jet particles (white region) through the hybrid composite laminates at each step increment

composite laminate with the AWJ stream. As the high velocity of AWJ tends to maintain the impact energy of the abrasive particles during the piecing processes, a large stagnation of the AWJ pressure on the bottom interface of the laminate surface was evidenced. Under these phenomena, the impact force of the high etch rate of AWJ exceeded the inter-laminar bond strength of the uncut thickness (Tan et al. 2017) which led to the initiation of delamination damage. On the other hand, it can be observed that the shearing action of the abrasive particles played an important role in the erosion mechanism by generating the kerf walls. Unfortunately, the abrasive nature and stiffness of the fiber would slow down the material removal rate and provide sufficient time for abrasive particles to generate crack initiation and propagation through shock loading of the abrasive particles, Fig. 12 iii–vi. The crack initiation took place ahead of the cutting front and along the side walls of the kerf. Following that, the hydrostatic pressurisation of the initiated micro-cracks pro-longed the delamination. Typically, this crack generation phenomenon can be classified as Mode I delamination damage (opening mode). The observations in the present study were in good agreement with Shanmugam et al. (2008), who produced an analytical model for delamination on graphite/epoxy composites in AWJ machining.

As the laminate weakened due to the onset of the delamination damage, the adjacent laminates undertook additional loads and subsequently endured damage initiation and evolution processes. Based on the present numerical model, this delamination damage phenomenon was severe when the composite material was progressively cut and, consequently, the shear-out damage mechanism occurred. Due to the continuous high stresses applied at the point of impact, the targeted laminate around the AWJ width was eroded, pushed forward and created a hole. Under the high-velocity impact of AWJ, the shear-out damage mechanism caused matrix cracking, debonding of the fiber-matrix interface and fiber peeling from the outer laminate on the top and bottom side of the laminate. The significant delamination around the impact point led to severe degradation of material mechanical strength. This is shown in Fig. 13. However, Fig. 12 vii and viii do not really illustrate the shear-out damage phenomenon. This might be due to the numerical coarse-grained abrasive particles that have significantly trimmed and worn away the targeted laminate along the jet width. In addition, the damage element was discarded immediately in order to prevent the large deformation model from generating the excessive incremental rotation of the elements in the element set. Another explanation might be the nature of the woven properties of the composite laminate which have a better impact strength than unidirectional composite laminate which relies on braided fiber orientation.

4 Summary

This study presented a numerical investigation into the high impact piercing response of hybrid carbon/glass composite laminates. The results of this study showed that:

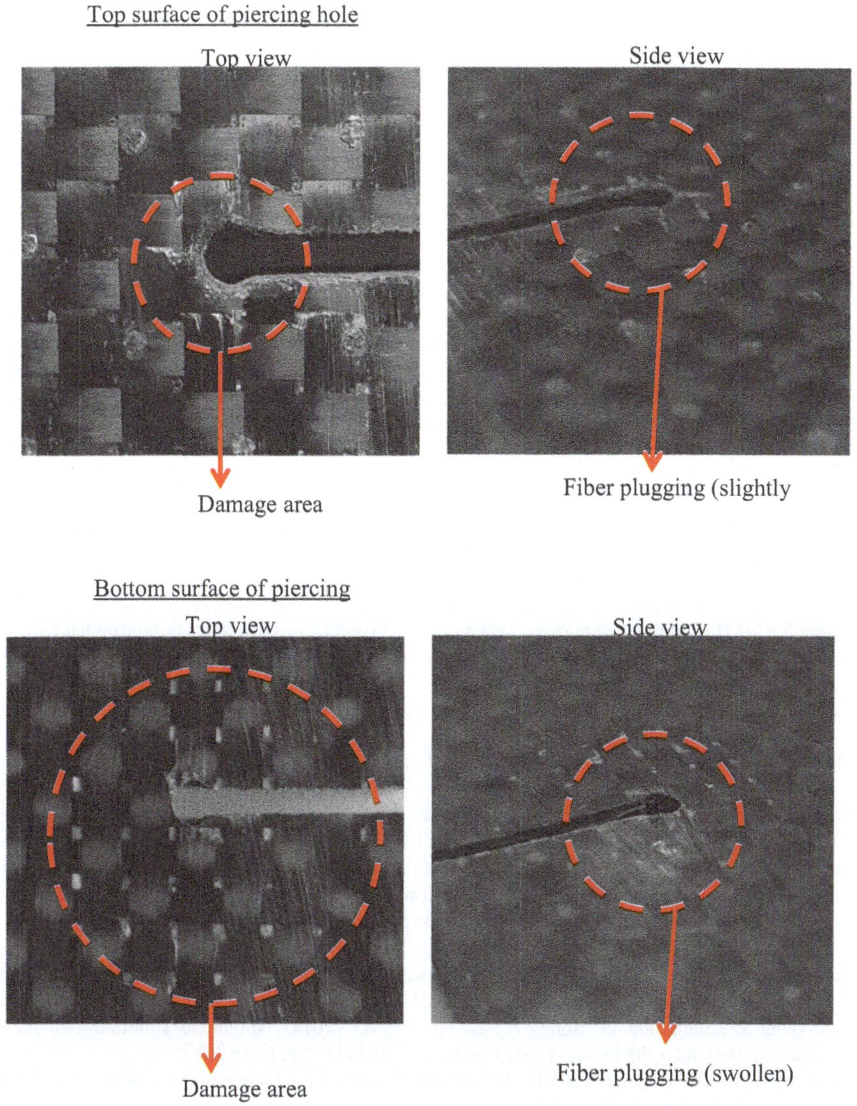

Fig. 13 Hole Piercing with AWJ: Shear-out damage mechanism occurred on both the top and bottom side of the composite and significant internal delamination around the impact point

- The model validations have a reasonable degree of accuracy for the top kerf width response.
- The laminate lay-up and architecture can affect the impact response of the composite laminates by changing the overall damage mechanism. The growth of delamination along the traverse direction compared with the unidirectional

composite can be prevented by involving a woven composite with 0°/90° fiber orientation.
- The interaction of a variety of damage mechanisms led to a more detrimental effect on the structure integrity and durability of the composite laminates. It is, therefore, necessary to understand the initiation and evolution of these damage mechanisms during the period of laminate failure.
- The major damage mechanisms during the piercing processes were delamination and, subsequently, shear-out mechanisms.

Acknowledgements The authors gratefully acknowledged the financial support of the Ministry of Science, Technology and Innovation (MOSTI) under the Science-Fund grant code UniMAP/RMIC/SF/06-01-15-SF0227/9005-00062(1) [(UniMAP Project Code: 9005-00062)] and Ministry of Education under Fundamental Research Grant code FRGS/1/2016/TK03/UNIMAP/02/7 (UniMAP Project Code: 9003-00605).

References

Anwar S (2013) Modelling of abrasive waterjet milled footprints. University of Nottingham

Anwar S, Axinte DA, Becker AA (2013) Finite element modelling of overlapping abrasive waterjet milled footprints. Wear 303(1–2):426–436

Azmi AI (2013) Chip formation studies in machining fibre reinforced polymer composites. Int J Mater Prod Technol 46(1):32–46

Chiu NSL (2015) Numerical and experimental study on composite structures under crushing loading. Monash University

Chiu LNS, Falzon BG, Boman R, Chen B, Yan W (2015) Finite element modelling of composite structures under crushing load. Compos Struct 131:215–228

Feng Y, Jianming W, Feihong L (2012) Numerical simulation of single particle acceleration process by SPH coupled FEM for abrasive waterjet cutting. Int J Adv Manuf Technol 59:193–200

Gingold RA, Monaghan JJ (1997) Smooth particle hydrodynamics: theory and application to non-spherical stars. R Astron Soc Mon Not 181:375–389

Gudimetla PV, Yarlagadda PK (2007) Finite element analysis of the interaction between an AWJ particle and a polycrystalline alumina ceramic. J Achiev Mater Manuf Eng 23(1)

Hongxiang J, Changlong D, Songyong L, Kuidong G (2014) Numerical simulation of rock fragmentation under the impact load of water jet. Hindawi Publ Corp 2014

Jianming W, Na G, Wenjun G (2010) Abrasive waterjet machining simulation by SPH method. Int J Adv Manuf Technol 50:227–234

Liu H, Falzon BG, Tan W (2018) Experimental and numerical studies on the impact response of damage-tolerant hybrid unidirectional/woven carbon-fibre reinforced composite laminates. Compos Part B 136:101–118

Lucy LB (1977) A numerical approach to the testing of the fission hypothesis. Astron J 82(12):1013–1024

Ma L, Hao Bao R, Mu Guo Y (2008) Waterjet penetration simulation by hybrid code of SPH and FEA. Int J Impact Eng 35(9):1035–1042

Mamiadaki K, Kestis T, Bilalis N, Antoniadis A (2007) A finite element-based model for pure waterjet process simulation. Int J Adv Manuf Technol 31:933–940

Naib SR (2015) Modeling the influence of water droplet impacts on steam turbine blade surfaces. University of Stuttgart

Phadnis VA, Silberschmidt VV (2018) Composites under dynamic loads at high velocities. Comprehensive composite materials II, vol 8, pp 262–285. Elsevier Ltd.
Phadnis VA, Pandya KS, Naik NK, Roy A, Silberschmidt VV (2013) Ballistic impact behaviour of woven fabric composite : finite element analysis and experiments. J Phys Conderence Ser 451:1742–6596
Puck A, Schurmann H (1998) Failure analysis of FRP laminates by means of physically based phenomenological models. Compos Sci Technol 58:1045–1067
Schwartzentruber J, Spelt JK, Papini M (2018) Modelling of delamination due to hydraulic shock when piercing anisotropic carbon-fibre laminates using an abrasive waterjet. Int J Mach Tools Manuf
Schwartzentruber J, Papini M, Spelt JK (2018b) Characterizing and modelling delamination of carbon-fibre epoxy laminates during abrasive waterjet cutting. Compos, Part A
Shanmugam DK, Nguyen T, Wang J (2008) A study of delamination on graphite/epoxy composites in abrasive waterjet machining. Compos Part A 39:923–929
Sheikh-Ahmad JY (2009) Machining of polymer composites
Srinivasu DS, Axinte DA, Shipway PH, Folkes J (2009) Influence of kinematic operating parameters on kerf geometry in abrasive waterjet machining of silicon carbide ceramics. Int J Mach Tools Manuf 49(14):1077–1088
Takaffoli M, Papini M (2012) Numerical simulation of solid particle impacts on Al6061-T6 Part II: materials removal mechanisms for impact of multiple angular particles. Wear 296:648–655
Tan CL, Wong I, Azmi AI, Muhammad N (2016) On improvement of mechanical properties of hybrid FRP composites via VARTM and rule of mixture. J Eng Appl Sci 11(9):2036–2043
Tan CL, Azmi AI, Muhammad N (2017) Critical thrust force for on-set delamination of hybrid FRP composite during Drilling Process. Key Eng Mater 740:111–117

Environmental Assessment of Composite Recycling for Machining Processes and Industries

Norshah Aizat Shuaib, Paul Tarisai Mativenga, Azwan Iskandar Azmi, and Hariz Zain

Abstract The most preferable and cost effective disposal route for glass fiber reinforced plastic (GFRP) waste is through landfill. However, this is not feasible as a long-term solution when taking into account strict regulations on landfill ban and loss of valuable materials. These pressures have driven the need for further study into composite recycling technology. However, environmental impact of end of life options, have yet to be thoroughly addressed. This chapter shows a simplified life cycle assessment method to assess possible recycling options and reuse applications for GFRP body-in-white automotive sheet moulding compound (SMC) structure. The results show that the high voltage fragmentation (HVF) recycling process has greater impact in all categories in comparison with the mechanical recycling method. This is as a result of high electricity energy demand and disposal of organic contaminated waste water in HVF process. The results also revealed that the global warming impact of a recycling process can be successfully reduced if the recyclate is used to replace high embodied energy virgin material. The methodology used in this chapter enables users to make an informed decision with respect to environmental implication of available end of life options for composite products. The methodology could be used for analysing environmental impact of other composite machining processes.

Keywords Composite recycling · Glass fiber reinforced composite · End of life · High voltage fragmentation · Mechanical recycling

---

N. A. Shuaib (✉) · A. I. Azmi · H. Zain
Faculty of Engineering Technology, Universiti Malaysia Perlis, 02100 Padang Besar, Malaysia
e-mail: norshahaizat@unimap.edu.my

P. T. Mativenga
School of Mechanical, Aerospace and Civil Engineering, University of Manchester, Manchester M13 9PL, UK

M. T. Hameed Sultan et al. (eds.), *Machining and Machinability of Fiber Reinforced Polymer Composites*, Composites Science and Technology,
https://doi.org/10.1007/978-981-33-4153-1_12

1 Introduction

Composite materials, particularly fiber reinforced plastics (FRP), have desirable mechanical properties such as high specific strength and fatigue resistance. These advantages have made this material highly desirable in mechanical demanding sectors such as, aerospace and automotive industries. Compared to equivalent metal counterparts, composite materials offer lightweight structures with comparable mechanical properties. In automotive sector, glass fiber reinforced plastics (GFRP) is commonly used in semi-structural applications in the form of sheet moulding compounds (SMCs) for vehicle parts such as front fender and body-in-white (BIW) structure (Palmer et al. 2009; Mayyas et al. 2012).

The increasing demand for composite materials has led to a global rising of composite waste. Recycling of thermoset based materials is challenging on account of its heterogeneous nature and because the matrix cannot be melted or remoulded. Currently, the most preferable route for disposal of GFRP waste is through landfill (Bains and Stokes 2013). However, tightening legislation on landfilling has caused this method to be impractical as a long-term solution. By 2025, landfilling of recyclable waste is no longer permitted in the European Union (EU) (European Commission 2014). EU End of life Vehicle Directive specifies that from 2015, 85% by weight of end of service vehicle needs to be recycled while the remaining 10 and 5% is for energy recovery and landfill respectively (European Commission 2000). With the increasing amount of FRP in modern vehicles, recycling of the material has to be considered in order to reach the 85% target. These legislation pressures and recycling challenges have intensified research and development into novel composite recycling technologies.

Composite recycling technologies can be separated into several main categories, namely thermal, mechanical, chemical, electrochemical, biotechnological and high voltage fragmentation methods. Mechanical recycling reduces size of composite waste into coarse and fine particle size fractions while the high voltage fragmentation method uses electrical pulses to disintegrate composite waste in an aqueous solution (Pickering 2006; Rouholamin et al. 2014). These two processes were found to have promising results in terms of energy demand and recyclate quality for GFRP recycling (Mativenga et al. 2016).

Reuse applications of composite recyclate have been demonstrated widely in literature. A study by Palmer et al. (2009), demonstrated that partial replacement of virgin fiber by fine and coarse fiber from mechanical recycling process can produce dough moulding compounds (DMCs) with comparable mechanical strengths with the virgin counterpart. In the construction sector, fine fraction from mechanical recyclates can be used to partially replace sand aggregates up to 20% in concrete mixtures (Correia et al. 2011). At an industry scale, Reprocover in Belgium manufactures railway cable duct covers completely from glass fiber and thermoset waste (Reprocover 2014).

By adding value to the recyclate and highlighting the resource benefits attained from recycling of GFRP waste, the marketability of glass fiber recyclate can be further improved. Positive environmental impact can be gained through consideration of credits obtained from recycling and energy recovery of the automotive waste. While impact of steel and aluminium recycling is well developed in literature, similar assessment for composite materials is still in need of attention. Likewise, environmental credit from the avoided production of virgin materials fulfilled by the reuse of composite recyclate is not yet extensively explored.

This chapter evaluates end of life scenarios for a glass fiber reinforced unsaturated polyester (GFRP) body-in-white (BIW) structure. A simplified life cycle assessment, which mainly focused on the recycling and reusing potential of the recyclate, was carried out. The main objectives were to assess environmental performance of mechanical and HVF recycling processes and to identify environmental credits for reusing the recyclate. The assessment incorporates a refined environmental dataset, developed from authors' previous works in Mativenga et al. (2016) and Shuaib and Mativenga (2016). The findings provide an extended waste management guide in collaboration with a robust environmental assessment. This could provide users with environmentally informed decisions on possible recycling options and associated applications for composite recyclate.

2 Methodology

Life cycle assessment (LCA) is a useful tool in evaluating environmental impact over a product's lifetime. LCA modelling was carried out using Simapro 8.1 software. The impact assessment method used was Environmental Product Declaration (EPD). All compulsory and optional impact categories in EPD were reported. The categories were acidification, eutrophication, global warming potential, photochemical oxidation, ozone layer depletion and abiotic depletion. Detailed analysis was carried out on global warming potential (GWP) as it is the most relevant impact category in reducing greenhouse gas emissions for the system under study. In this study, the LCA method was adapted to focus on the environmental burden and credits of the end of life phase of automotive component, specifically the sheet moulding compound (SMC) structure. The assessment also focused on possible environmental credits gained from recycling and material recovery initiatives. The end of life options considered in this study were mechanical recycling and high voltage fragmentation (HVF), as alternatives to the traditional landfill disposal. Potential avoided products associated with each recycling method were based on the recyclate's size and condition.

2.1 Goal and Scope

The LCA goal was to evaluate the life cycle impact of a body-in-white (BIW) structure of a passenger car. The functional unit in this study was defined as 123 kg of SMC structure made from glass fiber reinforced unsaturated polyester, as reported in

Table 1 Composition of sheet moulding compound (SMC), based on Witik et al. (2011)

Component	Weight percentage (%)
Unsaturated polyester	10
Styrene	10
Glass fiber	50
Sodium silicate	30

Mayyas et al. (2012). The system boundary was drawn around the composite BIW structure of the vehicle, leaving aside other components such as glass windows and metal compartments.

2.2 *Material Inventory*

A life cycle inventory was constructed to quantify environmental input and outputs of a system. The preparation of inventory analysis and life cycle impact assessment were carried out using the Simapro 8.1 software. SMC composition was taken from Witik et al. (2011), shown in Table 1.

2.3 *Energy and Environmental Signature Inventory*

Datasets for materials and electrical energy usage were retrieved from Ecoinvent 3 and the European Life Cycle Database (ELCD). However, not all data is available in the databases, in particular the resource data for recycling scenarios. The unavailable data was retrieved from reputable scientific literature and added to the Simapro software. The inventory items and the associated data sources are included in Table 2.

Data on landfilling of municipal waste was taken from Ecoinvent 3 database. Electricity demand for shredding as a pre-recycling and pre-landfill stage was taken as 0.09 MJ/kg (Witik et al. 2011). For mechanical recycling, the process was considered to be using a Wittmann MAS1 granulator at 30 kg/h processing rate, with energy demand of 0.37 MJ/kg (Shuaib and Mativenga 2016). The dust extractor used has a rated fan power of 1.5 kW, which gives energy consumption of 0.18 MJ/kg for 30 kg/h processing rate. The HVF method used a total of 1500 numbers of electrical pulses, and the energy demand was taken as 60 MJ/kg (Mativenga et al. 2016). Energy demand for the sieving or classification process for the recyclate was taken as 0.125 MJ/kg, as reported in Asmatulu et al. (2013). A system diagram for the end of life scenarios is shown in Figs. 1, 2 and 3.

Detailed size classification of glass fiber composite recyclate from an industrial scale process was reported in Palmer et al. (2009). The recyclate from the mechanical method was categorised into coarse and fine fractions, which were 28% and 72% by weight respectively. The fibrous fraction (42%), was assumed to be reusable by

Table 2 Data sources for life cycle inventory items

Life cycle inventory items	Source
Dry chopped strand glass fiber mat	ELCD
Unsaturated polyester resin	Ecoinvent 3
Sodium silicate	Ecoinvent 3
Sheet moulding compound (SMC) manufacturing process	Mayyas et al. (2012) and Song et al. (2009)
Transportation using 32 metric tonnes lorry	Ecoinvent 3
Landfilling of municipal solid waste	Ecoinvent 3
Shredding (pre-recycling stage)	Witik et al. (2011)
Mechanical recycling	Shuaib and Mativenga (2016)
High voltage fragmentation recycling	Mativenga et al. (2016)
Tap water	Ecoinvent unit
Waste water (organic contaminated) treatment	ELCD
Sieving (recyclate size classification)	Asmatulu et al. (2013)
Calcium carbonate	ELCD
Concrete	Ecoinvent 3
Sand aggregates	Ecoinvent 3
High density polyethylene (HDPE)	Ecoinvent 3
Synthetic rubber	Ecoinvent 3
Polyurethane foam	Ecoinvent 3
Polypropylene	Ecoinvent 3

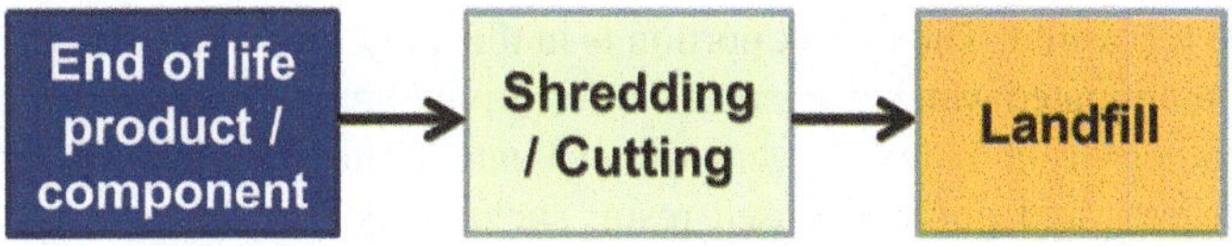

Fig. 1 System diagram for landfill as end of life option

partially mixing with virgin fiber in dough moulding compounds (DMCs) applications. The coarse fraction was not directly reusable and can be either sent to landfill or reprocessing. The powdered fraction (30%), was assumed to be replacing calcium carbonate as fillers in cement products. Full substitution of concrete from the mechanical process recyclates was based on a practical example by a company named

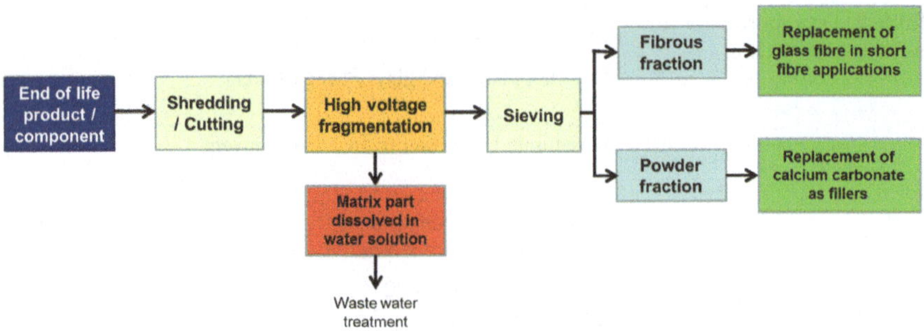

Fig. 2 System diagram for high voltage fragmentation (HVF) recycling as end of life option with potential reuse applications

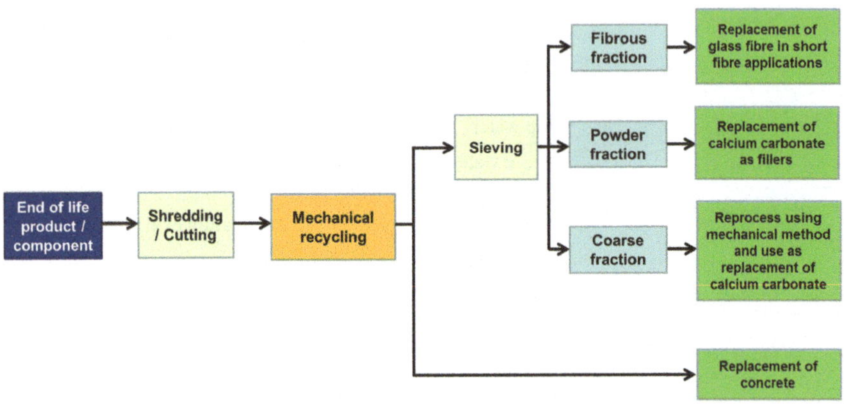

Fig. 3 System diagram for mechanical recycling as end of life option with potential reuse applications

Reprocover in Belgium (Reprocover 2014). For the recyclate obtained from the high voltage fragmentation (HVF) method, only the inorganic parts of the composite structure can be recovered. The matrix portion is in the aqueous solution and disposed as organic contaminated water. As part of the sensitivity analysis, different scenarios of replacing other virgin materials were also evaluated. The materials considered were sand aggregates and polymer based resin, such as, polyurethane foam, synthetic rubber, unsaturated polyester and high-density polyethylene. All data used for these raw materials were sourced from the ELCD and Ecoinvent 3 databases.

2.4 Assumptions and Limitations

The body-in-white (BIW) composite structure was assumed to be fully glass fiber reinforced unsaturated polyester with sodium silicate fillers and manufactured through an advanced sheet moulding compound process. BIW structures made from various SMC formulations were used in the sensitivity analysis. The structures were assumed to have comparable mechanical strength and weight despite the differences in terms of fiber weight.

Data for landfilling of composite material is scarcely found in literature and life cycle databases; hence, data for landfilling of municipal waste was used for disposing composite material in landfill. For HVF recycling process, the tap water consumption was assumed to be 3.3 L for 0.2 kg of waste processed. Embodied energy associated with consumables in recycling processes, such as granulator cutter and electrode, was not considered. The source of electricity was assumed to be based on UK conditions. Apart from that, the transport distance was assumed to be 100 kms on a 32 metric tonnes lorry from the waste collection centre to recycling or landfill site.

Despite low recycling uptake of composite waste (Halliwell 2006; Bains and Stokes 2013), this study assumed 100% recycling and recyclate recovery rate (no loss during material transfer). It has been assumed that replacement of virgin material by the recyclate in new applications will give comparable mechanical properties with their virgin counterparts.

The study uses the end of life recycling allocation method. This method gives the recycling credit at the end of a product's life with the assumption that the recycling avoids production of virgin material (Johnson et al. 2013). The benefit of recycling is attributed to the original product i.e. the GFRP BIW structure.

3 Results and Discussion

3.1 Environmental Impacts of Recycling Processes

Environmental burden of mechanical and HVF recycling processes are reported in Table 3. The HVF process has high environmental burden in all impact categories on account of its higher energy demand (60 MJ/kg), in comparison to only 0.37 MJ/kg for the mechanical recycling process. The environmental impact of mechanical recycling process is only from the electricity energy demand. For HVF process, the electricity energy demand, usage of tap water and disposal of untreated waste water needs to be considered. In general, the HVF process has about 100 times greater impact than the mechanical recycling process. The EPD impact categories for the HVF method was dominated by the electricity energy demand, around 86–99%. In the eutrophication category, the waste water accounts for about 14%. Future improvements should focus on minimising process direct energy demand. This can be executed through optimisation of process parameters and utilising machine basic energy demand by operating at the maximum capacity.

Table 3 Environmental product declaration (EPD) results for mechanical and high voltage fragmentation recycling processes

Impact category	Acidification (kg SO_{2eq})	Eutrophication (kg PO_{4eq})	Global warming potential (kg CO_{2eq})	Photochemical oxidation (kg C_2H_{4eq})	Ozone layer depletion (kg CFC-11_{eq})	Abiotic depletion (kg Sb_{eq})
High voltage fragmentation (SELFRAG, 1500 pulses)	4.36	1.50	1.30×10^3	1.76×10^{-1}	3.18×10^{-5}	2.86×10^{-4}
Mechanical recycling (MAS1 granulator, 30 kg/h)	4.56×10^{-2}	1.55×10^{-2}	1.35×10^1	1.83×10^{-3}	3.36×10^{-7}	3.04×10^{-6}

3.2 *Environmental Credits from Virgin Material Substitution*

Environmental credits from recycling are necessary to minimise a product's entire life cycle impact. The main benefits of choosing recycling as a disposal scenario are landfill avoidance and averting the need to produce virgin products. The possible scenarios for mechanical recycling and embodied energy of production for virgin material are shown in Table 4 and Table 5 respectively.

Table 4 Possible end of life scenarios using mechanical recycling and landfill

Scenario	Disposal technique	Allocation of recyclates
A	Mechanical recycling	• 28% of coarse fraction sent to landfill • 42% replacement of virgin glass fiber in dough moulding compounds (DMC) • 30% replacement of calcium carbonate as fillers
B		• 28% of coarse fraction reprocessed and replaces calcium carbonate as fillers • 42% replacement of virgin glass fiber in dough moulding compounds (DMC) • 30% replacement of calcium carbonate as fillers
C		• 100% replacement of concrete
D	Landfill	–

Table 5 Embodied energy of production for virgin material taken from Ecoinvent 3 database

Virgin material	Embodied energy of production (MJ/kg)
Glass fiber	45
Calcium carbonate	0.68
Concrete	0.84

Comparison of global warming potential (GWP) between traditional landfill disposal and mechanical recycling scenarios is shown in Fig. 4. The positive environmental credits allocated from avoided virgin material production, reduces the environmental burden of the mechanical recycling process. Figure 4 also indicates significant environmental benefits of reusing fibrous fraction in the replacement of virgin glass fiber in short fiber applications such as DMCs compared to the replacement of calcium carbonate as fillers. This is due to low embodied energy of virgin calcium carbonate.

The assessment for the HVF process was undertaken by considering three SMC formulations with different weight fractions of glass fiber (30, 40 and 50%). The result is shown in Fig. 5. At the maximum glass fiber recovery (50% by weight), the total environmental impact of the HVF process is still significantly higher than that of landfill. Replacement of virgin calcium carbonate, as filler only, has an environmental credit of 1.46 kg CO_{2eq}. The impact of tap water consumption and disposal of organic contaminated water is 0.64 kg CO_{2eq} (0.04% from total energy footprint of HVF) and 169.11 kg CO_{2eq} (11.77%) respectively. Future research on developing an industrial scale HVF process is recommended to reduce the process specific energy demand. The retrieved organic component can reduce the high environmental burden of the HVF recycling process through applications such as chemical and energy feedstock. Treatment of the contaminated or waste water to retrieve back organic components is suggested. However, the refining steps may escalate the energy demand and produce by-product contaminations. All mechanical recycling scenarios show negative values of GWP impact, indicating environmental benefits. From Fig. 5, landfill and mechanical recycling are the preferable end of life options in terms of GWP impact.

It is clearly apparent in Figs. 4 and 5 that the best strategy to utilise the composite recyclate is virgin fiber replacement. Replacement of calcium carbonate was found to

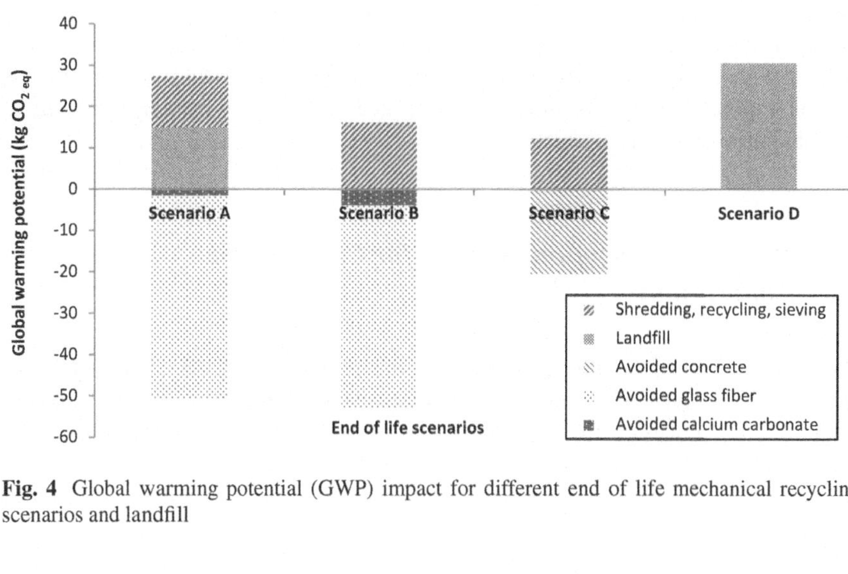

Fig. 4 Global warming potential (GWP) impact for different end of life mechanical recycling scenarios and landfill

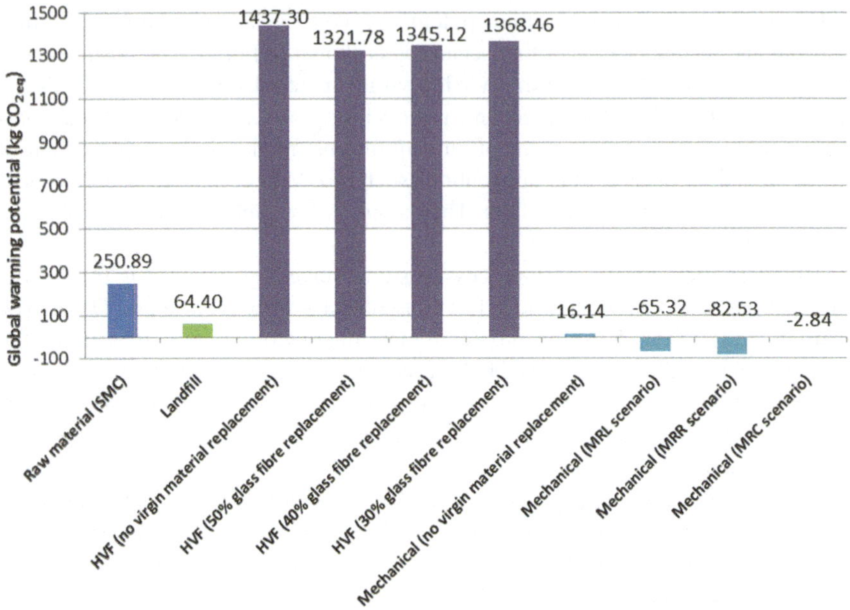

Fig. 5 Comparison of global warming potential (GWP) impact between SMC raw material production and different end of life scenarios

have low environmental credits. In addition, cheap virgin calcium carbonate (priced at around £0.10 per kg (Granta Design Limited 2015)), demonstrates that the replacement is not economically feasible. An advantage of HVF method is that fibers recovered have longer mean fiber length and are cleaner compared to those recovered from the mechanical process (Mativenga et al. 2016). However, low processing scale of the HVF method led to high energy demand per kilogram of processed waste. For the mechanical method, design of the granulator (such as the machine clearance gap) and process parameters (granulator screen size), should be considered to generate more fibrous fraction. Energy demand of the two processes can be utilised by operating at the maximum machine capacity.

3.3 *Recyclate Allocation Strategy for Cross Sector Applications*

The study on positive impact of avoiding the production of virgin material was extended to evaluate other potential applications. The breakeven material loading for a recycling process with a 10 MJ/kg energy footprint was evaluated in Fig. 6. The figure shows a relationship between material breakeven loading versus its embodied energy per kg. The breakeven loading indicates the minimum percentage of material

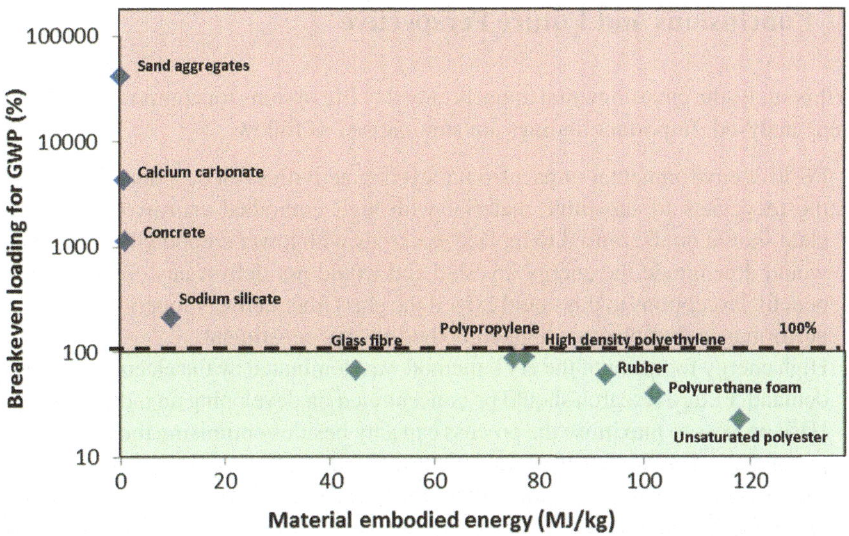

Fig. 6 Material breakeven loading for global warming potential for a recycling process with 10 MJ/kg energy footprint (y-axis is in logarithmic scale)

substitution required in reuse applications to provide environmental credits that is sufficient to totally eliminate global warming potential (GWP) impact. In general, material with high embodied energy will require low breakeven material loading.

Figure 6 exhibits the advantage of replacing polymer based materials such as high-density polyethylene, polyurethane foam, synthetic rubber, polypropylene and unsaturated polyester. However, substitution of these materials with organic product recovered from chemical recycling method is still under development and not commercially feasible (Yamada et al. 2010). For a thermal recycling method, the matrix component is broken down into smaller molecules and the recovery of the matrix is impossible. The fibrous fraction of GFRP recyclate should not be used to substitute sand aggregates, calcium carbonate or concrete, unless they functionally replace a constituent with higher embodied energy or render new innovative product performance.

From literature, glass fiber composite recyclate from mechanical method can be incorporated into injection moulded thermoplastic products to enhance mechanical strength. Such application can indirectly replace the usage of virgin polymer material, whilst also improving mechanical performance of the injection moulded products. From Fig. 6, it is apparent that replacement material with an embodied energy of less than about 40 MJ/kg, is not sufficient enough to offer environmental benefits for a recycling process with an energy demand of 10 MJ/kg. Future research on matrix recovery and reducing usage of polymer based materials by incorporating recyclate are paramount to ensure long term sustainability of composite recycling initiatives.

4 Conclusions and Future Perspective

In this study, the environmental impacts of end of life options for composite recyclate were analysed. Important findings are summarised as follow:

- Positive environmental impact from recycling activities can be gained by reusing the recyclates to substitute material with high embodied energy. GFRP recyclate should not be reused to replace materials with lower embodied energy. This would downgrade the energy invested and would not deliver any environmental benefit. Exceptions to this could exist if the glass fiber delivers superior functional performance or replaces a high embodied energy constituent.
- High energy footprint of the HVF method was dominated by the electricity energy demand. Future research should be concentrated on developing an industrial scale HVF process to maximise the process capacity besides optimising the operational parameters. The impact of organic contaminated waste water was around 12% of the global warming potential burden. The impact can be reduced by reusing the embedded organic products. However, the recovery steps may escalate the overall environmental footprint demand even further.
- The analysis of potential virgin material replacement suggests that replacement of polymer based virgin material has the lowest material breakeven loading for global warming potential impact. However, current technology for recovering matrix components of composite waste is still immature and more research is needed.
- As the current landfill uptake for composite waste is still high, the waste management strategy has to be changed in the near future. Recovery of material, as a result of landfill avoidance, can contribute to environmental credits from the use of recyclates in potential cross sector applications.

Acknowledgements The authors acknowledge the UK Engineering and Physical Sciences Research Council (EPSRC), grant EP/K026348/1, Efficient X-Sector use of HeterogeneoUs MatErials in Manufacturing (EXHUME).

References

Asmatulu E, Twomey J, Overcash M (2013) Recycling of fiber-reinforced composites and direct structural composite recycling concept. J Compos Mater

Bains M, Stokes E (2013) Developing a resource efficiency action plan for the composites sector. URS Netcompos

Correia JR, Almeida NM, Figueira JR (2011) Recycling of FRP composites: reusing fine GFRP waste in concrete mixtures. J Clean Prod 19:1745–1753

European Commission (2000) Directive on end of life vehicles (2000/53/EC). Off J Eur Communities 34–55

European Commission (2014) Communication from the commission to the European Parliament, The Council, The European Economic and Social Committee and the Committee of the Regions

Granta Design Limited (2015) CES EduPack software

Halliwell S (2006) End of life options for composite waste recycle, reuse or dispose. https://www.compositesuk.co.uk/LinkClick.aspx?fileticket=f3y8cNT6pIg%3D&tabid=111&mid=550. Accessed Dec 2013

Mativenga PT, Shuaib NA, Howarth J, Pestalozzi F, Woidasky J (2016) High voltage fragmentation and mechanical recycling of glass fibre thermoset composite. CIRP Ann Manuf Technol

Mayyas AT, Qattawi A, Mayyas AR, Omar MA (2012) Life cycle assessment-based selection for a sustainable lightweight body-in-white design. Energy 39:412–425

Palmer J, Ghita OR, Savage L, Evans KE (2009) Successful closed-loop recycling of thermoset composites. Compos a Appl Sci Manuf 40:490–498

Pickering SJ (2006) Recycling technologies for thermoset composite materials—current status. Compos Part A Appl Sci Manuf 37:1206–1215

Reprocover (2014) Production of cable duct covers. https://www.reprocover.com/en/node/618. Accessed July 2015

Rouholamin D, Shyng YT, Savage L, Ghita OR (2014) A comparative study into mechanical performance of glass fibre recovered through mechanical grinding and high voltage pulse power fragmentation. In: 16th European conference on composite materials, Seville

Shuaib NA, Mativenga PT (2016) Energy demand in mechanical recycling of glass fibre reinforced thermoset plastic composites. J Clean Prod 120:198–206

Song YS, Youn JR, Gutowski TG (2009) Life cycle energy analysis of fiber-reinforced composites. Compos Part A Appl Sci Manuf 40:1257–1265

Witik RA, Payet J, Michaud V, Ludwig C, Månson J-AE (2011) Assessing the life cycle costs and environmental performance of lightweight materials in automobile applications. Compos Part A Appl Sci Manuf 42:1694–1709

Yamada K, Tomonaga F, Kamimura A (2010) Improved preparation of recycled polymers in chemical recycling of fiber-reinforced plastics and molding of test product using recycled polymers. J Mater Cycles Waste Manag 12:271–274

The manufacturer's authorised representative in the EU is Springer Nature Customer Service Centre GmbH, Europaplatz 3, 69115 Heidelberg, Germany. If you have any concerns regarding our products, please contact ProductSafety@springernature.com

Printed and bound by CPI Group (UK) Ltd, Croydon, CR0 4YY
30/09/2026
02225289-0009